Nanomaterials and Nanoarchitectures

NATO Science for Peace and Security Series

This Series presents the results of scientific meetings supported under the NATO Programme: Science for Peace and Security (SPS).

The NATO SPS Programme supports meetings in the following Key Priority areas: (1) Defence Against Terrorism; (2) Countering other Threats to Security and (3) NATO, Partner and Mediterranean Dialogue Country Priorities. The types of meeting supported are generally "Advanced Study Institutes" and "Advanced Research Workshops". The NATO SPS Series collects together the results of these meetings. The meetings are co-organized by scientists from NATO countries and scientists from NATO's "Partner" or "Mediterranean Dialogue" countries. The observations and recommendations made at the meetings, as well as the contents of the volumes in the Series, reflect those of participants and contributors only; they should not necessarily be regarded as reflecting NATO views or policy.

Advanced Study Institutes (ASI) are high-level tutorial courses to convey the latest developments in a subject to an advanced-level audience

Advanced Research Workshops (ARW) are expert meetings where an intense but informal exchange of views at the frontiers of a subject aims at identifying directions for future action

Following a transformation of the programme in 2006 the Series has been re-named and re-organised. Recent volumes on topics not related to security, which result from meetings supported under the programme earlier, may be found in the NATO Science Series.

The Series is published by IOS Press, Amsterdam, and Springer, Dordrecht, in conjunction with the NATO Emerging Security Challenges Division.

Sub-Series

A.	Chemistry and Biology	Springer
B.	Physics and Biophysics	Springer
C.	Environmental Security	Springer
D.	Information and Communication Security	IOS Press
E.	Human and Societal Dynamics	IOS Press

http://www.nato.int/science
http://www.springer.com
http://www.iospress.nl

Series C: Environmental Security

Nanomaterials and Nanoarchitectures

A Complex Review of Current Hot Topics and their Applications

edited by

M. Bardosova
Tyndall National Institute UCC, Cork, Ireland

and

T. Wagner
Faculty of Chemical Technology, University of Pardubice, Pardubice, Czech Republic

Univerzita Pardubice Tyndall National Institute Springer

Published in Cooperation with NATO Emerging Security Challenges Division

Proceedings of the NATO Advanced Study Institute on Nanomaterials and Nanoarchitectures – A Complex Review of Current Hot Topics and their Applications in Photovoltaics, Plasmonics, Environmental and Security Areas
Cork, Ireland
30 June – 7 July 2013

Library of Congress Control Number: 2015946315

ISBN 978-94-017-9937-9 (PB)
ISBN 978-94-017-9920-1 (HB)
ISBN 978-94-017-9921-8 (e-book)
DOI 10.1007/978-94-017-9921-8

Published by Springer,
P.O. Box 17, 3300 AA Dordrecht, The Netherlands.

www.springer.com

Printed on acid-free paper

Naturae enim non imperatur nisi parendo
Nature can only be commanded by obeying her
Francis Bacon*, Novum Organum I, 129*

Preface

Science as 'knowledge' and technology as 'the practical use of that knowledge' have continually modified human existence, yet in the last two centuries they began to influence human development with growing ubiquity. The predecessor of modern science 'natural philosophy' evolved gradually, and its subject gave rise to different branches of science. Even at present, the ongoing progress brings about more narrowing specialisations, disciplines and subdisciplines. Nanoelectronics, nanomechanics, nanophotonics and nanoionics are examples of sciences that have evolved recently. At the same time, many interesting phenomena occur at the boundaries involving two or more scientific fields. This is the reason why a number of interdisciplinary scientific fields have emerged during the twentieth century. Increasing complexity of scientific problems requires the creation of large research teams with researchers having different backgrounds and skills. Contemporary science is perceived as a special brand of information about the world and is practised by a distinct group of highly trained individuals and pursued through a unique method. When science seeks a goal towards practical utility, it is called applied science. The anticipated applicability of the research outcome is the crucial aspect that influences the outcome of research proposals because the so-called 'exploratory' or 'curiosity-driven' research gets low priority when the funding is decided.

These general statements are also true for research and its applications in the field of nanotechnology, nanomaterials and nanoarchitectures. In this case, 'nanomaterials' are building blocks, 'nanoarchitecture' represents the design and 'nanotechnology' the means to produce the device, whereas material, device and function are nearly inseparable. The term 'nanotechnology' remains to a certain extent ambiguous. In fact, there are two definitions of it. One, more visionary and with a wee bit of sci-fi flavouring, originates from the works of Erich Drexler; the other was established by the National Nanotechnology Initiative and is currently accepted by the majority of scientists. The word 'nanotechnology' was used for the first time by Norio Toniguchi (1912–1999) at an international conference in Tokyo in 1974. Yet it was Eric Drexler who (unaware of Taniguchi's work) has nurtured the term in his paper [1] and later in books [2, 3] and by his vision inspired experts with different backgrounds, scientists and engineers to study materials and phenomena at the nanoscale. Richard Feyman's (1918–1988) lecture at Caltech (1959) 'There is plenty of room at the bottom', which is now considered the seminal event in the history of nanotechnology, had a negligible influence till the early 1990s [4]. The idea was rediscovered and advanced by Drexler who used Feynman's concept, citing it in his book and adding his own vision of a multitude of tiny robots (molecular assemblers) that could move molecules so quickly and position them so precisely that they could produce almost any substance out of ordinary ingredients. The self-replicating machine is a design that is capable of reproducing itself autonomously using raw materials found in the environment, thus exhibiting self-replication found in nature. In computing, a similar concept was elaborated by John von Neumann in the 1950s [5]. Drexler's concept of nanotechnology referred to the technological goal of precisely manipulating atoms and molecules for the fabrication of macroscale products. Research in this area, now called 'molecular nanotechnology' or 'molecular manufacturing', in which mechanosynthesis is used, continues. In mechanosynthesis individual molecules are positioned close together so that stronger chemical attractions can overcome weaker ones in a controlled way, depositing or removing atoms as desired. Successful experiments involving manipulating individual atoms have been reported. The first of them was the placement of xenon atoms so that they formed letters IBM onto a copper surface with the tip of a scanning tunnelling microscope in 1988. Despite this and other partial successes, the proof of the concept involving a molecular assembler or atomically precise manufacturing (APM) does not exist at the moment. Drexler is currently an academic visitor at the 'Programme on the Impacts of Future Technology' at Oxford University in the UK. In 2013 he published a new book [6] in which he predicts that the Golden Era for humanity is around the corner. He states that using APM would make it possible to produce virtually any product from materials widely available (such as carbon nanotubes) and accomplish this production close to the places where it is needed. Three-dimensional printing illustrates this principle on macroscale. While his scientific ideas remain visionary and will materialise in due course, albeit maybe in a modified form, his presumptions of societal impacts seem unrealistic.

A generally accepted description of nanotechnology established by the NNI[1] defines nanotechnology as the manipulation of matter with at least one dimension sized from 1 to 100 nm. This means that a particular technological goal present in Drexler's concept was replaced by a research category inclusive of all types of research and technologies that deal with special properties of matter that occur below the given size threshold. For Drexler's *revolutionary* description of nanotechnology to be widely accepted, a paradigm shift would be necessary. A paradigm shift is a change of basic assumptions – a profound change in a fundamental model. Scientists currently do not 'encounter anomalies that cannot be explained by the universally accepted paradigm within which scientific progress

[1] US government R&D National Nanotechnology Initiative formed in 2000

has been made',[2] i.e. the condition for a scientific revolution formulated in [7] has not been fulfilled. This and the lack of a proof that molecular assemblers could work the way Drexler envisioned were the strong underlying argument during the heated discussion between Richard Smalley (1943–2005), the chemist who shared the 1996 Nobel Prize for discovering fullerenes, and Drexler. Their exchange took place in 2001–2003 and had the form of a series of journal articles and open letters. The final disputation 'Point-Counterpoint' [8] was published in the *Chemical and Engineering News*.

Research on nanoscale is highly specific. It has much promise and at the same time carries many hazards. To date, the development of nanotechnology has focused on the novel materials, properties of which can be predicted with computer simulation and modelling. They represent a critical dimension of nanotechnology because their composition is linked with the relative functions and devices. Currently, zero-dimensional materials (such as quantum dots), one-dimensional materials (nanowires) and two-dimensional materials (thin films) are studied. Nanomaterials are pushing the boundaries of physical laws to achieve novel technologies, using new methods to organise individual nanostructures into higher-order architectures, such as self-assembly.

Nanoscience and nanotechnology entered our everyday lives aggressively and on massive scale when the commercialisation of research began early, maybe too early. Currently, there are more than 1,600 manufacturer-identified nanotech products publicly available [9] – and the number is growing. The speed with which new products appear on the market (3–4 per week) is astonishing. Most applications so far are still limited to passive nanomaterials – using silver nanoparticles as antibacterial agent, nanoparticle-based sunscreens, stain-resistant textiles, food packaging and even as additives in food itself, such as TiO_2 in yogurts and chocolates [10]. Serious health concerns have been raised in connection with using nanoparticles and nanofibres – inhaling leads to pulmonary diseases; silver nanoparticles released from garments kill useful bacteria in the environment; TiO_2 nanoparticles widely used in many industries have been linked with DNA and chromosome damage in lab mice; etc. [11, 12].

Here is an incomplete list of scientific disciplines, which are using nanomaterials and nanotechnologies and products emerging from their research: biomedicine (implants, diagnostic procedures, biosensors, drugs and therapeutic methods, tissue engineering), photonics (3D, holographic, OLED, QD and screenless displays), electronics (electronic nose, e-textiles, memristors, molecular electronics, spintronics), energy (biofuels, batteries), IT (artificial intelligence, data storage, optical computing, RFID), manufacturing (3D printing, assemblers), materials science (research on fullerene, graphene, high-temperature superconductivity and superfluidity, metamaterials, QDs) and also military, neuroscience, robotics and transport, with applications in new disciplines emerging.

[2] Citing from Thomas Kuhn's book [7]

Thanks to nanoscience it is still possible to continue the exponential growth of the number of transistors in a dense integrated circuit predicted by Moore's law. The hardware advancements allow the production of smaller, yet more efficient computers, consumer electronic appliances and also contribute to further progress in informatics and artificial intelligence. In July 2014 a chatbot[3] named Eugene Goostman [13] reportedly passed the Turing test [14] by convincingly imitating a 13-year-old boy. One could argue, whether *he* did learn from the interaction and demonstrate problem-solving skills or whether *he* was only *mimicking* understanding. This is a small step advancing the area of artificial intelligence (AI). Research in this field depends to a big extent on steadily improving hardware, i.e. increasing computational power of computers, which in turn depends on nanotechnologies. As with other fields, there were optimistic prognoses in the past as well as setbacks. At present, positivism prevails and the timing was set up for the expected emergence of the so-called strong-AI, which would combine all human skills and exceed human abilities. Such outcome inspires both hopes and fears. For many years, scientists and philosophers have been expressing the view that the ever-accelerating progress of technology and changes in the mode of human life will ultimately lead to some essential *technological singularity* beyond which 'human affairs, as we know them, could not continue' [5, 15].

There is a growing concern that the less developed countries might not have access to infrastructure, funding and human resources to support nanotechnology, and as a result certain goods such as water purification, solar energy and advancements in medicine might not be available for them. Producers in developing countries could also be disadvantaged by the replacement of natural products (rubber, cotton, coffee) by developments in nanotechnology. These natural products are important export crops for developing countries, and their substitution with industrial nano-products could negatively affect the economies of developing countries and make them even poorer.

Some nanotechnology procedures and products influence environment in the negative sense (increased toxicity and reactivity), but there are also ways in which nanotechnology can help the environment (e.g. remediation, prevention, green manufacturing).

The military is one of the largest and primary drivers of new technologies throughout history, resulting in significant changes to society and international relations. Nanotechnology is no exception to this trend since the military aggressively pursues research in this field by supporting projects aimed at making lightweight body armour, fabricating ultrahydrophobic surfaces by using chemical vapour deposition technique (CVD), designing negative refractive index metamaterials to make objects invisible and fabricating chemical and biological sensors. Research projects are also underway leading to the enhancement of physical and cognitive abilities. Some of these will in time find way for civilian use.

[3] A computer program designed to simulate conversation with human users

From this short account, it is clear that nanotechnology, more than other areas of contemporary research, can have immense consequences reaching far into the future and could influence the very essence of humanity. Researchers need to be aware of possible ethical issues, environmental and health hazards and positive/negative economical effects of their work. Without thinking about them in the broad context, we will only be soulless cogs in the machine with intent and function unknown to us. Having this in mind, we decided to organise a summer school for young scientists active in nanosciences, at which they could get a complex review of current hot topics in the field and the applications in photovoltaics, plasmonics and environmental and security areas as well as discuss nanosciences in historical and societal context. We applied for and got support for this idea from the NATO Science for Peace Division.

The school was one of dissemination events of the FP7 project entitled 'Photonic Applications of Nanoparticle Assemblies and Systems' (acronym PHANTASY). The members of the consortium are Tyndall National Institute, Cork (Ireland); University of Erlangen–Nurnberg (Germany); University of Pardubice (Czech Republic); Ioffe Institute (Russia); and NIMS Tsukuba, Japan. The objective of the joint research programme is to improve the present level of technology in order to produce advanced structures that include colloids, nanoparticles, oxides and metals and to gain a fundamental understanding of their properties with the aim of using them for manipulation of light on the nanoscale.

The individual chapters in this book are based on lectures and round-table discussions presented at the NATO Advanced Study Institute 'Nanomaterials and Nanoarchitectures', which took place at Cork in July 2013. Chapter 1 (Fudouzi) describes biomimetics, which is the imitation of the models, systems and elements of nature for the purpose of solving human problems. Biomimetics has given rise to new technologies inspired by biological solutions at macro- and nanoscales. Chapter 2 (Aroca et al.) analyses ultrasensitive detection techniques using plasmon-enhanced spectroscopy, surface-enhanced Raman scattering (SERS), surface-enhanced fluorescence (SEF) and shell-isolated nanoparticle-enhanced fluorescence (SHINEF). Chapter 3 (Ariga et al.) demonstrates a new concept in nanotechnology, which allows manual control of nano-/molecular phenomena and functions by macroscopic mechanical force such as hand motions. Chapter 4 (Luby et al.) is based on the round-table discussions where Prof Luby acted as the moderator. It places nanoscience and nanotechnology in both historical and societal contexts. Chapters 5, 6 and 7 (Lvov et al., Pemble and Bardosova, Frumar et al.) outline properties of promising nanomaterials and related composites. Lvov et al. describe halloysite clay nanotube composites and their applications. Pemble and Bardosova discuss colloidal photonic crystals and the use of atomic layer deposition to infill them in order to tune the refractive index contrast in the material. Frumar et al. review and discuss crystalline, nanocrystalline and glassy and amorphous chalcogenides and their thin films and fibres, their preparation, structure, properties, changes and applications in optics, optoelectronics and electronics, data storage and sensors with accent on recent data and progress. Chapter 8 (Giesen et al.) describes plasmonic sensing strategies and concludes that plasmonic structures may lead towards the detection of chemical and catalytic events down to the single molecule level. Chapter 9 (Romanov) analyses light propagation in two photonic–plasmonic architectures. Chapter 10 (Oliveira et al.) describes the preparation and properties of solid organised films and their applications in sensing and biosensing.

Cork, Ireland — Maria Bardosova
Pardubice, Czech Republic — Tomas Wagner

References

[1] Drexler KE (1981) Molecular engineering: an approach to the development of general capabilities for molecular manipulation. Proc Natl Acad Sci 78(9):5275–5278
[2] Drexler KE (1986) Engines of creation. The coming era of nanotechnology, Donbleday. ISBN 0-385-19973-2
[3] Drexler KE (1992) Nanosystems: molecular machinery, manufacturing and computation. Wiley. ISBN 0-471-575474
[4] Toumey Ch (2008) Reading Feynman into nanotechnology: a text for a new science. Techne 12(3 Fall 2008):133–168
[5] Ulam S (1958) Tribute to John von Neumann. Bull Amer Math Soc 64:1–49
[6] Drexler KE (2013) Radical abundance: how a revolution in nanotechnology will change civilisation. PublicAffairs. ISBN 978-1-61039-113-9
[7] Kuhn TS (1962) The structure of scientific revolutions. University of Chicago Press. ISBN 9780226458113
[8] Drexler KE, Smalley RE (2003) Point-counterpoint. Nanotechnology. Chem Eng News 81(48):37–42
[9] http://www.nanotechproject.org/cpi/. Retrieved 28 Sept 2014
[10] Tiny ingredients. Big risks. Nanomaterials rapidly entering food and farming. Report written by Ian Illuminato. Friends of the Earth. May 2014 (The findings have been disputed by the producers affected and their statements can be accessed via [9])
[11] Trouiller B, Reliene R, Westbrook A, Solaimani A, Schiestl RH (2009) Titanium dioxide nanoparticles induce DNA damage and genetic instability in vivo in mice. Cancer Res 69(22):8784–8789
[12] Magdolenova Z, Collins A, Kumar A, Dhawan A, Stone V, Dusinska M (2014) Mechanisms of genotoxicity. A review of in vitro and in vivo studies with engineered nanoparticles. Nanotoxicology 8(3):233–278
[13] http://www.kurzweilai.net/chatbot-eugene-goostman-passes-turing-test-warwick-claims. Retrieved 28 Sept 2014
[14] Turing AN (1950) Computing machinery and intelligence. Mind 59:433–360
[15] www.independent.co.uk/news/science/stephen-hawking-transcendence-looks-at-the-implications-of-artificial-intelligence–but-are-we-taking-ai-seriously-enough-9313474.html. The Independent, 1 May 2014. Retrieved 1 Sept 2014

Contents

Chapter 1
Colloidal Photonic Crystal Films: Fabrication and Tunable Structural Color and Applications

Hiroshi Fudouzi

Abstract Colloidal photonic crystals have been attracting much attention due to their novel use as 3D photonic crystals and tunable structural color. The tunable structural color by swelling and strain is demonstrated on examples of opal composites. In addition, a high quality opal film coating process is reported.

1.1 Introduction

1.1.1 Opals and Structural Color

Opal gemstones have been utilized in jewelry making since ancient Greek period [1]. Figure 1.1a shows a natural opal gemstone illuminating due to iridescent structural color. The coloration mechanism was investigated with a scanning electron microscope in the middle of 1960s. A mineralogist in Australia, Sanders reported that natural opal gemstones consist of ordered array structures of mono-dispersed silica particles [2]. Afterwards silica particles were produced by the Stöber method [3]. The method produces mono-dispersed spherical amorphous silica particles from hydrolysis and condensation reactions of tetraethoxysilane. The sedimentation of mono-dispersed silica particles over several months forms an ordered array structure. Figure 1.1b shows a piece of synthetic silica opal gemstone produced by the Kyocera Company. Both natural and synthetic opal gemstones show iridescent colors. The color caused by colorless silica colloids is known as structural color. When the array of periodic length is approximately half of the wavelength of visible light, a specific wavelength of light selectively reflects due to Bragg diffraction. In addition, iridescence or rainbow colors arise from the incident angle due to the random orientation of the domain microstructure of silica particle arrays.

H. Fudouzi (✉)
Photonic Materials Unit, National Institute for Materials Science, Tsukuba, Japan
e-mail: FUDOUZI.Hiroshi@nims.go.jp

M. Bardosova, T. Wagner (eds.), *Nanomaterials and Nanoarchitectures*, NATO Science for Peace and Security Series C: Environmental Security, DOI 10.1007/978-94-017-9921-8_1

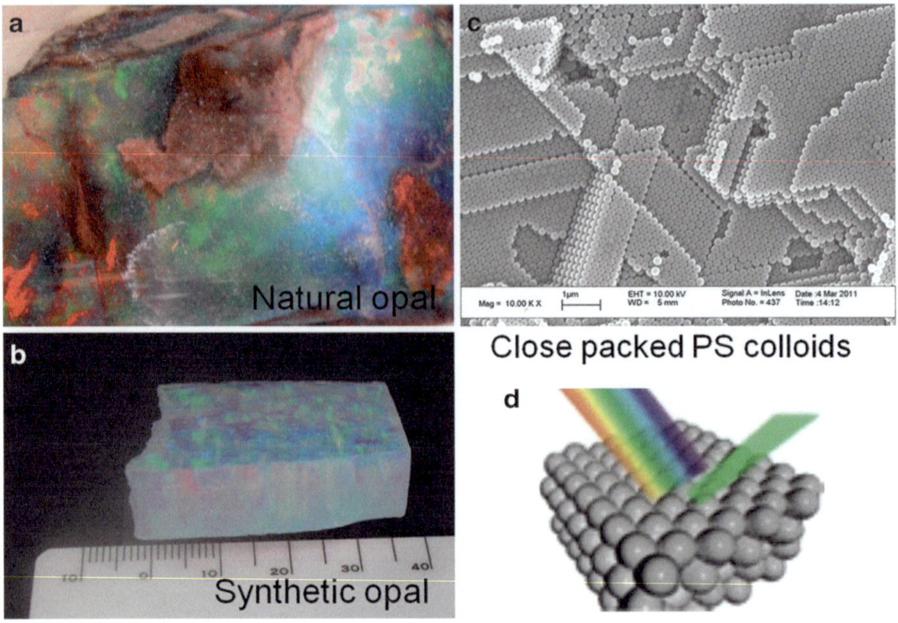

Fig. 1.1 Structural color in opals caused by Bragg diffraction of visible light. (**a**) Natural opal gemstone displayed at the Ibaraki Nature Museum, Japan, (**b**) A synthetsis silica opal produced by the Kyocera, (**c**) SEM image of cubic close packed, ccp, polystyrene spheres of 200 nm diameter. The ccp (111) planes are parallel to the substrate. (**d**) Illustration of Bragg diffraction of only a specific wavelength light from incident white light

Since the late 1990, synthetic opals and related structures have attracted much attention as three dimensional photonic crystals and monochromatic structural color materials. More detail is given in review papers and books [4–8]. In this chapter, we shall start from high quality opal films with crystal planes oriented on the substrate. Figure 1.1c shows a scanning electron microscope, SEM, image of close-packed polystyrene, PS, particles on a silicon wafer. Here, mono-dispersed 200 nm PS particles form a three dimensional periodic submicron structure of cubic close packing, ccp. From the microstructure in the SEM image, the ccp (111) plane of the particle array is parallel to the substrate. Figure 1.1a, d show structural color due to the diffraction of visible wavelength light from the stacking of ccp (111) planes on the substrate in vertical direction. The periodic ccp (111) planes cause the monochromatic structural color of the high quality opal film.

The structural color depends on the refractive index, tilting angle and distance between the ccp (111) planes. The reflected wavelength λ is expressed by combining Bragg's equation with Snell's law [9] as below:

$$\lambda = 2d_{111}\sqrt{\left(n_{\mathrm{eff}}^2 - \sin^2\theta\right)} \tag{1.1}$$

Here d_{111} is the distance between the ccp (111) planes, n_{eff} is the average refractive index of the opal film and θ is the angle of incidence.

1.1.2 Opal Films as Photonic Crystals

From a viewpoint of bottom-up nanotechnology, nanostructured colloidal crystals can be fabricated via a simple and low cost process. A large number of research papers on opal photonic crystals have been published in the past decade [12–18]. In theory, an inverse opal structure with high refractive index contrast, larger than 2.8, has a full photonic band gap as a 3D photonic crystal, PhC [10]. A milestone achievement was the design of an inverse silicon opal structure with a full band gap by Norris's group in 2001 [11]. However, there are many technological issues that must be addressed in order to apply opal films for 3D PhC devices compared with other top down approaches, such as EB lithography. Recently, colloidal crystals have attracted increasing attention due to the structural color phenomenon, Bragg diffraction peak and tuning stop band of 1D photonic crystals.

Figure 1.2 shows the photonic band structure of cubic close packed silica particles in air. The PBG structure of the opal film was calculated by using the Band SOLVE (Rsoft Corp, USA) based on Plane Wave Expansion algorithm. The reflectivity of the opal film was recorded by using an optical-fiber spectrometer (Ocean Optics, USA). The probing light was set normal to the surface of the opal films. As shown in Fig. 1.2a, two band gaps appear from the (111) and (100) planes corresponding to L point and X point, respectively. Unfortunately, there is no full photonic band gap to prohibit the presence of light in all directions. In this chapter, we will focus on opal films as 1D photonics crystals, especially the Bragg diffraction peak of ccp (111) direction or stop band at L point, because it is well known that the ccp (111) plane in opal films is preferentially oriented compared with the (100) or (110) planes.

Figure 1.2b shows the first Brillouin zone ΓL direction of the opal film. The measured Bragg peak of the silica opal film is in reasonable agreement with the theoretical stop band gap. The stop band gap at L point is caused by Bragg diffraction from the ccp (111) planes. From a photonic crystal viewpoint, the

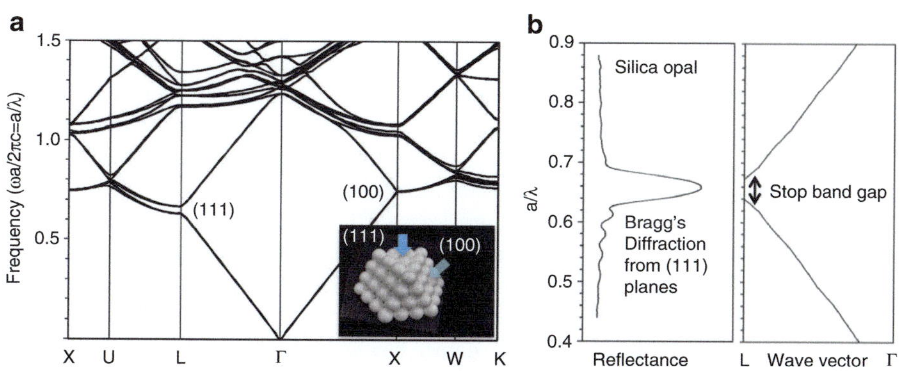

Fig. 1.2 Photonic band structure of a silica opal film. (**a**) Photonic band diagram for cubic close packed silica particles in air, (**b**) Comparison of the theoretical calculation near ΓL direction and experimental Bragg diffraction from (111) colloidal planes of the silica opal film

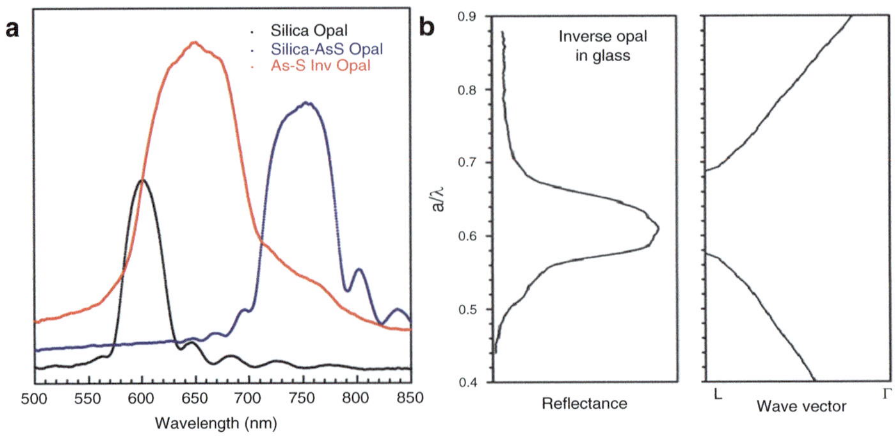

Fig. 1.3 Photonic crystal film made of inverse opal structure chalcogenide (As-S) glass. (**a**) Bragg diffraction peaks from silica opal, silica opal with As-S glass and As-S inverse opal films. (**b**) Comparison of the inverse opal As-S structure theoretical calculation near ΓL direction and experimental Bragg diffraction

structural color of a silica opal is mainly caused by this stop band gap. Furthermore, opal films and related nano-structures show tunable structural color and tunable stop bands, which will be described in detail later. When there is low contrast of refractive index, the Bragg diffraction peak becomes sharp, i.e. the stop band gap width narrows. High intensity and sharp peaks are desirable for sensing indicator applications and vivid structural color.

The stop band gap is designed by material and structure. Figure 1.3 shows the expanded stop band gap of a nano-structured material based on an opal film. Here chalcogenide glass, AsS, was chosen as a high refractive index material. In this case $n_{As30S70}$ is about 2.26. Figure 1.3a shows Bragg diffraction peaks from silica opal, silica opal with AsS and AsS inverse opal [19]. The silica opal film assembled with mono-dispersed silica particles of 280 nm diameters. Silica-AsS opal film has the void spaces filled with $As_{30}S_{70}$ glass. The inverse opal As-S structure was achieved by etching of unprotected silica particles in HF acid solution. The Bragg peak maxima were found near 602 nm for silica opal film, near 769 nm for silica-AsS opal film and near 658 nm for As-S inverse opal structure. The peak position depends on the average refractive index. In contrast, the width of the peak is influenced by the contrast of refractive index and the crystal lattice structure.

Figure 1.3b shows the first Brillouin zone ΓL direction of the inverse opal structure made of chalcogenide glass and the measured Bragg diffraction peak. Both are in reasonable agreement with the stop band gap in the theoretical calculation. The stop band gap is wider than that of silica opal in air as shown in Fig. 1.2b. The ccp (111) planes of AsS inverse opal structure have a more superior function as a 1D photonic crystal. A wider stop band gap is applicable for band pass optical filters, dielectric mirrors, resonant cavities and waveguides.

1.2 Fabrication of High Quality Opal Photonic Crystal Films

1.2.1 Opal Films Fabricated by Convective Self-Assembly

A key issue for industry is the scaling-up of the processing of high-quality colloidal crystal films. From a viewpoint of bottom-up nanotechnology, nanostructured colloidal crystals can be fabricated in a simple and low cost process. Simple preparation techniques based on convective self-assembly have been developed and widely used in this research field. The most widely used processes are evaporative driven self-assembly as shown in Fig. 1.4. The basic and fundamental mechanism of the process was investigated in the early 1990s [20]. Convective flow and capillary forces play an important role in the formation of high-quality colloidal crystal films. Self-assembly processes include a withdrawal method [21] as shown in Fig. 1.4a and a vertical deposition method (the so-called Colvin method [22] as shown in Fig. 1.4b. Both methods are based on the evaporative self-assembly of colloidal particles as shown in Fig. 1.4c. Convective flow and capillary forces play an important role in the formation of high-quality colloidal crystal films [23]. Recently, reducing cracks in opal films for large areas was reported for fabricating inverse opal films [24]. In this field, the evaporative self-assembly process is regarded as a standard procedure; colloidal particles are self-assembled to form ccp (111) colloidal crystal films in mild experimental conditions.

From a technological application viewpoint, one of the key issues is the development of a coating technique for the deposition of a high-quality colloidal crystal film onto a solid substrate at low cost for large areas. Evaporative self-assembly of

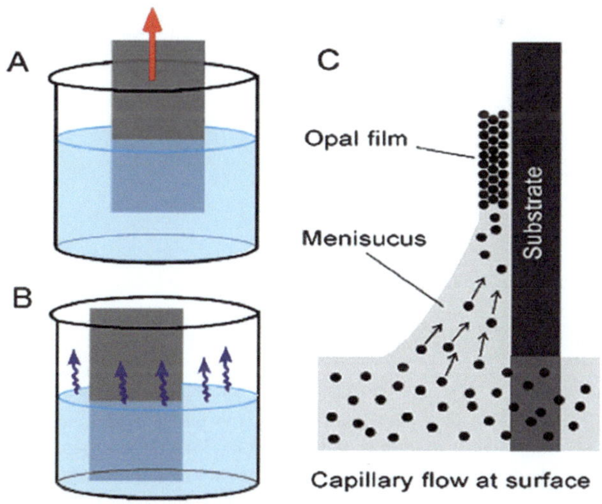

Fig. 1.4 Opal films coating process by evaporative self-assembly, (**a**) Withdrawal of a substrate with hydrophilic surface from colloidal suspension, (**b**) Vertical deposition on a substrate by drying colloidal suspension, (**c**) Mechanism of self-assembly by capillary flow at the meniscus

colloidal particles is a strong candidate as a technology typical coating technique for colloidal crystal films [25]. However, alternative self-assembly approaches were proposed. In the next section, we show an opal film formation method based on the phase transition from a concentrated colloidal suspension.

1.2.2 Opal Film Growth Under Silicone Oil

Figure 1.5 shows a phenomenon of opal film growth from aqueous PS colloidal suspension under a silicone liquid cover layer [9]. The mechanism of opal film formation is a little different from a conventional evaporative self-assembly. Polydimethyl silicone oil with a low viscosity of 10 cSt is used as the cover layer. the silicone oil has low density of 0.936 g/cm^3 and low surface tension of 20.1 mN/m. Silicone oil is also hydrophobic and chemically and thermally stable. Wettability of the liquid surface is expressed by the spreading coefficient, S. In the case of silicone

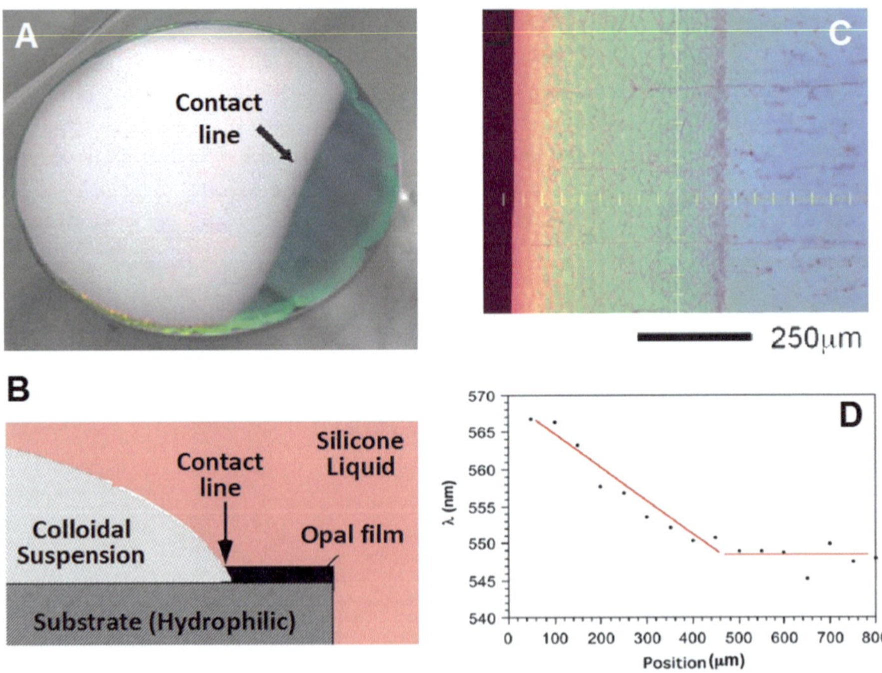

Fig. 1.5 Crystal growth of an opal film from suspension under a silicone liquid layer. (**a**) The color change from *milky white* to *green* at the contact line on a silicon wafer. (**b**) A cross sectional illustration of opal film formation. (**c**) Optical microscope image near the contact line and crystalline growth area. (**d**) Bragg diffraction peak position plotted against the distance from the contact line

oil on pure water, S is larger than 0. This means the silicone oil perfectly wets the water surface, same as PS colloidal suspension.

Figure 1.5a shows a film made by this method of 202 nm PS on a silicon wafer [25]. The milky white area is the colloidal suspension and the green area is opal film. The colloidal crystal growth occurs at a contact line indicated by the arrow. Figure 1.5b shows the cross sectional image of the contact line under crystallization. Control of the interface growth rate, afforded by the silicone oil immersion method, is an important aspect in forming high quality opal films. By detecting Bragg diffraction, it is estimated that the phase transition from non-crystalline to crystalline occurred crossing at the contact line. To understand the process of crystal growth at the contact line, we investigated microstructure at the contact line using an optical microscope equipped with a spectroscopy system [26].

Figure 1.5c shows the optical microstructure image at the contact line during crystal growth of the opal film [27]. The transition from non-crystalline to crystalline phase is observed within a region between the disordered colloidal suspension and the colloidal crystal film. Within this region, that spans a distance of about 400 μm, the lattice of the colloidal crystal reduces until it transitions to a close-packed structure. In addition, steps, terraces and triangular microstructures are also observed in this region between the suspension (black area) and the opal (green area). The structural color continually changes from red to green. Figure 1.5d shows the peak position of Bragg diffraction as a function of position [27]. The peak position shifts ca 20 nm towards shorter wavelengths with an increasing distance from the contact line. A non-close packed colloidal crystal is formed at first, which is then changed to a close packed structure.

A low cost and large area opal film coating is important for mass production in wide applications. The growth speed at the interface decides the rapidity of growth of the colloidal crystal. The growth speed at the interface, in the middle stage of the opal formation was estimated as 0.55 μm/s. However, the speed was not constant. In the final stage, the speed reduced and the opal film became thicker. For improving the opal film formation, we have proposed a vertical crystal growth method based on the under silicone oil method.

1.2.3 High Quality Opal Film Formation by the Silicone Oil Method

Figure 1.6 shows a vertical coating system enabling growth of opal film under a silicone liquid covering layer. A high quality opal film was coated on a polyethylene terephthalate, PET, sheet with a thickness of 50 μm in thickness. The surface of the PET sheet is modified from hydrophobic to hydrophilic to be wetted with PS colloidal suspension. The PET sheet is mounted on a supporting substrate and immersed in a container of PS colloidal suspension as shown in Fig. 1.6a. The

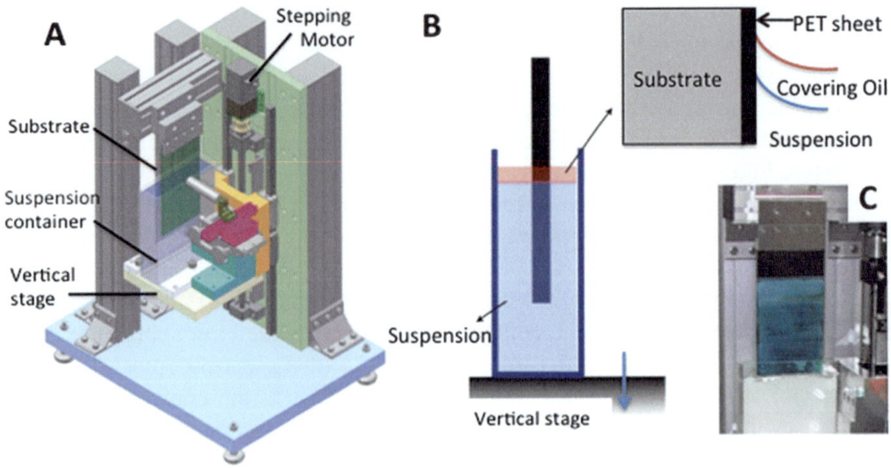

Fig. 1.6 Vertical coating opal films by crystal growth under a silicone liquid layer. (**a**) Schematic drawing of the equipment. The substrate and colloidal suspension container are fixed to the lifting stage. (**b**) Scheme and photo of the opal film coating system. (**c**) Black PET sheet coated with opal film made of 200 nm PS particles

container fixed on a vertical stage moves slowly down along the vertical stage, with a typical speed of 1 μm/s at room temperature. The detailed experimental setup is shown in Fig. 1.6b. PS colloidal suspension is covered with silicone oil in the container. Thus an opal film is grown under the silicone oil layer. Figure 1.6c shows an opal film deposited on a black PET sheet mounted on a glass substrate, 200 mm by 100 mm.

Using this vertical system, the opal film coats all area uniformly. Figure 1.7 shows the opal film coated on PET sheet after fixing treatment. The opal film shown in Fig. 1.6c consists of close packed colloidal particles without additional bonding between them. Because it is fragile, the voids between the close packed PS particles were filled with polydimethylsiloxane elastomer, Sylgard 184, Dow Corning. The PET sheet was then peeled off the supporting glass substrate. Figure 1.7a shows a bright and uniform green structural color indicating a high quality opal film with over 150 cm^2 area on the black colored PET sheet. The opal film is strongly adhered to the upper surface of PET sheet. The opal film with PET sheet shows robust and flexible properties in engineering applications described in the final section. Figure 1.7b shows a cross sectional image of the opal film by SEM. The cubic close packed (111) planes form an array of 31 layers. By measurement of Bragg peaks at multiple locations, we confirmed that the diffraction peaks had almost identical wavelengths and intensities independent of position, i.e. the coated opal film exhibits high quality and uniform film thickness. For use in engineering applications, the meter scale opal film is required for scalable mass production process. Thus we are also developing the roll-to-roll coating process for industrial use in further research.

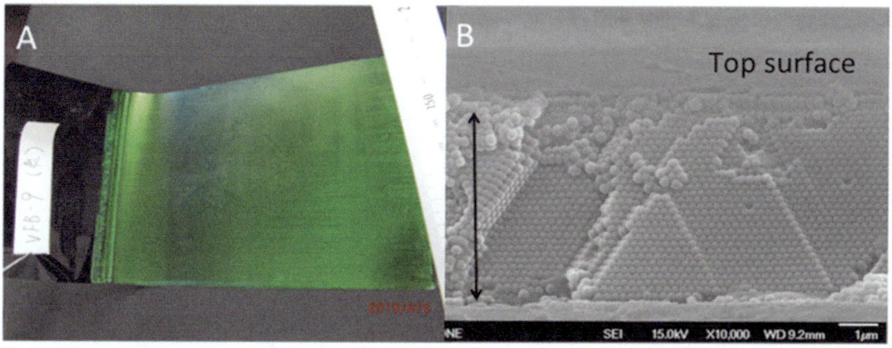

Fig. 1.7 High quality opal film coated by vertical coating equipment shown in Fig. 1.6. (**a**) *Green* structural color opal film coated on the *black* colored PET sheet of 100 mm in width. (**b**) A cross sectional SEM image of the opal film made of 31 stacking layers of 200 nm PS particles

1.3 Opal Photonic Crystals with Tunable Color

1.3.1 Soft Materials Based on Colloidal Crystals

One of the hot topics in colloidal crystal research is tuning structural color or tunable stop band gap [18]. Figure 1.8 shows a classification of colloidal photonic crystals into three different types, i.e. opal composite, inverse opal and traditional colloidal crystal. Figure 1.8a shows the opal composite made of cubic close packed particles embedded in an elastomer: illustration and an SEM image of opal composite film. Many pioneering papers were reported on opal composites of ccp (111) planes focusing on tuning the internal space of d_{111} [28–32]. Using opal film as a template, an inverse opal structure can be produced which exhibits porous morphology and tunable color as shown in Fig. 1.8b. Inverse opal hydrogels have potential applications in chemical sensors [33, 34] and elastomer inverse opal films can be used for mechanical fingerprinting [35].

Figure 1.8c shows a colloidal crystal embedded in a gel or elastomer. Diffraction of light by ordered suspensions is well known [36]. Non-close-packed colloidal crystals are formed in deionized water and exhibit iridescence, i.e. structural color. In the early 1990s, Asher et al. reported intelligent or smart sensing applications by tuning the lattice of colloidal crystals [37]. The lattice spacing of the colloidal crystal is tuned by the temperature-induced phase transition in the hydrogel. As a result, the diffracted wavelength from the colloidal crystal is thermally tunable across the entire visible spectrum. The color of hydrogel colloidal crystals is also sensitive to pH and ion concentration [38, 39], and colloidal crystals embedded in hydrogels can serve as mechanical sensors to measure strain due to uniaxial stretching or compression [40]. There have been a wide variety of studies reported on colloidal crystals of soft materials; electrically tunable structural color [41], tunable laser

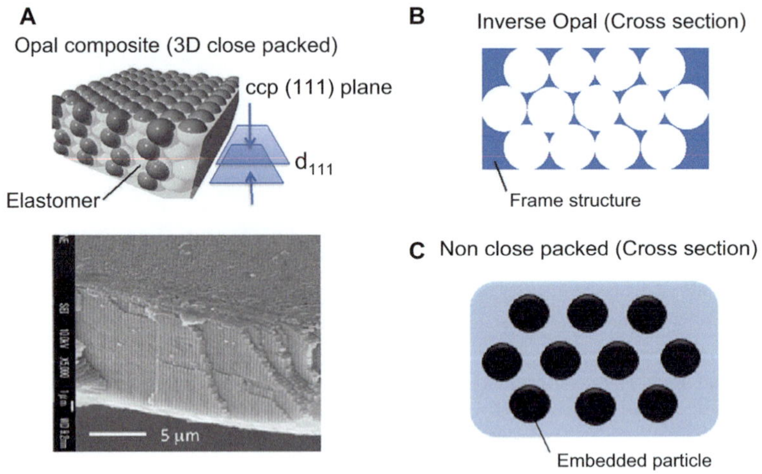

Fig. 1.8 Three structure types of colloid-based soft materials, which exhibit tunable structural color by diffracting visible light: (**a**) opal composite made of 3D close-packed colloidal spheres bonded by an elastomer, (**b**) inverse opal made of a soft-material frame structure – here opal acts as a template for the soft material, (**c**) non-close-packed colloidal crystal embedded in a soft material, typically a hydrogel or elastomer

emission [42], structural color printing [43] and tuning and fixing structural color by magnetic field [44, 45].

In the next section, the tuning of structural color of the opal composite film shown in Fig. 1.8a is demonstrated by two types of stimuli. The soft material film consists of closely packed PS particles and a polydimethylsiloxane (PDMS, Sylgard 184) elastomer. The expansion or compression, e.g. due to swelling of liquids [46] or due to mechanical strain by external force [47], reversibly tunes the spacing between the ccp (111) planes, d_{111}, in opal composite films.

1.3.2 Tuning Structural Color by Swelling

Figure 1.9 shows how an opal composite film made of 175 nm PS particles changes its color after being immersed in a hydrophobic liquid, octane [48]. The opal film in dry condition has the lattice spacing d_1. In octane liquid, PDMS swells without dissolving and the lattice spacing expands to d_2. This expansion shifts the Bragg diffraction peak to a longer wavelength due to increasing the volume of elastomer by absorbing octane. As shown in the photograph, the coloration of the opal film exhibits blue color at the dry area and red color at the wetting area. Evaporating the octane recovers the initial blue color. This wavelength shift depends on the swelling liquid, and can be changed continuously, e.g. by using silicone oils. The

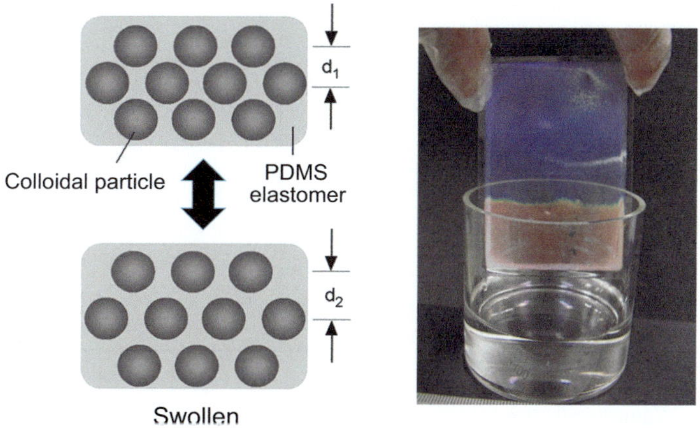

Fig. 1.9 Reversible structural color changing by swelling phenomena. A glass substrate coated with a soft opal film was dipped in a hydrophobic liquid, octane. In the photograph (*right*), the dry area has the initial *blue* color, and the wet area is *red*. The color is reverted from *red* to *blue* by evaporating the octane. The model (*left*) shows how the spacing between colloidal particles is changed by the swollen PDMS elastomer from d_1 to d_2 (Copyright 2011, National Institute for Materials Science, IOP Publishing Ltd.)

shift decreases with the molecular weight of silicone oil, such that it can be used in a new type of colorimetric detection [46].

The structural color varies according to the solvents' dependence on the swelling ability of the PDMS elastomer. Quantitative analysis of the swelling phenomena enables to measure the peak shift of Bragg diffraction. In our previous work, an opal composite became an indicator for the swelling PDMS elastomer with the components of mixture of two liquids [49]. The peak shifts as a function of the mixing ratio of the solvents, i.e., methanol, ethanol, and propanol. Here, the peak position is proportional to the solvent concentration.

Figure 1.10a shows the peak shift for swelling with the components of the mixture of isopropyl alcohol and a small amount of pure water, less than 1.0 vol.%. The Bragg diffraction peak gradually shifts to lower wavelengths from increasing pure water ratio 0–1.0 vol.% in isopropyl alcohol. Because the hydrophobic PDMS elastomer doesn't swell with pure water, the swollen volume PDMS with IPA will reduce according to the quantity of water. Figure 1.10b shows the peak wavelength as a function of water concentration in isopropyl alcohol matrix liquid. A commercially available optical fiber spectrometer easily and quickly detects the volume concentrations of water of the order of 0.1 vol.% in isopropyl alcohol. The peak shift is about 8 nm for 1 vol.% change, and the linear relationship suggests the swelling of PDMS elastomer can be used to detect the presence of small concentrations of water in IPA. The opal composite has the potential to be used in a rapid analysis that employs a portable optical fiber spectrometer.

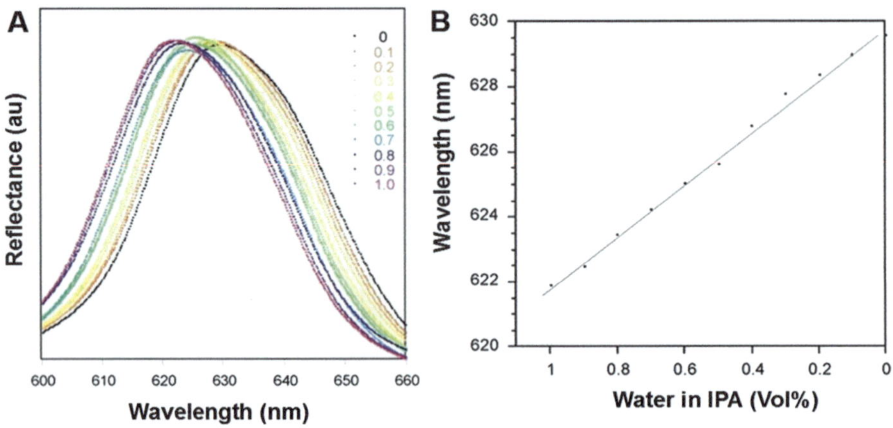

Fig. 1.10 Swelling PDMS elastomer with the components of mixture of isopropyl alcohol, IPA, and a small amount of pure water. (**a**) Reflectance spectroscopy of liquids in the IPA mixture. (**b**) Peak position as a function of the water volume concentration from 0 % to 1.0 %

1.3.3 Tuning Structural Color by Strain

Figure 1.11 shows the tunable and reversible structural color change by elastic deformation of soft opal film coated on a rubber sheet [48]. The structural color change from red to green was reversible and repeatable by applying and releasing mechanical stress.

By reflectance measurements, the Bragg diffraction peak shifted from 630 to 580 nm upon stretching. The stress in horizontal direction is expressed as $\sigma_x = E\varepsilon_x = E(\Delta L/L_0)$, where E is the Young's modulus, L_0 is initial length and $\Delta L = L - L_0$ is elongation. The horizontal stretching induces strain in the vertical direction, which can be expressed using the Poisson's ratio υ as $\varepsilon_z = -\upsilon\ (\sigma_x/E) = -\upsilon\ (\Delta L/L_0)$. The lattice spacing d is proportional to ε_z; therefore the change in Bragg diffraction peak, $\Delta\lambda$, shifts with the elongation of the rubber sheet as below.

$$\Delta\lambda \propto -\ \upsilon\ \ (\Delta L/L_0) \tag{1.2}$$

From the Eq. 1.2, the strain of the opal composite film was obtained from the peak of the Bragg diffraction.

We refer to a soft opal film fabricated on a rubber sheet capable of elastic deformation as "Photonic rubber sheet". The photonic rubber sheet is made of opal composite film situated on a supporting rubber sheet, not limited to PDMS elastomer. In addition, the specifics of photonic rubber sheet is that structural color can be observed by the naked eye but the Bragg diffraction peak can be also measured by spectroscopy. The photonic rubber sheets have practical applications as structural color indicators of stress and a new type of tension gauge. A shift of

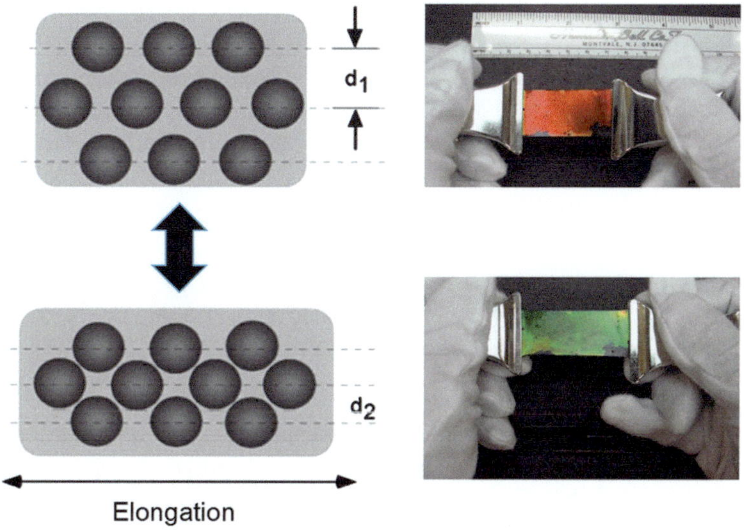

Fig. 1.11 Reversible structural color changing by mechanical deformation. The model (*left*) shows the reduction of the spacing between the ccp (111) planes from d_1 to d_2. The photographs (*right*) show a soft opal film on a PDMS rubber sheet changing color from *red* to *green* upon stretching. The structural color is restored by releasing the strain (Copyright 2011, National Institute for Materials Science, IOP Publishing Ltd.)

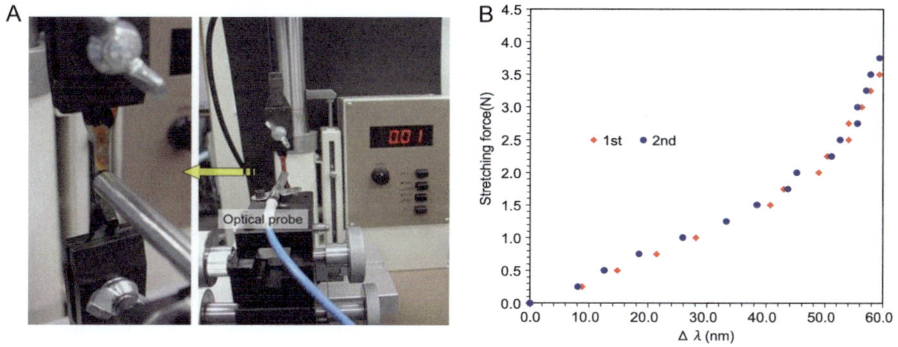

Fig. 1.12 Stress sensing of the photonic rubber sheet. (**a**) Photograph of a specimen on a tension machine. (**b**) Graph showing the relationship between the peak shift, $\Delta\lambda$, and stretching force for two measurements (Copyright 2006, Society of Photo-Optical Instrumentation Engineers)

Bragg peak indicates the tensile stress. Figure 1.12 shows the calibration of a stress sensor using the photonic rubber sheet [50].

A photonic rubber sheet was set on a tensile testing machine, shown in Fig. 1.12a. The optical fiber probe was placed at right angles. The incident light was aligned perpendicular to the ccp (111) plane of the opal composite in the photonic rubber sheet. The Bragg diffraction peak was located at 613 nm. Figure 1.12b shows the relationship between the shift of peak position measured initially and the position after the stress was applied *versus* the stretching force. The repetitive test was investigated between the original state and the stretched state. Based on this relationship, the photonic rubber sheet can indicate the tensile force without using the tension machine.

1.4 Potential Applications of Structural Color

1.4.1 Color Tunable Fiber Fabric

Polymer opals consisting of core-shell colloidal particles are amongst the candidates for this market. Baumberg et al. have been developing polymer opal fibers and opal composite sheets [51, 52]. Polymer opals consist of a hard polystyrene core, coated with a thin polymer layer containing allyl-methacrylate as a grafting agent, and a soft poly-ethyl acrylate outer shell. The polymer opal fibers are fabricated using the manufacturing technology of synthetic fibers. The fibers exhibit structural color with a spectrum stretch-tunable across the visible region. These structural color changes are caused by the decrease of the inter-planar distance during stretching. As the fibers become increasingly stretched, the diffraction peak is shifted to lower wavelengths. This process has the potential for easy scale-up for industrial mass production.

1.4.2 Structural Color for Printing and Displays

Structural color is applied in imaging technology. The concept of "Photonic ink or P-ink" was demonstrated as a prototype display device [53, 54]. The matrix sheet was made of a silica opal composite with polyferrocenylsilane, PFS organic solvent gel and cationic iron sites. This PFS organic gel is swollen with redox-controlled polarity of PFS chains by applied electric field. The structural color change was reversible, rapid (within sub-second) and of broad wavelength (the whole visible region). A different concept of "M-ink" was proposed using magnetic responsive colloidal magnetic particles of Fe_3O_4 [45]. This color changed due to the distance of colloidal particles' chains in a magnetic field. Using M-ink on a flexible polymer sheet, a prototype magnetic recording was reported. Magnetic tunable structural color shows a potential to overcome the limitations of the conventional magnetic recording technology.

1.4.3 Imaging Local Strain of Deformed Metal Plates

A joint team consisting of researchers from the National Institute for Materials Science, the Public Works Research Institute and the University of Hiroshima has been investigating a simple and low cost method to visualize strain damage for health monitoring of civil infrastructures, such as bridges, towers, buildings, tunnels, dams and highway roads. In a preliminary study, the team demonstrated a new visualization technique of strain deformation of metal plates and detecting strain during tensile tests [55] as described below.

The aluminum plates were set up on a tensile test system. Then a uniaxial strain was applied to the specimens. Figure 1.13a shows the deformed aluminum plate and local reflectance spectroscopy after tensile test. Structural color changes reveal a local strain distribution of deformation of the metal plate. The change of structural color was caused by uniaxial elongation. The photograph shows a plastic deformation at the neck area with strain concentrating near the dumbbell center. Figure 1.13b shows Bragg diffraction peak at the green area. In contrast, Figure 1.13c shows the peak at the red area. The peak position is shifted from 623 to 548 nm by elongation of the aluminum plate. This result demonstrates a local strain image of metal plate by changes in the opal composite film. This approach could be useful as a new strain gauge having a visual indicator to detect

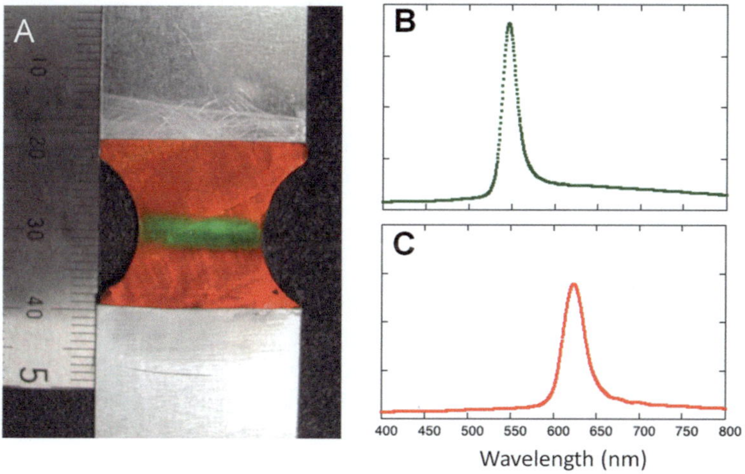

Fig. 1.13 Strain imaging of a deformed metal plate. (**a**) Photograph of opal composite film coated on PET sheet. Change of structural color by uniaxial elongation of aluminum plate. The structural color changed from *red* to *green* color at the center of dumbbell shape area, due to concentrating strain. (**b**) Reflectance spectroscopy at deformed area of *green* color. Diffraction peak was located at 548 nm. (**c**) Diffraction peak at *red* color area was located at 623 nm (Copyright 2012, Society of Photo-Optical Instrumentation Engineers)

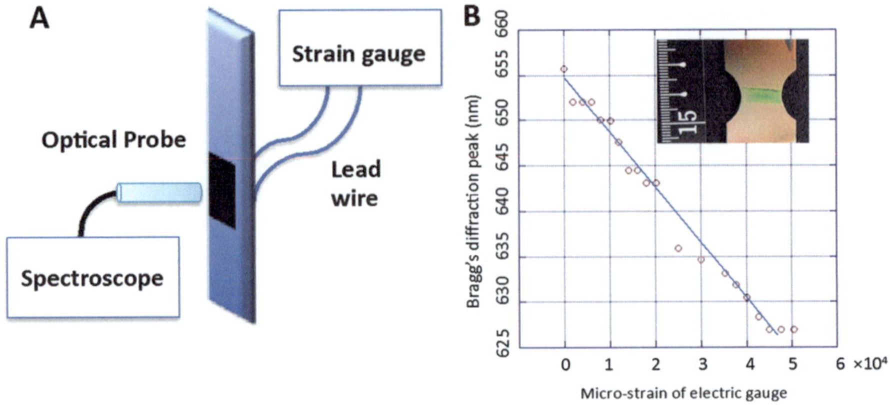

Fig. 1.14 Uniaxial elongation due to plastic deformation of metal plate revealed by a structural color strain indicator. (**a**) Setup showing the use of two types of strain sensors, an optical probe spectroscope and an electric strain gauge at the backside of the plate. (**b**) The relationship between micro-strain and Bragg diffraction peak shows a linear inverse proportion. The inserted photo shows a deformed aluminum plate after the tensile test (Copyright 2012, Society of Photo-Optical Instrumentation Engineers)

mechanical deformation. The opal composite film was coated onto a PET sheet as shown in Fig. 1.7a. The PET sheet with opal composite film was bonded to the flat surface of an aluminum plate with a cyanoacrylate adhesive. The black pigmented polyethylene terephthalate, PET sheet plays two functions - as a supporting substrate of flexible colloidal opal composite film, and an absorbing layer for transmitted light. This method is a simple and practical way to coat an opal composite film onto target specimens in civil engineering applications.

Figure 1.14 shows the peak shift of Bragg diffraction in an opal composite film used to measure the elongation of a metal plate during tensile test. Using conventional electric strain gauge, the structural color strain indicator was investigated for comparison and to examine its potential as a strain gauge. As shown in Fig. 1.14a, two types of gauges were adhered to each surface of aluminum. By applying uniaxial strain to the aluminum plate, peak shift and micro-strain were measured at the same time. Figure 1.14b shows a plot of Bragg diffraction peak by this new method and micro strain by the conventional method. This result suggests that the correlation seen was relatively stable.

1.5 Conclusions

Colloidal photonic crystal films show response to external stimuli, such as swelling and mechanical strain. Films with tunable structural color have potential applications as new sensing materials and devices. Colloidal crystal films made of 3D arrayed polystyrene particles were infilled with an elastomer, polydimethylsiloxane (PDMS). Colloidal photonic crystal films were grown under silicone oil. High quality opal films were formed on glass and silicone substrates, black color rubber and PET sheets. We demonstrated the change of structural color by swelling and by strain. The change of structural color on the rubber sheet by stretching is reversible and repeatable. In contrast, the color change by plastic deformation of PET sheet is irreversible. The strain of the metal deformation is visualized as a change of the structural color. One application is monitoring of structural changes in civil engineering.

Acknowledgments This work is financially supported by AMATA metal works foundation, IKETANI science foundation, KUMAGAI foundation for science and technology and Grants-in-Aid for Scientific Research, Kaken-hi Kiban C18560682, B23360313 and B26289139 from The Japan Society for the Promotion of Science, JSPS and a feasibility study in Adaptable and Seamless Technology Transfer Program 2011 through Target-driven R&D in Japan Science and Technology Agency, JST.

References

1. Eckert AW (1997) The world of opals. Wiley, New York
2. Sanders JV (1968) Diffraction of light by opals. Acta Cryst A24:427
3. Stöber W, Fink A, Bohn E (1968) Controlled growth of monodisperse silica spheres in the micron size range. J Colloid Interface Sci 26:62
4. Xia Y, Gates B, Yin Y, Lu Y (2000) Monodispersed colloidal spheres: old materials with new applications. Adv Mater 12:693
5. López C (2003) Materials aspects of photonic crystals. Adv Mater 15:1679
6. Busch K, Lölkes S, Wehrspohn RB, Föll H (eds) (2006) Photonic crystals: advances in design, fabrication, and characterization. Wiley, Weinheim, New York
7. Ozin GA, Arsenault AC (2008) Nanochemistry, 2nd edn. RCS Publishing, Cambridge, New York
8. Marlow F, Muldarisnur, Sharifi P, Brinkmann R, Mendive C (2009) Opals: status and prospects. Angew Chem Int Ed 48:6212
9. Fudouzi H (2004) Fabricating high-quality opal films with uniform structure over a large area. J Colloid Interface Sci 275:277
10. Vlasov YA, Bo XZ, Sturm JC, Norris DJ (2001) On-chip natural assembly of silicon photonic bandgap crystals. Nature 414:289
11. Joannopoulos JD, Johnson SG, Winn JN, Meade RD (2008) Photonic crystals: molding the flow of light, vol 2. Princeton University Press, Princeton, New York
12. Harun-ur-rashie M, Seki T, Takeoka Y (2009) Structural colored gels for tunable soft photonic crystals. Chem Rec 9:87
13. Furumi S, Fudouzi H, Sawada T (2010) Self-organized colloidal crystals for photonics and laser applications. Laser Photonics Rev 4:205
14. Aguirre CI, Reguera E, Stein A (2010) Tunable colors in opals and inverse opal photonic crystals. Adv Funct Mater 20:2565
15. Zhao Y, Zhao X, Gu Z (2010) Photonic crystals in bioassays. Adv Funct Mater 20:2970
16. Kim SH, Lee SY, Yang SM, Yi GR (2011) Self-assembled colloidal structures for photonics. NPG Asia Mater 3:25
17. Galisteo-Lõpez JF, Ibisate M, Sapienza R, Froufe-Pérez LS, Blanco Ú, Lõpez C (2011) Self-assembled photonic structures. Adv Mater 23:30

18. Yadong Y (ed) (2013) Responsive photonic nanostructures: smart nanoscale optical materials. RSC Publishing, Cambridge, New York
19. Kohoutek T, Orava J, Sawada T, Fudouzi H (2011) Inverse opal photonic crystal of chalcogenide glass by solution processing. J Colloid Interface Sci 353:454
20. Denkov ND, Velev OD, Kralchevsky PA, Ivanov IB, Yoshimura Y, Nagayama K (1992) Mechanism of formation of two-dimensional crystals from latex particles on substrates. Langmuir 8:3183
21. Gu ZZ, Fujishima A, Sato O (2002) Fabrication of high-quality opal films with controllable thickness. Chem Mater 14:760
22. Jiang P, Bertone JF, Hwang KS, Colvin VL (1999) Single-crystal colloidal multilayers of controlled thickness. Chem Mater 11:2132
23. Velikov KP, Velev OD (2006) Novel materials derived from particles assembled on liquid surfaces. In: Binks BP, Horozov TS (eds) Colloidal particles at liquid interfaces. Cambridge University Press, New York
24. Hatton B, Mishchenko L, Davis S, Sandhage KH, Aizenberg J (2010) Assembly of large-area, highly ordered, crack-free inverse opal films. Proc Natl Acad Sci U S A 107:10354
25. Innocenzi P, Malfatti L, Paolo Falcaro P (2013) Water droplets to nanotechnology: a journey through self-assembly. RSC Publishing, Cambridge, New York
26. Fudouzi H (2007) Novel coating method for artificial opal films and its process analysis. Colloids Surf A 311:11
27. Fudouzi H, Sawada T, Brush L (2010) Concentrated crystallization on closely packed colloidal crystal films from aqueous suspension under silicone oil. MRS Proc 1273:MM08-06
28. Yoshino K, Kawagishi Y, Ozaki M, Kose A (1999) Mechanical tuning of the optical properties of plastic opal as a photonic crystal. Jpn J Appl Phys Part 2 38:L786
29. Arsenault AC, Miguez H, Kitaev V, Ozin GA, Manners I (2003) A polychromic, fast response metallopolymer gel photonic crystal with solvent and redox tunability: a step towards photonic ink (P-Ink). Adv Mater 15:503
30. Fudouzi H, Xia YN (2003) Photonic papers and inks: color writing with colorless materials. Adv Mater 15:892
31. Rugge A, Ford WT, Tolbert SH (2003) From a colloidal crystal to an interconnected colloidal array: a mechanism for a spontaneous rearrangement. Langmuir 19:7852
32. Furumi S, Fudouzi H, Miyazaki HT, Sakka Y (2007) Flexible polymer colloidal-crystal lasers with a light-emitting planar defect. Adv Mater 19:2067
33. Takeoka Y, Watanabe M (2003) Template synthesis and optical properties of chameleonic poly(N-isopropylacrylamide) gels using closest-packed self-assembled colloidal silica crystals. Adv Mater 15:199
34. Lee YJ, Braun PV (2003) Tunable inverse opal hydrogel pH sensors. Adv Mater 15:563
35. Arsenault AC, Clark TJ, Von Freymann G, Cademartiri L, Sapienza R, Bertolotti J, Vekris E, Wong S, Kitaev V, Manners I, Wang RZ, John S, Wiersma D, Ozin GA (2006) From colour fingerprinting to the control of photoluminescence in elastic photonic crystals. Nat Mater 5:179
36. Hiltner PA, Krieger IM (1969) Diffraction of light by ordered suspensions. J Phys Chem 73:2386
37. Weissman JM, Sunkara HB, Tse AS, Asher SA (1996) Thermally switchable periodicities and diffraction from mesoscopically ordered materials. Science 274:959
38. Holtz JH, Asher SA (1997) Polymerized colloidal crystal hydrogel films as intelligent chemical sensing materials. Nature 389:829
39. Hu Z, Lu X, Gao J (2001) Hydrogel opals. Adv Mater 13:1708
40. Iwayama Y, Yamanaka J, Takiguchi Y, Takasaka M, Ito K, Shinohara K, Sawada T, Yonese M (2003) Optically tunable gelled photonic crystal covering almost the entire visible light wavelength region. Langmuir 19:977
41. Xia J, Ying Y, Foulger SH (2005) Electric-field-induced rejection-wavelength tuning of photonic-bandgap composites. Adv Mater 17:2463
42. Lawrence JR, Ying Y, Jiang P, Foulger SH (2006) Dynamic tuning of organic lasers with colloidal crystals. Adv Mater 18:300

43. Jiang P, Smith DW Jr, Ballato JM, Foulger SH (2005) Multicolor pattern generation in photonic bandgap composites. Adv Mater 17:179
44. Kim H, Ge J, Kim J, Choi S, Lee H, Lee H, Park W, Yin Y, Kwon S (2009) Structural colour printing using a magnetically tunable and lithographically fixable photonic crystal. Nat Photonics 3:534
45. Ge J, Hu Y, Yin Y (2007) Highly tunable superparamagnetic colloidal photonic crystals. Ange Chem Int Ed 46:7428
46. Fudouzi H, Xia Y (2003) Colloidal crystals with tunable colors and their use as photonic papers. Langmuir 19:9653
47. Fudouzi H, Sawada T (2006) Photonic rubber sheets with tunable color by elastic deformation. Langmuir 22:1365
48. Fudouzi H (2011) Tunable structural color in organisms and photonic materials for design of bioinspired materials. Sci Tech Adv Mat 12:064704–1
49. Fudouzi H, Sawada T (2007) Photonic crystal sensing for components of a liquid mixture with an optical fiber spectrometer. Proc SPIE 6767:676704–1
50. Fudouzi H, Sawada T (2006) Tuning stop band of soft opal film by deformation for strain sensing applications. Proc SPIE 6369:63690D-1
51. Wohlleben W, Bartls FW, Altmann S, Leyrer RJ, Finlayson CE, Spahn P, Snoswell DRE, Yates G, Kontogeorgos A, Haines AI, Hellmann GP, Baumberg JJ (2011) 3D bulk ordering in macroscopic solid opaline films by edge-induced rotational shearing. Adv Mater 23:1540
52. Finlayson CE, Goddard C, Papachristodoulou E, Snoswell DRE, Kontogeorgos A, Spahn P, Hellmann GP, Hess O, Baumberg JJ (2011) Ordering in stretch-tunable polymeric opal fibers. Opt Express 19:3144
52. Arsenault AC, Puzzo DP, Manners I, Ozin GA (2007) Photonic crystal full color displays. Nat Photonics 1:468
54. Puzzo DP, Arsenault AC, Manners I, Ozin GA (2009) Electroactive inverse opal: a single material for all colors. Angew Chem Int Ed 48:943
55. Fudouzi H, Sawada T, Tanaka Y, Ario I, Hyakutake T, Nishizaki I (2012) Smart photonic coating as a new visualization technique of strain deformation of metal plates. Proc SPIE 8345:8351S1

Chapter 2
Plasmonics and Ultrasensitive Detection

Ricardo F. Aroca

Abstract An approach for experimental design in plasmon-enhanced spectroscopy is discussed based on its basic elements: electromagnetic radiation, adsorbed molecule and the metal nanostructure. Optimization of the plasmon enhancement may be achieved by tuning the electromagnetic radiation to take advantage of resonances with molecules and nanostructures. For instance, when the excitation is in resonance with a molecular electronic transition, resonance Raman scattering is observed providing a very efficient scattering with cross section for vibrational transitions several orders of magnitude higher than normal Raman. Tuning the excitation of the nanostructure might depend on the degree of aggregation, or the properties of two and three dimensional array of fabricated nanostructures. Several examples of surface enhanced Raman scattering (SERS), SERS and surface enhanced fluorescence using shell isolated nanoparticles are presented. The experimental results illustrate the remarkable optical properties of metal nanoparticles which are governed by the excitation of localized surface plasmon resonances producing local enhancement of the electromagnetic field. However, each experiment is unique and requires a selection of the setting for each one of the three elements that would lead to the most efficient plasmon enhancement.

2.1 Introduction

The development of ultrasensitive detection techniques using plasmonics involves three fundamental components: the molecular system to be detected, a radiative plasmon excitation and the impinging electromagnetic (EM) radiation. When only the molecule and the EM radiation are involved we have the broad field of spectroscopy. According to IUPAC spectroscopy is: "*The study of physical systems by the electromagnetic radiation with which they interact or that they produce*", and "*Spectrometry is the measurement of such radiations as a means of obtaining information about the systems and their components*". When the third

R.F. Aroca (✉)
Department of Chemistry and Biochemistry, University of Windsor, Windsor, Canada
e-mail: raroca1@cogeco.ca

M. Bardosova, T. Wagner (eds.), *Nanomaterials and Nanoarchitectures*,
NATO Science for Peace and Security Series C: Environmental Security,
DOI 10.1007/978-94-017-9921-8_2

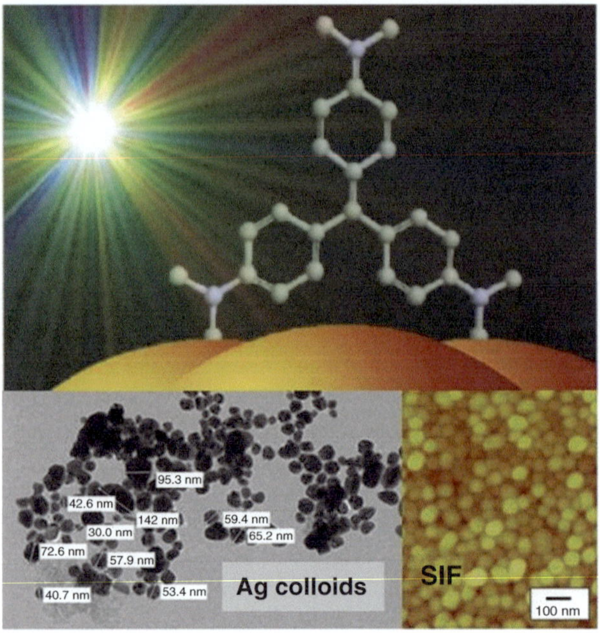

Fig. 2.1 The three components of plasmon enhanced spectroscopy

component – plasmon excitation – is added a new field is unwrapped: plasmon enhanced spectroscopy (PES). In this new field the role of the nanostructure that supports localized surface plasmon resonances (LSPR) [1] is to amplify the intensity of the frequencies that are characteristic of molecular systems. The enhancement of optical signal started with the discovery of surface enhanced Raman scattering (SERS) [2–4], although the first observation of amplification was reported earlier [5].

The trinity of plasmon enhanced spectroscopy is illustrated in Fig. 2.1 [6], that also shows two of the most common nanostructures used in plasmon enhanced spectroscopy: silver colloids (transmission electron microscopy) and the atomic force microscopy of silver island film (SIF).

Although the role of the surface plasmon resonances in the amplification of optical signals was pointed out very early on the development of SERS [4], the field grew with a focus on the enhancing nanostructure after the independent reports in 1997 of Kneipp et al. [7] and Nie and Emory [8], on the detection of the vibrational spectrum of a single molecule. The new focus was, and continues to be, on the fabrication and control of optical properties of metallic nanostructures for plasmonic enhancement [9] with the emphasis in the development of SERS substrates [10]. Fluorescence has been the reference for ultrasensitive detection due to large cross section and the clear detection of emission signals [11]. At room temperature cross section of the order of 10^{-17} cm^2/molecule is measured for aromatic molecules such as Rhodamine 6G. Single molecule spectroscopy is a well-developed multidisciplinary field [12] and fluorescence excitation provides

Table 2.1 Normal Raman scattering cross sections

Molecule	Vibrational wavenumber (cm^{-1})	Laser line (nm)	Cross section
Benzene liquid	992	514.5	3×10^{-29} cm^2sr^{-1} molecule^{-1}
Benzene gas	992	514.5	0.7×10^{-29} cm^2sr^{-1} molecule^{-1}
Cyclohexane liquid	1,267	488	0.5×10^{-29} cm^2sr^{-1} molecule^{-1}
CH_3Cl gas	725	514.5	0.2×10^{-29} cm^2sr^{-1} molecule^{-1}
CCl_4 gas	459	514.5	0.5×10^{-29} cm^2sr^{-1} molecule^{-1}
N_2 gas	2,331	514.5	0.4×10^{-30} cm^2sr^{-1} molecule^{-1}

Table 2.2 Resonance Raman cross sections

Molecule	Vibrational wavenumber (cm^{-1})	Laser line (nm)	Cross section
Nile Blue in water [16]	595	514.5	3.6×10^{-26} cm^2sr^{-1}molecule^{-1}
Nile Blue in ethanol [16]	1,645	514.5	1.2×10^{-25} cm^2sr^{-1}molecule^{-1}
Rhodamine 6G [17]	1,356	532	2.6×10^{-23} cm^2molecule^{-1}
Rhodamine 6G [17]	1,647	532	2.0×10^{-23} cm^2molecule^{-1}
Rhodamine 6G [17]	604	532	4.1×10^{-23} cm^2molecule^{-1}

superior signal-to-noise if the emission is collected efficiently [13]. However, the information rich vibrational spectrum collected from inelastic scattering, or Raman [14] effect, is very inefficient optical process, and typical Raman cross sections are in the 10^{-29} cm^2/molecule. Here we give a few examples in Table 2.1 taken from McCreery's book [15].

However, when the excitation is in resonance with electronic transition the resonance Raman scattering (RRS) is much more efficient and the cross section for vibrational transitions are several orders of magnitude higher as can be seen in Table 2.2.

The maximum absorption of Rhodamine 6G has been measured with a cross section $\sigma = 2.08 \times 10^{-16}$ cm^2molecule^{-1} at 530 nm [18]. Therefore, the cross section of highly fluorescent dyes would be in the 10^{-17} cm^2 molecule^{-1} region that we take as reference for single molecule detection (SMD). The normal Raman effect would need an enhancement of $\sim 10^{12}$ for SMD while the resonance Raman would only need an enhancement factor (EF) of $\sim 10^7$. In contrast, the remarkable optical properties of metal nanoparticles which are governed by the excitation of localized surface plasmon resonances (LSPR) producing local enhancement of the electromagnetic field by the plasmonic response are characterized by large absorption and scattering cross sections, orders of magnitude better than the best absorbing dyes. This is illustrated in Fig. 2.2, where the scattering and absorption of a silver sphere of 30 nm radius in water (refractive index $n = 1.33$) and the absorption for a corresponding sphere of gold are shown. The scattering cross sections are of the order of 10^{-10} cm^2. For particles small compared with

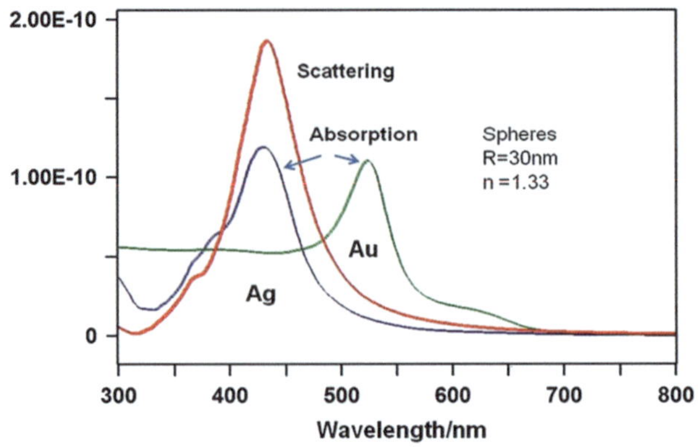

Fig. 2.2 Scattering and absorption of a silver sphere of 30 nm radius in water and the absorption for a sphere of gold

wavelength of excitation the corresponding scattering and absorption cross section can be calculated from (Bohren and Huffman [19], Chap. 5): $C_{sca} = \frac{8\pi}{3}k^4a^6\left|\frac{\varepsilon-\varepsilon_m}{\varepsilon+2\varepsilon_m}\right|^2$ and $C_{abs} = k\text{Im}\,[\alpha] = 4\pi ka^3\ \text{Im}\left[\frac{\varepsilon-\varepsilon_m}{\varepsilon+2\varepsilon_m}\right]$, where ε_m is the dielectric constant of the medium and a is the radius of the sphere. Notably, the efficiency of absorption, scaling as a^3, may be dominant for small nanoparticles. However, the scattering efficiency increases very rapidly with size, scales as a^6, and will dominate for larger scatterers. The latter, $C_{sca} > C_{abs}$ can be seen for a silver nanoparticle of $a = 30$ nm in Fig. 2.2. For particles small compared with the wavelength of the impinging radiation, $2a << \lambda$, the plasmon mode has the character of an electric dipole, which rapidly dephases (femtoseconds [20]). The LSPRs are collective resonances of the conduction electrons of metallic nanostructures driven by the electromagnetic field, the spectral positions of these resonances are strongly dependent on dielectric function of the material, particle size and shape [21]. It should be pointed out that the full width at half maximum (FWHM) of the absorption and scattering in Fig. 2.2 is about 75 nm, which in that part of the electromagnetic spectrum translates into more than 4,000 cm^{-1}, a wavenumber range that can encompass the full vibrational Raman spectrum of any polyatomic molecule. Considering the giant scattering cross section of the collective electron oscillations in LSPR, a plasmon-molecule coupling [22] through the near field with the plasmon resonance could provide the opportunity for observing molecular frequencies elastically scattered with the scattering efficiency of the nanostructure. The latter helps to understand Moskovits' working definition [23] of SERS or plasmon-enhanced Raman scattering: "*As it is currently understood SERS is primarily a phenomenon associated with the enhancement of the electromagnetic field surrounding small metal (or other) objects optically excited near an intense and sharp (high Q), dipolar resonance such as a surface plasmon polariton. The enhanced re-radiated dipolar fields excite the adsorbate, and, if the resulting molecular radiation remains at or near resonance with the* *enhancing object, the scattered radiation will again be enhanced (hence the most intense SERS is really frequency-shifted elastic scattering by the metal). Under appropriate circumstances the field enhancement will scale as E^4 where E is the local optical field*".

In practise, the maximum Raman enhancement occurs when the plasmon resonance extinction band of the metal nanostructure encompasses the excitation frequency and vibrational Raman spectrum [24]. Similar correlations are seen in surface enhanced fluorescence (SEF) [25]. Harnessing the enhanced and highly confined local fields in metallic nanostructures for plasmon enhanced spectroscopy, is by no means limited to Raman and fluorescence, plasmon supported spectroscopic effects include all linear and non-linear optical signals [26]. Finally, the LSPR providing the strong local fields depends on the dielectric function of the metal $\varepsilon(\omega)$. The optimal values of $\varepsilon(\omega)$ require a negative real part of the complex dielectric function with small damping, i.e. small imaginary part. Notably, silver and gold nanostructures fulfil $\varepsilon(\omega)$ requirements in the visible region of the electromagnetic spectrum and are the most common materials in plasmon enhanced spectroscopy. Notwithstanding the fact that plasmonic enhancement of other metals is actively pursued [27].

2.2 The Enhancement Factor (EF)

Probably, the most challenging task for plasmon-enhanced practitioners has been and still is: precise control and quantification of the enhancement effect. Correspondingly, there is extensive literature eon enhancement factors discussing definitions of the SERS EFs in the context of SERS applications [28]. More recent comprehensive reviews are also available [29, 30]. Experimentally, it has been clearly determined that aggregated nanoparticles are better enhancers than isolated nanoparticles. In colloidal SERS is common practise to induce aggregation to boost the enhancement factor [31]. In theory, intense electromagnetic fields and large electromagnetic enhancements are created when light illuminates metallic nanostructures with nanoscale gaps and fractals [32]. When light is concentrated on the nanometer scale it produces what is called a 'hotspot'. Recently, Cang et al. [33] were able to image the fluorescence enhancement profile of single hotspot of 15 nm. Klienman et al. [30] propose the following working definition for a hotspot: "*A junction or close interaction of two or more plasmonic objects where at least one object has a small radius of curvature on the nm scale. This structural motif has the ability to concentrate an incident electromagnetic field and effectively amplify the near field between and around the nanostructures.* The experimental confirmation of the hotspot permits one to distinguish two different regimes in plasmon enhanced Raman scattering. First, there is **average SERS EF** which is characteristic of well-defined SERS substrate, and is reproducible for any given molecular system. Second, there is a single molecule detection **regime in SERS,** where the active site empowering a SMD is a hot spot. These high EF sites are

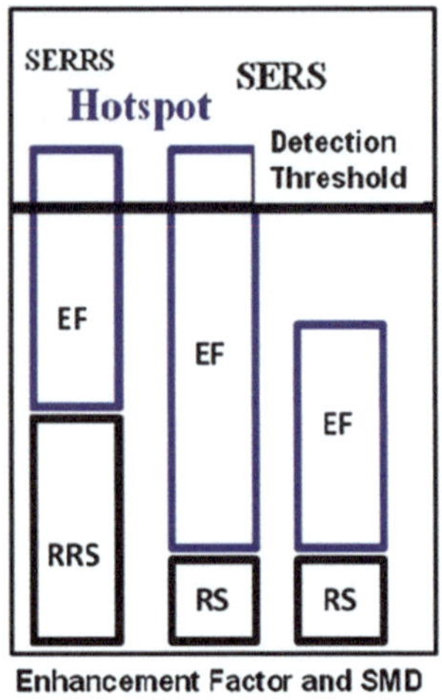

Fig. 2.3 Enhancement factor illustration

unique and hard to reproduce. The experiment requires that the SERS or SERRS cross section should high enough to pass the threshold of SMD as illustrated in Fig. 2.3. In a recent review Kleinman et al. [30] have compiled a table of EFs for variety plasmonic nanostructures that is reproduced here in Table 2.3.

It can be seen that there is a wide range of EF values. The main message from the accumulated experience is that a careful selection the appropriate enhancing structure for the problem at hand in plasmon enhanced spectroscopy, is the key for successful applications.

2.3 Surface-Enhanced Raman Scattering and the Challenge of Spectral Interpretation

Normal Raman Scattering (RS) and Resonance Raman Scattering (RRS) In spectroscopy, the word **molecule** refers to any stable system of nuclei and electrons. For each electronic state, there is a set of molecular vibrational modes (molecular fingerprints) that characterizes the stable electronic state of the molecule. Since most systems are found under normal conditions in their ground electronic state; the molecular vibrations are the fundamental set of characteristic frequencies that allow the unique identification of molecules. When the incident radiation producing the inelastic scattering is far below the frequency of the first excited singlet electronic state we term the inelastic scattering normal Raman scattering. The scattered light is generated by oscillating electric dipoles p induced by the electric field of incident electromagnetic radiation: $p = \alpha E$. Where α - is the molecular polarizability tensor

Table 2.3 Characteristics EF values for SERS substrates reported in the literature [30]

SERS substrate	EF	Gap in nm
Non-hot spot substrates		
Island films	10^4	10–100
Periodic particle array	10^7	10–100
Nanorods	10^3–10^7	NA
Nanoshells	$\sim 10^6$	NA
Fabricated substrates		
MFONs	10^3–10^{11}	0.1–10
Lase-shell NPs	$\sim 10^2$	1–10
Fs laser etching	$<10^9$	NR
E-beam lithography	10^3–10^5	1–100
Templated nanowires	10^3–10^3	5–100
Bowtie electromigration	10^8	0.1–10
Spheres on E-beam posts	10^8	0–1
Assembled structures		
Aggregated colloids	10^6–10^{10}	0–10
DNA hybridization	10^8–10^{12}	1–100
Embedded polymers	$\sim 10^5$	10–100
SiO_2 encapsulated cores	10^6–10^8	0–10
Particles in wells	10^9	0–10
Nanofingers	10^{10}–10^{11}	0–10
TERS	10^4–10^9	0–100
SHINERS	10^5–10^8	1–10

Acronyms: *MFONs* metal film over nanospheres, *TERS* tip-enhanced Raman scattering, *SHINERS* shell-isolated nanoparticle enhanced Raman scattering

and $E = E_0 \cos(\omega_0 t)$', is the incident field. Each component of the induced is given by: $\begin{bmatrix} p_x \\ p_y \\ p_z \end{bmatrix} = \begin{pmatrix} \alpha_{xx} & \alpha_{xy} & \alpha_{xz} \\ \alpha_{yx} & \alpha_{yy} & \alpha_{yz} \\ \alpha_{zx} & \alpha_{zy} & \alpha_{zz} \end{pmatrix} \begin{bmatrix} E_x \\ E_y \\ E_z \end{bmatrix}$

Since excited states are not involved, the classical Placzek's polarizability theory control the intensity for normal Raman Scattering (RS) [14]. Near the equilibrium, the change of each component of the polarizability tensor with the normal coordinate $Q_i = Q_{i0} \cos(\omega_i t)$ can be written in a Taylor series:

$$\alpha = \alpha_0 + \alpha_0^1 Q_1 + \cdots + \frac{1}{2}\alpha_0^{11} Q_1^2 + \cdots + \alpha_0^{12} Q_1 Q_2 + \cdots + \frac{1}{6}\alpha_0^{111} Q_1^3 + \cdots$$

Where, $\alpha_0^1 = \left(\frac{\partial \alpha}{\partial Q_1}\right)_0$; $\alpha_0^{12} = \left(\frac{\partial^2 \alpha}{\partial Q_1 \partial Q_2}\right)_0$, *and* $Q_1, \ldots, Q_{3N-6}\,(Q_{3N-5})$

N – is the number of nuclei.

Including, in the induced dipole moment only the first derivatives $\alpha' = \left(\frac{\partial \alpha}{\partial Q}\right)_0$,

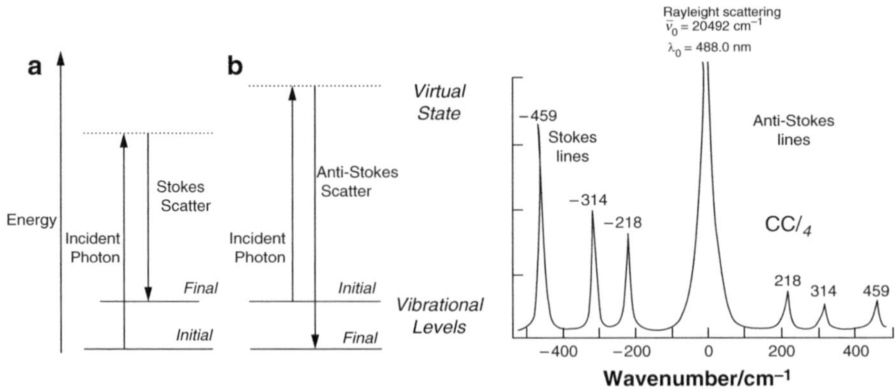

Fig. 2.4 Inelastic Raman scattering; Stokes and anti-Stokes spectral lines

$$p = \left[\alpha_0 + \alpha' Q_{i0} \cos\left(\omega_i t\right)\right] E_0 \cos\left(\omega_0 t\right),$$

Using the trigonometric identity, $\cos a \times \cos b = \frac{1}{2}\left[\cos\left(a+b\right) + \cos\left(a-b\right)\right]$, we arrive at

$$\begin{array}{l} p = \alpha_0 E_0 \cos\left(\omega_0 t\right) + \dfrac{1}{2}\alpha' E_0 Q_{i0} \cos\left(\omega_0 - \omega_i\right) t + \dfrac{1}{2}\alpha' E_0 Q_{i0} \cos\left(\omega_0 + \omega_i\right) t \\ \quad \text{Rayleigh} \qquad + \ \text{Stokes} \qquad\qquad\qquad\qquad + \ \text{Anti-Stokes} \end{array}$$

Given rise to the Rayleigh line (elastic scattering) and the Stokes and anti-Stokes Raman, as illustrated in Fig. 2.4. However, this simple approach is no longer valid as soon as the excitation energy is in resonance with an excited electronic state. The observed Resonance Raman scattering, in general, is governed by the KHD equation [14]. The polarizability elements are given by:

$$\left(\alpha_{\rho\sigma}\right)_{GF} = \frac{1}{\hbar}\sum_n \left(\frac{\langle i\left|\mu_\rho\right|n\rangle\ \langle n\left|\mu_\sigma\right|f\rangle}{\omega_0 - \omega_{nf} - i\Gamma_n} - \frac{\langle i\left|\mu_\sigma\right|n\rangle\ \langle n\left|\mu_\rho\right|f\rangle}{\omega_0 - \omega_{ni} - i\Gamma_n}\right)$$

Where i, n, and f designate the initial, intermediate, and final states, respectively. For RRS, ω_0 is close to ω_{ni}, so the second term dominates the equation. In practice, although RS and RRS report the same fundamental vibrational frequencies, the spectra could be drastically different in their relative intensities, and the RRS may also show overtone and combinations with strong relative intensity. To illustrate this difference, the SERRS spectrum of a perylene tetracarboxylic derivative [34] is shown in Fig. 2.5, where the laser line at 514.5 nm is in full resonance with the electronic absorption and the 633 nm outside the absorption band. The spectra were recorded from a sample fabricated by coating a silver island film with a single Langmuir-Blodgett (LB) [35] monolayer of the target molecule. These are averaged SERS (and SERRS) spectra with about 10^6 molecules in the field of view

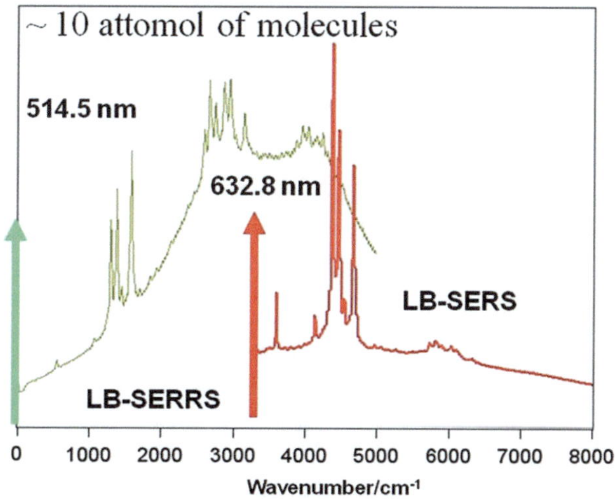

Fig. 2.5 SERS and SERRS spectra of Langmuir-Blodgett [35] of a single monolayer on silver island film

of the Raman microscope. Apart from the differences in relative intensities, the LB-SERRS spectrum shows strong first and second overtones and combinations, which are normally not observed in the RS spectra. The background observed at 514.5 nm excitation is due to what's left fluorescence, heavily quenched by the silver island film.

2.4 SERS and SERRS Spectral Interpretation

The concept of SERS is such a powerful force that has been inappropriately stretched out to include Raman spectra of molecules obtained without plasmon enhancement, or molecules adsorbed onto materials that do not support surface plasmon resonances. The significance of some misconceptions regarding the origin of the enhancement in SERS has been addressed [36]. The discovery of SERS has contributed to the development of plasmonics, an emerging research field based on the manipulation of light at the nanoscale using noble metal nanostructures. Plasmonic structures, due to their strongly enhanced optical near-fields, provide new tools for molecular sensing applications, which are sensitive to the local optical environment. The two main tasks in SERS research are then: (1) Plasmonic nanostructure; fabrication and characterization of noble metal nanostructures for specific light manipulation and ultrasensitive detection, and (2) proper interpretation of the observed enhanced spectra.

The plasmonic properties can be illustrated using the simple model of a metal sphere of radius a and dielectric function $\varepsilon(\omega)$, which can radiate a dipolar field:

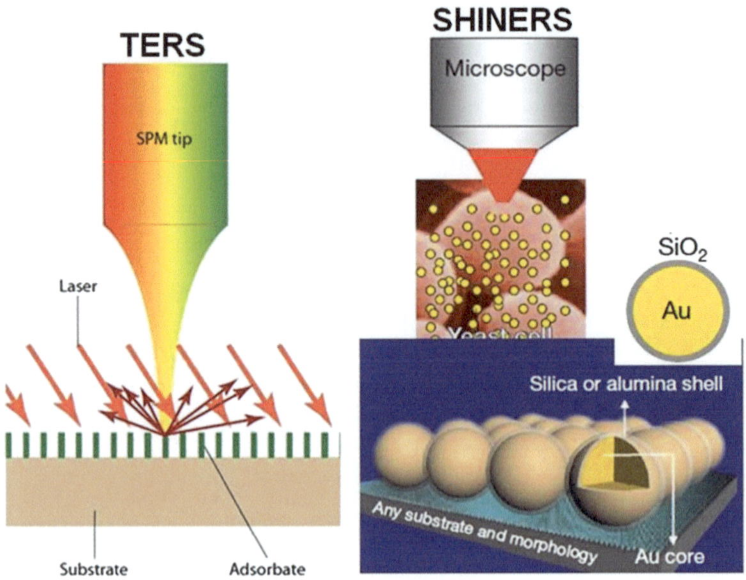

Fig. 2.6 Enhanced Raman spectroscopy techniques: SHINERS and TERS. Adapted from references [40] and [39]

$E_{LSPR} = \frac{\varepsilon(\omega)-\varepsilon_m}{\varepsilon(\omega)+2\varepsilon_m}\left(\frac{a}{a+d}\right)^3 E_0$, d- is the metal-molecule separation distance. The plasmon resonance peaks at $\varepsilon(\omega_{LSPR}) + 2\varepsilon_m = 0$, with a small imaginary part of $\varepsilon(\omega)$, provides large enhancement thanks to the factor $g = \frac{\varepsilon(\omega)-\varepsilon_m}{\varepsilon(\omega)+2\varepsilon_m}$ [37, 38].

The distance dependence $\left(\frac{a}{a+d}\right)^3$, is a unique property of the plasmon enhancement that is carefully controlled in two very important SERS techniques: Shell-isolated nanoparticle-enhanced Raman spectroscopy(SHINERS) [39], and tip-enhanced Raman spectroscopy (TERS) [40] illustrated in Fig. 2.6, adapted from references [40] and [39].

Tip-enhanced Raman scattering offers the additional advantage of high spatial resolution, and the fact that using TERS one can record spectra on any adsorbate/substrate configuration.

The new SHINERS technique is based on the fabrication of coated nanoparticles with a very thin layer of an oxide (such as SiO_2). These shell-isolated enhancing nanoparticles are spread onto a sample to obtain SERS that now will be called SHINERS. Therefore, the plasmonic enhancer may be sprayed over any potential target for enhanced Raman scattering. In addition, the magnitude of the tangential and perpendicular components of the electric field at the surface of the sphere change with the excitation frequency, and this problem will be discussed in connection with the spatial orientation of the adsorbed molecule [37].

2.5 The Observed SERS Spectra

The induced dipole on the molecular system is: $p = \alpha E_0$, and any change in the polarizability "α" may induce changes in the polarizability derivative tensor, and they will be reflected on the inelastic scattering spectrum. In general, the spectral properties of the species adsorbed onto the metal nanoparticles may change on account of the "chemical or physical" interactions with the nanostructure [41], with other molecules, molecular orientation at surface and, correspondingly, the plasmon enhanced SERS spectrum will report not only on the target molecule; but it will contain information that can shed light on these molecule-interactions. Since good Raman scatters are highly polarisable molecules, their polarizability and, correspondingly, their polarizability derivatives are sensitive to the environment (intermolecular and molecule surface interactions) that may change the Raman band location (wavenumber), the relative intensity, or the full width at half maximum. The net result is that the SERS spectrum may be quite different from the reference Raman spectrum.

Therefore, a number of determining factors should be considered in the interpretation of observed SERS.

1. SERS Spectra of Chemically Adsorbed Molecules This is the obvious case when the molecule forms a new chemical bond with the metal on the surface of the nanostructure.

A very clear example is found in the SERS spectra of thiophenol or benzenethiol shown in Fig. 2.7. The reference RS shows the characteristic stretching vibrational wavenumber of the S–H $(2,570\ \mathrm{cm}^{-1})$, C–H $(3,060\ \mathrm{cm}^{-1})$ and C = C $(1,584\ \mathrm{cm}^{-1})$. However, in the SERS spectrum recorded on silver colloids, the

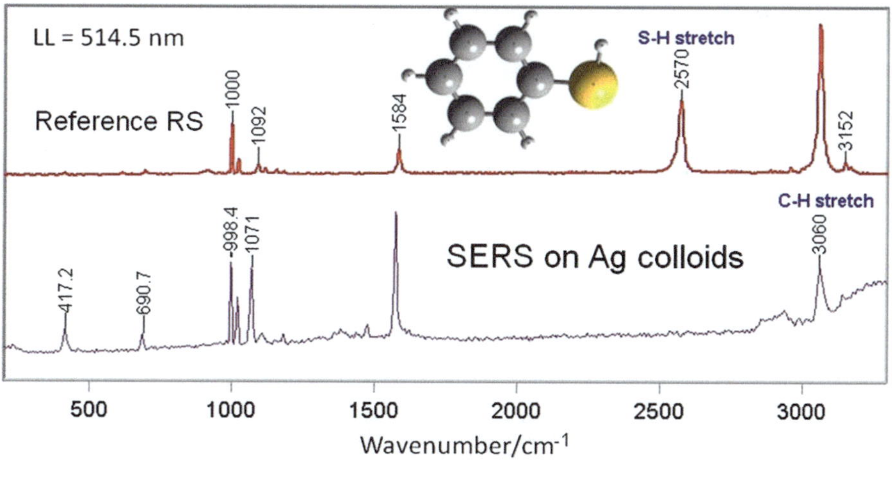

Fig. 2.7 Raman spectrum and SERS spectrum of thiophenol on Ag colloids

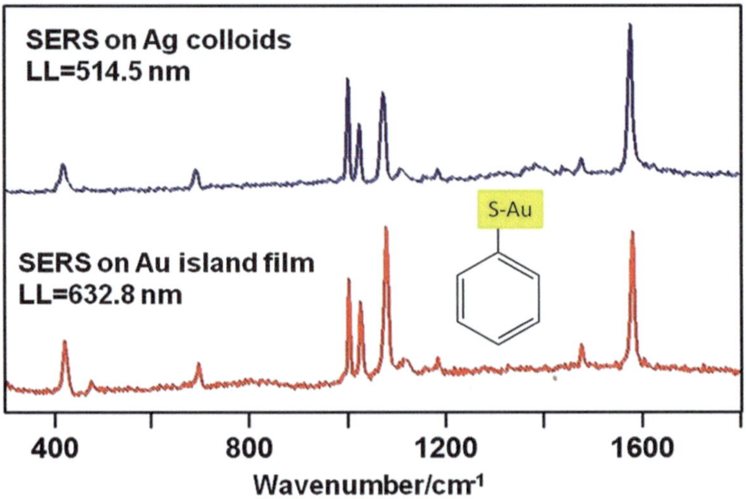

Fig. 2.8 Average SERS spectra of thiophenol excited with two different laser lines: 514.5 nm for Ag colloids and 632.8 nm for gold substrates

S – H vibrational wavenumber is absent due to the formation of a new Ag–S chemical bond and a surface complex. The same surface reaction is observed on gold colloids a gold island films. The average SERS spectra show the same characteristic vibrational frequencies, as can be seen in Fig. 2.8, for SERS spectra excited with two different laser lines: 514.5 nm for Ag colloids and 632.8 nm for gold substrates. These average SERS spectra are obtained reproducibly on both, metal colloids or metal island films of silver or gold. The same spectrum is observed with excitation at 785 nm, indicating minimum dispersion in the 500–785 nm region. It can also be concluded that the two complexes have the same structure, and given the broad spectral region tested it can be safely assumed that the silver and gold complexes do not have electronic resonances in that region.

If an excitation frequency is close to or in-resonance with an electronic state, there is either pre-resonance RS or resonance Raman scattering, changing the relative intensities in the vibrational spectrum and adding absolute intensity to the plasmon enhancement [42]. Therefore, for chemisorbed species, the vibrational spectrum observed in SERS is that of the complex formed at the surface, not the original molecule [23]. On the other hand chemisorbed species enjoy maximum plasmon enhancement since the molecule-metal separation is $d = 0$. The latter is sometime known as the "first layer effect".

2. The Cases of Double Resonances Case 1. Adsorbed molecules with excited electronic states in resonance with the exciting electromagnetic radiation. When the impinging EM is also in resonance with the plasmon extinction spectrum, we call the observed spectra SERRS (surface enhanced resonance Raman scattering, as in Fig. 2.5), and there is ample literature on the subject [23, 24]. SERRS cross sections are the largest, since they benefit from the multiplicative effect of RRS (see Table 2.2) and plasmon enhancement. These chromophores can also be used as SERRS probes (labeled molecules) for single molecule detection of molecules with intrinsically low cross sections, such as phospholipids [43].

Case 2. In the SERS experiment, the overall effect of the molecule-metal interaction may be a change in the molecular electronic structure creating resonances within the adsorbed molecule itself that could lead to SERRS spectra.

Case 3. Charge transfer. It was first observed in electrochemical SERS experiments, where electronic resonances could be obtained by varying the applied potential. "*The observed wavelength dependence of these resonances showed that charge transfer (CT) between the molecule and the metal conduction band was responsible for this effect. This charge transfer can occur in either direction depending on the relative energies of the metal Fermi level and the HOMO and LUMO levels of the adsorbed molecule, and we have shown experimentally that the CT does take place in both directions, that is, metal cluster-to-molecule or molecule-to-metal cluster, depending on the nature of the molecule.*" [41] Therefore, when the excitation wavelength is in the region of a charge-transfer and the plasmon extinction band, a particular type of SERRS is observed. The review of Lombardi and Birke [41] examine this case in details, and a good example can be found in our own work on SERRS spectra of Fe(II) phthalocyanine (FePc) adsorbed onto a Ag electrode [44].

3. Molecular Orientation at the Surface of the Enhancing Nanostructure Let's consider a spatially oriented molecule at the surface of an enhancing nanostructure. The dynamics of the interaction of light with the adsorbate leads to a pattern of Raman intensities determined by "surface selection rules"[45]. Surface selection rules encompass the symmetry properties of the dipole transitions and the modification of the intensities due to the components of the local electric field vector at the surface. In a metal sphere, a tangential and normal component is present at the surface as can be seen in Fig. 2.9. The magnitude of the tangential and the normal

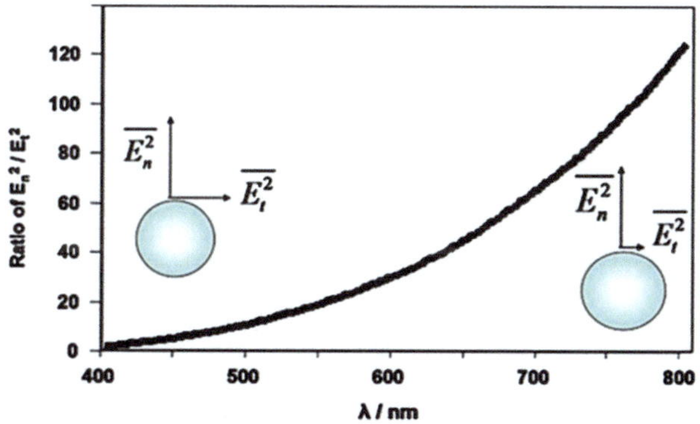

Fig. 2.9 Electric field tangential and normal components on a silver sphere

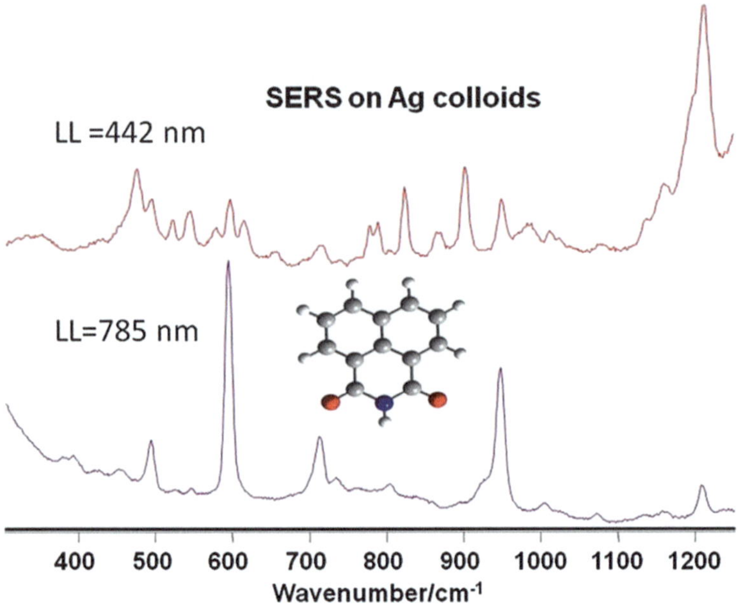

Fig. 2.10 The SERS spectra of naphthalimide on silver colloids recorded with two laser lines

components of the electric vector changes with the wavelength of the incident EM field, and the normal component dominates at longer wavelength. Consequently, the component of the transition dipole normal to the surface is excited more strongly than the tangential component. In the limit of high reflectivity the tangential field is almost completely absent. Therefore, for molecules anchored on nanostructures Raman intensities are further modulated by the spatial orientation of the local electric field interacting with the polarizability derivative tensor. Determining the symmetry point group of the adsorbed molecule allows the interpretation of vibrational modes observed with different relative intensities in SERS spectra recorded with different laser lines. The SERS spectra of naphtalimide on silver colloids recorded with two laser lines (LL = 442 nm and 785 nm) are shown in Fig. 2.10, clearly illustrating the effect of the component of the electric field at the surface.

Selecting the appropriate experimental conditions and with the help of group symmetry, molecular orientation can be determined using SERS and the surface selection rules. In addition, SERS spectral interpretation can be supported by quantum mechanical computations of the most likely molecular structure of the adsorbate.

In summary, it may be advantageous for the practitioner to focus first on the search of the appropriate plasmonic structure (to provide the enhancement) that can be fabricated and calculated independently, based on the optical properties of the metal. Understanding the role of plasmonics helps the practitioners to concentrate on the development of nanostructures for specific applications. Then obtain the

Raman spectra and vibrational assignment of the reference molecule. Examine and optimize the adsorption conditions to attain the optimum SERS spectrum. Carry out the vibrational interpretation of the SERS spectra of the molecule adsorbed onto the plasmonic substrate. The ultrasensitive vibrational SERS can provide information about the nature of metal-molecule interaction (such as surface complex formation) and molecular orientation. There are also polarization techniques that can enhance the structural information.

2.6 Surface-Enhanced Fluorescence (SEF) and Shell-Isolated Nanoparticle Enhanced Fluorescence (SHINEF)

Fluorescence Spontaneous emission of radiation ***(luminescence)*** from an excited molecular entity with the formation of a molecular entity of the same spin ***multiplicity (IUPAC)***. Luminescence is formally separated into fluorescence and phosphorescence, Fluorescence is emission of light from excited singlet states. Phosphorescence is emission of light from excited triplet states. Fluorescence typically occurs from aromatic molecules (fluorophores). A characteristic of fluorescence is that the fluorescence emission is independent of the excitation wavelength. On the contrary, Raman spectrum follows the excitation wavelength. It is only required that the excitation light has enough energy to reach the excited states, but at higher energies the emission spectrum remains the same. This is because upon relaxation, all excess energy is dissipated rapidly leaving the fluorophore in the lowest vibrational state of S_1 (lowest singlet state) and the emission and relaxation proceeds from that state only (*Kasha's rule*) [11].

Experimentally, it is observed that most fluorophores follow the so-called mirror image rule, shown in Fig. 2.11, where the emission spectrum is a mirror image of the absorption spectrum. It is always observed that the energy of the emission is typically less than that of absorption (Stokes shift). This is because when the molecule relaxes and emits, it does not necessarily return to the lowest energy state, but to some excited ground vibrational state. The main properties of a fluorophore are the fluorescence lifetime and the quantum yield. Quantum yield Q is the ratio of emitted photons to absorbed photons. Far from saturation of the excited state, the fluorescence emission rate I_F is given by: $I_F = I_0 Q$, where I_0 is the excitation rate. The quantum yield can be defined in terms of the radiative decay rate Γ, and the non-radiative processes represented by decay rate k_{nr}; $Q = \frac{\Gamma}{\Gamma + k_{nr}}$

Statistically, the lifetime represents the life expectation of the excited molecule, i.e., the average time the molecule spends in the excited state. Fluorescence lifetimes are found to be of a few nanoseconds. The lifetime τ is given by; $\tau = \frac{1}{\Gamma + k_{nr}}$.

It should be pointed out that the fluorescence emission is also a sensitive function of external parameters accounting for the local environment of the molecule. For instance, for molecules near metal surfaces, fluorescence lifetime changes due to modifications of both the radiative and the non-radiative decay rates [46], that results in strong quenching of the fluorescence emission.

$$Q = \frac{radiating\ molecules}{excited\ molecules} = quantum\ yield$$

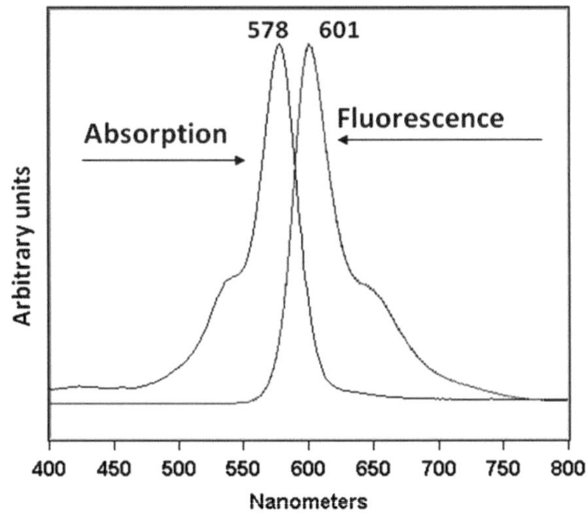

Fig. 2.11 Absorption spectrum with mirror image emission spectrum

2.7 Shell-Isolated Nanoparticle Enhanced Fluorescence (SHINEF)

Single-molecule fluorescence detection can be achieved with fluorophores of high quantum yield [47] thanks to their intrinsic large cross sections. Notably, the detection of single fluorophore was first achieved at cryogenic temperatures [13], almost a decade before the SMD SERS was reported [7, 8]. The absorption cross section of a dye molecule at low temperatures is many orders of magnitude larger than at room temperature. In addition, photodecomposition channels can be suppressed at cryogenic conditions, which makes the dye molecules much more photo stable than at ambient temperatures [12]. New technologies in the field of high-sensitivity detectors and high-quality optics have made possible the single-molecule fluorescence detection at ambient conditions. These technological developments also benefit SERS spectroscopy. The ability of metal nanostructures to enhance optical fields can enhance the molecular fluorescence expanding its applications, for instance, in bio sensing and imaging.

Plasmon or surface enhanced fluorescence [48] of molecules near metallic nanostructures is the result of two competing processes [49]: the amplification of the emission due to the enhanced local field and non-radiative decay due to radiantionless energy transfer to the metal. It should be mentioned that same physical phenomenon was renamed (in 2002) as metal enhanced fluorescence

(MEF) [50]. There are several important properties of SEF that deserved a thorough discussion: the metal-molecule distance dependence, the relationship between the plasmon and the emission frequency, the role hot spots, the role of nanoparticle size and shape and the role of the molecular quantum yield.

2.8 Distance Dependence of the Enhancement

Since the work of Wokaun et al. [49] in 1983, several research groups have demonstrated the unique distance dependence of SEF on silver and gold nanostructures [51], as shown in Fig. 2.12. With the simple model of the sphere of radius *a,* and the fluorophore at a distance *d,* the two competing effects give the trend seen in Fig. 2.12. The enhanced local field distance dependence is given by, indicating a slow decay as *d* increases. At short distances from a metal surface, there is $\frac{1}{d^3}$ dependence for the relaxation, i.e., an enhancement of the ***nonradiative decay*** of excitations [46] leading to fluorescence quenching. After a critical distance, the trend follows the surface average of the near field intensity enhancement, $\langle |E|^2 \rangle$ as calculated directly using Mie theory. The $\langle |E|^2 \rangle$ enhancement dependence of SEF has been demonstrated experimentally [52] using shell-isolated nanoparticle enhanced fluorescence [53]. In a nutshell, let $\mathbf{E}_0$ be the incident field and, after plasmon excitation, the enhancement factor $|E| = \left|\frac{E_{loc}}{E_0}\right|$ is provided by enhanced local field. Since Raman benefits from the enhancement of both the incident field and the scattered field, it results in a Raman enhancement proportional to the fourth power [37, 54]; $|E|^4 = \left|\frac{E_{loc}}{E_0}\right|^4$ However, surface enhanced fluorescence can only benefit from the enhanced local filed and should be proportional to $|E|^2 = \left|\frac{E_{loc}}{E_0}\right|^2$,

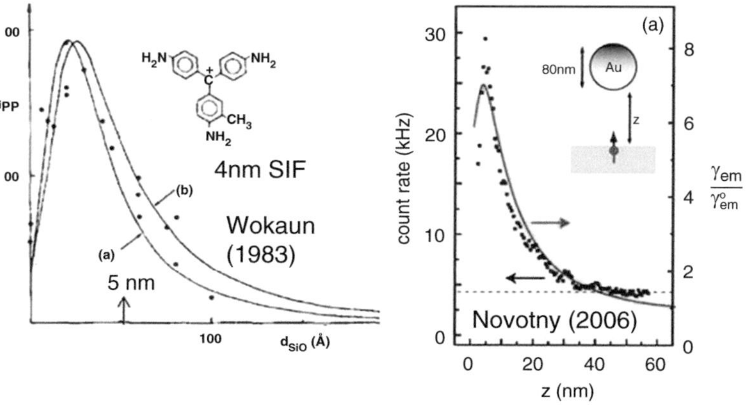

Fig. 2.12 Distance dependence of SEF on silver and gold nanostructures

and correspondingly, with EFs much modest compared to SERS. After the critical distance in SEF (maximum in Fig. 2.12), the Raman and SEF enhancement factors should follow the $\frac{|E|^4}{|E|^2}$ relationship for a molecule sitting at a distance d from the surface of the enhancing nanostructure. To illustrate the evaluation of the $\frac{|E|^4}{|E|^2}$ factor, we choose the ethyl violet (EV) molecule, which has a low quantum yield, allowing the Raman and the fluorescence to be observed in the same spectrum by selecting the appropriate laser line. Using shell-isolated nanoparticles (SHIN), a separation between the target molecule and the metal surface is guarantee, and SHINEF is collected for molecules located outside the zone of strong quenching. In addition, experiments are carried out in solution; thereby, we are measuring reproducible average properties. As a result, with Ag-SHINs, the plasmon enhanced vibrational Raman is observed "unperturbed" from the reference spectrum in solution, and there are not dramatic changes in the lifetime.

The experimental results presented in Fig. 2.13, confirm that the plasmon-enhanced fluorescence is proportional to the square of the local field enhancement while the plasmon-enhanced Raman scattering is given by the fourth power of the local field enhancement. What is unique about these far field measurements is that the vibrational SERS (SHINERS) [39] and the enhanced fluorescence (SHINEF) [53] signal come from the same EV molecule sitting at a distance d from the metal core. Therefore, the enhancements can be directly compared, as it is done in Fig. 2.13 to calculate the enhanced local field from both spectra.

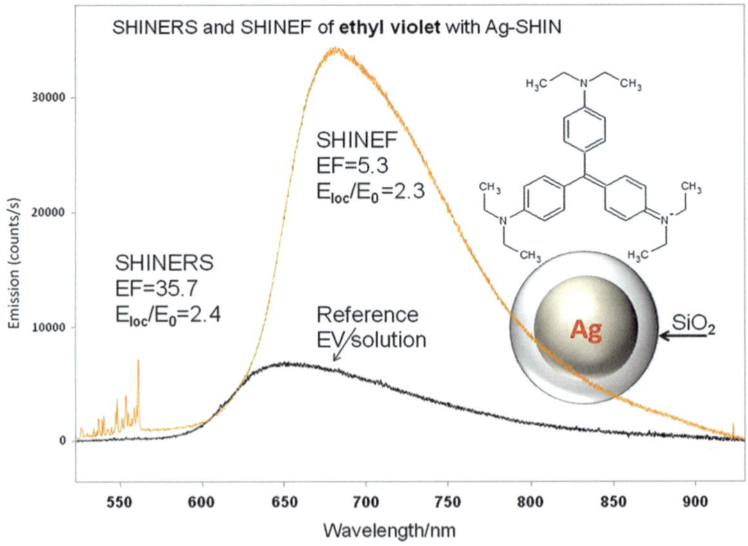

Fig. 2.13 SHINERS and SHINEF of ethyl violet on Ag-SHINs

2.9 Tuning the Plasmon Extinction and the Molecular Emission

Several groups have explored the correlation between the plasmon resonance frequency and the emission frequency of the molecule. Since, the full with at half maximum is very large, it helps to overlapped with a fluorescence band nearby, as can be seen in Fig. 2.14 for an Au-SHIN and two dyes; R18 (octadecyl RhodamineB) and a PTCD derivative (bis(n-butylimido)perylene).

Tam et al. [55] have examined the role of the nanoparticle plasmon resonance energy and nanoparticle scattering cross section on the fluorescence enhancement of adjacent indocyanine green (ICG) dye molecules, and found out that "*From this observation, it appears that the fluorescence enhancement of a nanoparticle substrate is optimal when the nanoparticle plasmon resonance is tuned to the emission wavelength of the molecular fluorophore*" [55]. Another manifestation of this close relationship is found in the so-called spectral profile modification that has experimentally demonstrated and support by a theoretical model by Le Ru et al. [56]: *"The spectral profile modification (SPM) that should accompany any plasmon-resonance-mediated enhancement or quenching of the fluorescence close to metal surfaces. We argue theoretically and demonstrate experimentally that under SEF conditions the plasmon resonances affect not only the fluorescence intensity (enhancement or quenching) but also its spectral profile, sometimes to a point where the original fluorescence spectrum may no longer be recognizable as such"*. Our own experiments using SHINEF also confirm the SPM effect in the observed enhanced fluorescence [57]. It simply means that the re-emission by the enhancing nanostructure may change the shape of the molecular fluorescence

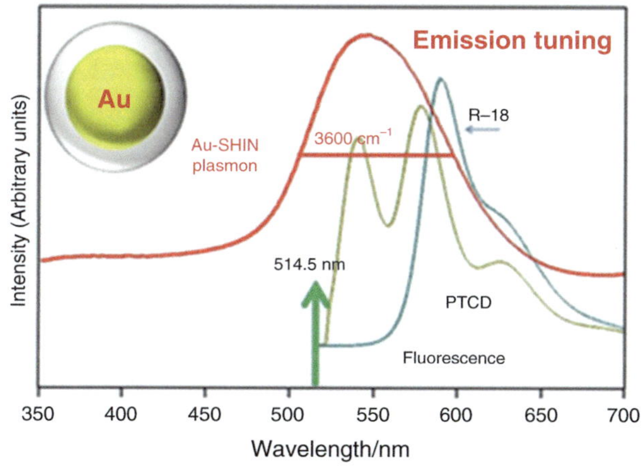

Fig. 2.14 Au-SHIN plasmon extinction and fluorescence emission of two dyes. For a Au-SHIN and two dyes; R18 (octadecyl rhodamine B) and a PTCD derivative (bis(n-butylimido)perylene)

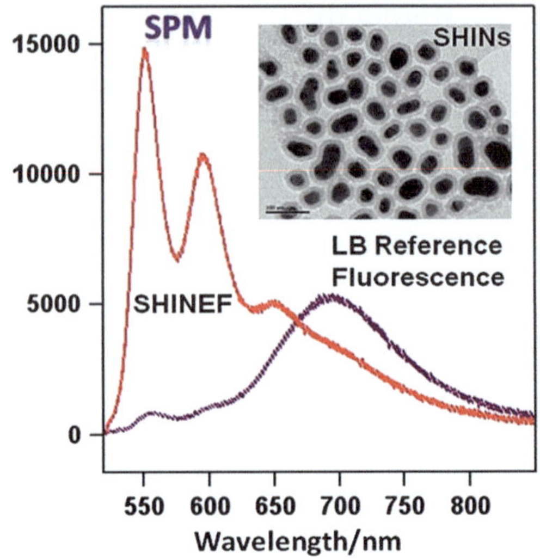

Fig. 2.15 SPM effect observed with shell-isolated nanoparticle enhanced fluorescence [57]

producing a modified band shape, as can be seen in Fig. 2.15. Here, a Langmuir-Blodgett monolayer of bis-(3,4 dichloro-benzylimido) perylene (PTCD), provides the reference fluorescence spectrum. Adding Au-SHINS rresults in enhanced fluorescence or SHINEF with a different band shape due to the fact that the Au-SHINs have plasmon resonance center at 535 nm, which is in tune with the monomer emission of PTCD, The band of PTCD around 700 nm is the excimer emission in the LB film. There are many more examples in the literature that confirm the correlation as, for example, Chen et al. [58], have reported: "*We use single-particle darkfield scattering and fluorescence microscopy to correlate the fluorescence intensity of the dyes with the localized surface plasmon resonance (LSPR) spectra of the individual metal nanoparticles to which they are attached. For each of three different dyes, we observe a strong correlation between the fluorescence intensity of the dye and the degree of spectral overlap with the plasmon resonance of the nanoparticle*".

2.10 The Role of the Molecular Quantum Yield

The role of the quantum yield of the molecule in the enhancement has been extensively discussed in the literature. Some research work has shown that molecules with low intrinsic quantum yield would get larger enhancement factor from introducing metallic nanoparticles in close proximity to fluorophore. Conversely, molecules with high intrinsic quantum yield may not benefit from the same enhancement. Indeed, theoretical models predict [59, 60] a simple scaling of the SEF enhancement factor with $\frac{1}{Q_{mol}}$, Q_{mol} being the non-modified quantum yield of the fluorophore. However, it should be pointed out that high enhancements are also possible for fluorophores with quantum yield near unity. In fact, SEF of Alexa Fluor 488 with a quantum efficiency Q = 0.92 has been reported [61]. In our own SHINEF work using LB monolayers of fluorophores we find similar EF for both, high and low quantum yield molecules [62].

2.11 Conclusions

Considering the basic elements of plasmon-enhanced spectroscopy; electromagnetic radiation, adsorbed molecule and metal nanostructure, provides an approach for experiment design and optimization. The EM radiation may be tuned into resonance with the adsorbate electronic states or new charge transfer transitions, and also in resonance with the plasmon absorption of the nanostructure [63]. The latter excitation tuning might depend on the degree of aggregation, or the two and three dimensional array of fabricated nanostructures. Control of the adsorption process, and an awareness of the potential outcome due to photo processes near the nanostructure, is required. In summary, each experiment is unique and requires a selection of the setting for each one of the three elements that would lead to the most efficient plasmon enhancement.

Acknowledgment Funding from the National Science and Engineering Research Council of Canada (NSERC) is gratefully acknowledged.

References

1. Maier SA (2007) Plasmonics: fundamentals and applications. Springer, New York
2. Albrecht MG, Creighton JA (1977) Anomalously intense Raman spectra of pyridine at a silver electrode. J Am Chem Soc 99(15):5215–5217. doi:10.1021/ja00457a071
3. Jeanmaire DL, Van Duyne RP (1977) Surface Raman spectroelectrochemistry. Part I. Heterocyclic, aromatic, and aliphatic amines adsorbed on the anodized silver electrode. J Electroanal Chem Interf Electrochem 84(1):1–20
4. Moskovits M (1978) Surface-roughness and enhanced intensity of Raman-scattering by molecules adsorbed on metals. J Chem Phys 69(9):4159–4161. doi:10.1063/1.437095
5. Fleischmann M, Hendra PJ, McQuillan AJ (1974) Raman-spectra of pyridine adsorbed at a silver electrode. Chem Phys Lett 26(2):163–166. doi:10.1016/0009-2614(74)85388-1
6. Pieczonka NPW, Aroca RF (2008) Single molecule analysis by surface-enhanced Raman scattering. Chem Soc Rev 37(5):946–954
7. Kneipp K, Wang Y, Kneip H, Perelman LT, Itzkan I, Dasari RR, Feld M (1997) Single molecule detection using surface-enhanced Raman scattering. Phys Rev Lett 78:1667–1670
8. Nie S, Emory SR (1997) Probing single molecules and single nanoparticles by surface-enhanced Raman scattering. Science (Washington, DC) 275(5303):1102–1106
9. Rycenga M, Cobley CM, Zeng J, Li WY, Moran CH, Zhang Q, Qin D, Xia YN (2011) Controlling the synthesis and assembly of silver nanostructures for plasmonic applications. Chem Rev 111(6):3669–3712. doi:10.1021/cr100275d

10. Fan M, Andrade GFS, Brolo AG (2011) A review on the fabrication of substrates for surface enhanced Raman spectroscopy and their applications in analytical chemistry. Anal Chim Acta 693(1–2):7–25. doi:10.1016/j.aca.2011.03.002
11. Lakowicz JR (2006) Principles of fluorescence spectroscopy, vol 1, 3rd edn. Springer, New York
12. Basche T, Moerner WE, Orrit M, Wild UP (eds) (1997) Single-molecule optical detection, imaging and spectroscopy. VCH, Weinheim/New York
13. Moerner WE (2002) A dozen years of single-molecule spectroscopy in physics, chemistry, and biophysics. J Phys Chem B 106(5):910–927. doi:10.1021/jp012992g
14. Long DA (2001) The Raman effect. Wiley, Chichester
15. McCreery RL (2000) Raman spectroscopy for chemical analysis, 3rd edn. Wiley, New York
16. Reigue A, Auguie B, Etchegoin PG, Le Ru EC (2013) CW measurements of resonance Raman profiles, line-widths, and cross-sections of fluorescent dyes: application to Nile Blue A in water and ethanol. J Raman Spectrosc 44(4):573–581. doi:10.1002/jrs.4233
17. Shim S, Stuart CM, Mathies RA (2008) Resonance Raman cross-sections and vibronic analysis of rhodamine 6G from broadband stimulated Raman spectroscopy. Chemphyschem 9(5):697–699. doi:10.1002/cphc.200700856
18. Gazeau MC, Wintgens V, Valat P, Kossanyi J, Doizi D, Salvetat G, Jaraudias J (1993) New experimental-method for determining the s(1)- s(n) absorption-spectrum of highly fluorescent dyes. Can J Phys 71(1–2):59–65
19. Bohren CF, Huffman DR (1983) Absorption and scattering of light by small particles. Wiley Interscience, New York
20. Hartland GV (2011) Optical studies of dynamics in Noble metal nanostructures. Chem Rev 111(6):3858–3887. doi:10.1021/cr1002547
21. Kelly KL, Coronado E, Zhao LL, Schatz GC (2003) The optical properties of metal nanoparticles: the influence of size, shape, and dielectric. J Phys Chem B 107:668–677
22. Chen HJ, Ming TA, Zhao L, Wang F, Sun LD, Wang JF, Yan CH (2010) Plasmon-molecule interactions. Nano Today 5(5):494–505. doi:10.1016/j.nantod.2010.08.009
23. Aroca R (2006) Surface-enhanced vibrational spectroscopy. Wiley, Chichester
24. Le Ru EC, Etchegoin PG (2009) Principles of surface enhanced Raman spectroscopy (and related plasmonic effects). Elsevier, Amsterdam
25. Ming T, Chen HJ, Jiang RB, Li Q, Wang JF (2012) Plasmon-controlled fluorescence: beyond the intensity enhancement. J Phys Chem Lett 3(2):191–202. doi:10.1021/jz201392k
26. Kneipp K (2007) Surface-enhanced Raman scattering. Phys Today 60(11):40–46. doi:10.1063/1.2812122
27. McMahon JM, Schatz GC, Gray SK (2013) Plasmonics in the ultraviolet with the poor metals Al, Ga, In, Sn, Tl, Pb, and Bi. Phys Chem Chem Phys 15(15):5415–5423. doi:10.1039/c3cp43856b
28. Le Ru EC, Blackie E, Meyer M, Etchegoin PG (2007) Surface enhanced Raman scattering enhancement factors: a comprehensive study. J Phys Chem C 111(37):13794–13803
29. Le Ru EC, Etchegoin PG (2013) Quantifying SERS enhancements. Mrs Bullet 38(8):631–640. doi:10.1557/mrs.2013.158
30. Kleinman SL, Frontiera RR, Henry A-I, Dieringer JA, Van Duyne RP (2013) Creating, characterizing, and controlling chemistry with SERS hot spots. Phys Chem Chem Phys 15(1):21–36. doi:10.1039/c2cp42598j
31. Kruszewski S, Cyrankiewicz M (2012) Aggregated silver sols as SERS substrates. Acta Phys Polonica A 121(1A):A68–A74
32. Stockman MI, Shalaev VM, Moskovits M, Botet R, George TF (1992) Enhanced Raman-scattering by fractal clusters – scale-invariant theory. Phys Rev B 46(5):2821–2830. doi:10.1103/PhysRevB.46.2821
33. Cang H, Labno A, Lu CG, Yin XB, Liu M, Gladden C, Liu YM, Zhang X (2011) Probing the electromagnetic field of a 15-nanometre hotspot by single molecule imaging. Nature 469(7330):385. doi:10.1038/nature09698

34. Alessio P, Constantino CJL, Aroca RF, Oliveira ON Jr (2011) Surface-enhanced Raman scattering: metal nanostructures coated with Langmuir-Blodgett films. J Chil Chem Soc 55(4):469–478
35. Petty MC (1996) Langmuir-Blodgett films: an introduction, 1st edn. Cambridge University Press, Cambridge, UK
36. Moskovits M (2013) Persistent misconceptions regarding SERS. Phys Chem Chem Phys 15(15):5301–5311. doi:10.1039/c2cp44030j
37. Moskovits M (1985) Surface-enhanced spectroscopy. Rev Mod Phys 57(3):783–826. doi:10.1103/RevModPhys.57.783
38. Kerker M, Wang DS, Chew H (1980) Surface enhanced Raman scattering (SERS) by molecules adsorbed at spherical particles: errata. Appl Optics 19:4159–4173
39. Li JF, Huang YF, Ding Y, Yang ZL, Li SB, Zhou XS, Fan FR, Zhang W, Zhou ZY, Wu DY, Ren B, Wang ZL, Tian ZQ (2010) Shell-isolated nanoparticle-enhanced Raman spectroscopy. Nature 464(7287):392–395
40. Pettinger B, Schambach P, Villagomez CJ, Scott N (2012) Tip-enhanced Raman spectroscopy: near-fields acting on a few molecules. In Johnson MA, Martinez TJ (eds) Annual review of physical chemistry, vol 63, pp 379–399. doi:10.1146/annurev-physchem-032511-143807
41. Lombardi JR, Birke RL (2009) A unified view of surface-enhanced Raman scattering. Acc Chem Res 42(6):734–742
42. Sonntag MD, Klingsporn JM, Garibay LK, Roberts JM, Dieringer JA, Seideman T, Scheidt KA, Jensen L, Schatz GC, Van Duyne RP (2012) Single-molecule tip-enhanced Raman spectroscopy. J Phys Chem C 116(1):478–483. doi:10.1021/jp209982h
43. Pieczonka NPW, Moula G, Aroca RF (2009) SERRS for single-molecule detection of dye-labeled phospholipids in Langmuir-Blodgett monolayers. Langmuir 25(19):11261–11264
44. Corio P, Rubim JC, Aroca R (1998) Contribution of the Herzberg-Teller mechanism to the surface-enhanced Raman scattering of iron phthalocyanine adsorbed on a silver electrode. Langmuir 14(15):4162–4168
45. Moskovits M (1982) Surface selection rules. J Chem Phys 77(9):4408–4416. doi:10.1063/1.444442
46. Chance RR, Prock A, Silbey R (1978) Molecular fluorescence and energy transfer near interfaces. In Advances in chemical physics. Wiley, pp 1–65. doi:10.1002/9780470142561.ch1
47. Xia T, Li N, Fang XH (2013) Single-molecule fluorescence imaging in living cells. Annu Rev Phys Chem 64(64):459–480. doi:10.1146/annurev-physchem-040412-110127
48. Gersten J, Nitzan A (1981) Spectroscopic properties of molecules interacting with small dielectric particles. J Chem Phys 75(3):1139–1152
49. Wokaun A, Lutz HP, King AP, Wild UP, Ernst RR (1983) Energy transfer in surface enhanced luminescence. J Chem Phys 79(1):509–514
50. Geddes CD, Lakowicz JR (2002) Metal-enhanced fluorescence. J Fluoresc 12(2):121–129. doi:10.1023/B:JOFL.0000031824.48401.5c
51. Anger P, Bharadwaj P, Novotny L (2006) Enhancement and quenching of single-molecule fluorescence. Phys Rev Lett 96(11):113002–113001 to 113002–113004
52. Guerrero AR, Zhang Y, Aroca RF (2012) Experimental confirmation of local field enhancement determining far-field measurements with shell-isolated silver nanoparticles. Small 8(19):2964–2967. doi:10.1002/smll.201200750
53. Guerrero AR, Aroca RF (2011) Surface-enhanced fluorescence with shell-isolated nanoparticles (SHINEF). Angew Chem Int Ed Engl 50(3):665–668
54. Le Ru EC, Etchegoin PG (2006) Rigorous justification of the |E|4 enhancement factor in surface enhanced Raman spectroscopy. Chem Phys Lett 423(1–3):63–66
55. Tam F, Goodrich GP, Johnson BR, Halas NJ (2007) Plasmonic enhancement of molecular fluorescence. Nano Lett 7(2):496–501. doi:10.1021/nl062901x
56. Le Ru EC, Etchegoin PG, Grand J, Felidj N, Aubard J, Levi G (2007) Mechanisms of spectral profile modification in surface-enhanced fluorescence. J Phys Chem C 111(44):16076–16079

57. Aroca RF, Teo GY, Mohan H, Guerrero AR, Albella P, Moreno F (2011) Plasmon-enhanced fluorescence and spectral modification in SHINEF. J Phys Chem C 115(42):20419–20424. doi:10.1021/jp205997u
58. Chen Y, Munechika K, Ginger DS (2007) Dependence of fluorescence intensity on the spectral overlap between fluorophores and plasmon resonant single silver nanoparticles. Nano Lett 7(3):690–696. doi:10.1021/nl062795z
59. Bardhan R, Grady NK, Halas NJ (2008) Nanoscale control of near-infrared fluorescence enhancement using Au nanoshells. Small 4(10):1716–1722. doi:10.1002/smll.200800405
60. Gill R, Le Ru EC (2011) Fluorescence enhancement at hot-spots: the case of Ag nanoparticle aggregates. Phys Chem Chem Phys 13(36):16366–16372. doi:10.1039/c1cp21008d
61. Xie F, Pang JS, Centeno A, Ryan MP, Riley DJ, Alford NM (2013) Nanoscale control of Ag nanostructures for plasmonic fluorescence enhancement of near-infrared dyes. Nano Res 6(7):496–510. doi:10.1007/s12274-013-0327-5
62. Mohan H (2013) Plasmon enhanced fluorescence (PEF) of high and low quantum yield molecules. University of Windsor, Windsor
63. McFarland AD, Young MA, Dieringer JA, Van Duyne RP (2005) Wavelength-scanned surface-enhanced Raman excitation spectroscopy. J Phys Chem B 109(22):11279–11285

Chapter 3
Functional Nanomaterials Prepared by Nanoarchitectonics-Based Supramolecular Assembly

Katsuhiko Ariga, Yusuke Yonamine, and Jonathan P. Hill

Abstract Recently, we have introduced the new concept of nanoarchitectonics to traditional supramolecular self-assembly to promote the development of the bottom-up approaches. The concept of nanoarchitectonics can be developed using synergic contributions of technical innovations such as atomic/molecular-level control, chemical nanofabrication, and self- and field-controlled organization. This concept might be applied to inorganic, organic, biochemical, and other systems, beyond the levels of substance, material, and system regardless of their dimensions i.e., macro, micro, or nano scale. In this chapter, recent examples of research on functional nanomaterials prepared by nanoarchitectonics-based supramolecular assembly are explained. These examples are classified according to their scale, (i) molecular-level nanoarchitectonics, (ii) microscale assembly, (iii) macroscopic materials with internal nanostructures.

3.1 Introduction

Development of human activities depends on the advancement of nanoscience and nanotechnology for fabrication of materials and systems containing microscale and nanoscale structural features. Various techniques such as lithographic methods provide top down approaches that can be used to construct integrated microstructures and there has been considerable success in the use of the resulting microstructures over the past few decades. For example, the widespread availability of cellular phones and mobile devices is a consequence of the great success of these methods. However, the current rate of miniaturization (governed by so-called Moore's law) of

K. Ariga (✉) • Y. Yonamine • J.P. Hill
World Premier International (WPI), Research Center for Materials Nanoarchitectonics (MANA), National Institute for Materials Science (NIMS), 1-1 Namiki, Tsukuba 305-0044, Japan

Core Research for Evolutional Science and Technology (CREST), Japan Science and Technology Agency (JST), 1-1 Namiki, Tsukuba 305-0044, Japan
e-mail: ARIGA.Katsuhiko@nims.go.jp

M. Bardosova, T. Wagner (eds.), *Nanomaterials and Nanoarchitectures*, NATO Science for Peace and Security Series C: Environmental Security, DOI 10.1007/978-94-017-9921-8_3

silicon memory technology is anticipated to encounter the physical limits of device dimensions imposed by ultra-violet, electron/ion beam and soft x-ray lithographic techniques in the very near future.

Therefore, materials' fabrications based on bottom-up-type self-assembly have become more important [1, 2]. Both formation and preparation of nanoscale objects rely strongly on self-assembly of unit molecules or nanomaterials. The bottom-up approaches based on supramolecular self-assembly techniques often provide structures required for sophisticated arrangement of their components leading to expression of subtle or complex functionalities. Recently, we have introduced the new concept of nanoarchitectonics [3–10] to traditional supramolecular self-assembly to promote the development of the bottom-up approaches. The concept of nanoarchitectonics can be developed using synergic contributions of technical innovations such as atomic/molecular-level control, chemical nanofabrication, and self- and field-controlled organization. This concept might be applied to inorganic, organic, biochemical, and other systems, beyond the levels of substance, material, and system regardless of their dimensions i.e., macro, micro, or nano scale.

In this chapter, recent examples of research on functional nanomaterials prepared by nanoarchitectonics-based supramolecular assembly are explained. These examples are classified according to their scale, (i) molecular-level nanoarchitectonics, (ii) microscale assembly, (iii) macroscopic materials with internal nanostructures. In addition, our innovative method for accessing molecular systems from the macroscopic length scale is mentioned in the final chapter.

3.2 Molecular-Level Nanoarchitectonics

The most basic supramolecular systems are molecular complexes that can be converted into regular molecular arrays by molecular-level nanoarchitectonics. Formation of porphyrin arrays in the two-dimensional plane is a good example. Hill et al. used porphyrin derivative, tetrakis(3,5-di-tert-butyl-4-hydroxyphenyl)porphyrin containing an essentially planar tetrapyrrole core [11]. Submonolayer coverage of this porphyrin derivative on a Cu(111) surface comprised surface-mobile hexagonally-packed domain islands. This hexagonally-packed structure underwent a transition to a square packed grid motif at ambient temperature (Fig. 3.1). The phase transition occurred on a timescale of seconds enabling its real-time observation by scanning tunneling microscope (STM). The molecules adopt non-planar conformations which optimize the van der Waals contacts between tert-butyl groups by formation of the square grid structure. On the other hand, two-electron oxidation induced formation of a non-aromatic derivative with enhanced stability of the non-planar conformation of the tetrapyrrole. These molecules adsorbed at Cu(111) surfaces with submonolayer coverage and exhibited only one type of surface-immobile hexagonal supramolecular structure at room temperature, and transition to the square phase was not observed.

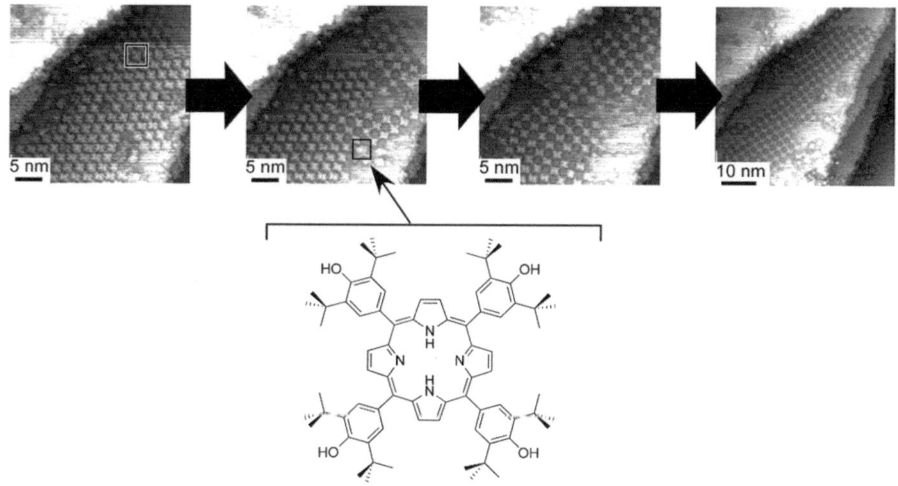

Fig. 3.1 Phase transition of porphyrin arrays from hexagonal to square packing

Hill et al. used porphyrin substituted with 3,5-dimethyl-4-hydroxyphenyl groups, (5,10,15,20-*tetrakis*(3,5-dimethyl-4-hydroxyphenyl)porphyrin), in order to construct hydrogen-bonded arrays [12]. The observed STM images suggested that the methyl groups are in contact with the surface and that hydrogen bonding as intermolecular interactions may occur at the *meso* substituents. Both macrocyclic and ring-open forms of a trimer, formed through hydrogen bonding, were commonly observed in STM images. In addition, a hexagonal structure was occasionally observed as a porous honeycomb-type structure, a Kagome structure, containing void areas within its lattice.

Formation of a two-dimensional pattern containing bright and dark sports in STM images of molecular arrays of 5,10,15,20-tetrakis(3,5-di-tert-butyl-4-hydroxyphcnyl)porphyrin on Λu(111) was also reported[13]. Formation of the bright oblong spots could be due to deformation of the porphyrin core through alternate rotation of phenyl rings. Consequent steric hindrance between the rotated phenyl substituents and the tetrapyrrole macrocycle induces alternate tilting of its pyrrole rings. The bright spots and dark spots of the molecular array on Au(111) surface were caused by the height differential. Such a structure might be useful in construction of nanoscale molecular two-digit information storage (0 or 1, dark or blight) where a single signal can be written within ca. 2×2 nm square.

Unlike solid surfaces, two-dimensional environments presented by gas-liquid interfaces are sometimes useful in structural fabrication based on dynamic molecular interactions. In particular, enhancement of hydrogen bonding efficiency at the air-water interface can be used in a novel nanofabrication method, that of two-dimensional molecular patterning [14, 15]. As shown in Fig. 3.2, molecular patterning was achieved through interaction between amphiphiles at the interface with an aqueous (subphase) template molecule having multiple binding sites,

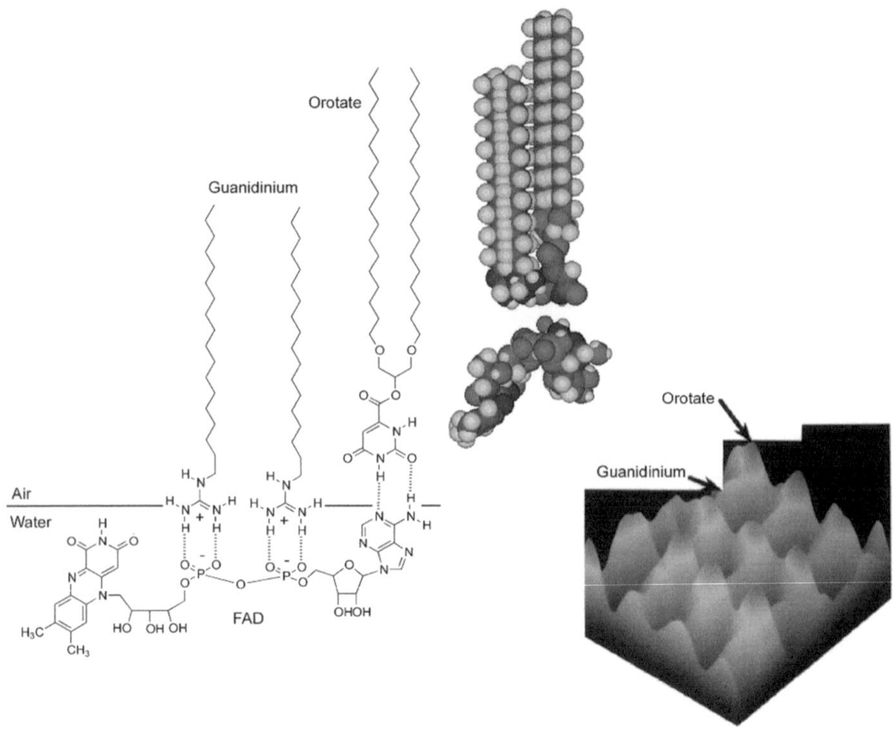

Fig. 3.2 Two-dimensional molecular patterning through molecular recognition at the air-water interface

flavin adenine dinucleotide (FAD) [16]. This template molecule recognizes the guanidinium and orotate amphiphiles. A periodic wavelike structure composed of two distinct peaks with several angstroms height difference was formed in the transferred monolayer.

Another approach for formation of controlled structures at the air-water interface was demonstrated by Oishi et al. through control of interactions among hydrophobic or/and hydrophilic parts of the components in mixed monolayers consisting of an alkyl guanidinium and fluorocarbon carboxylic acids [17]. Formation of a nanoscopic domain structure with control of the domain size and shape from the nanometer to micrometer scales was achieved by changing the pressure, composition, temperature, pH, or ionic strength of the subphase. In these systems, the ionic and/or hydrogen bonding attraction competes with unfavorable contact between hydrocarbon and fluorocarbon. The balance between these respectively energetically advantageous and disadvantageous interactions determines the domain sizes in two-dimensional plane.

3.3 Microscopic-Level Assembly

Self-assembly processes of various materials including organic molecules and inorganic substances provide various functional structures. For example, Sathish et al. demonstrated preparation of two-dimensional C_{60} fullerene microcrystals with hexagonal and rhombic forms, as well as microrod formation (Fig. 3.3) [18]. These structures can be prepared by simple self-assembly processes, and the shapes of microobjects were controlled only by selection of solvents as the assembly media. Uniformly shaped hexagonal and rhombic structures were respectively obtained by precipitation at the liquid-liquid interfaces of iso-propyl alcohol/CCl_4 and tert-butyl alcohol/toluene interfaces. Hexagonal nanosheets possessed a hexagonal crystalline lattice while the rhombic nanosheets have a mixed-crystalline lattice comprised of face-centered cubic and hexagonal structures.

Shrestha et al. reported studies on nanofabrication involving highly crystalline cubic-shaped crystals of a C_{60}-$AgNO_3$ organometallic heteronanostructure [19]. These crystals could be converted into uniquely structured formations of well-oriented assemblies of C_{60} nanoneedles, known as bucky cubes (Fig. 3.4). The unique feature of these assemblies depends on the orientation of the individual needlelike crystals on the orientation of the different crystal planes of the original crystals. This unique structured bucky cube has potential for the study of its electronic and optical properties.

Similarly, fullerene crystals with porous nanostructures, so called mesoporous fullerene, can be prepared by liquid-liquid interfacial precipitation (Fig. 3.5) [20]. The resulting mesoporous C_{60} crystals have a bimodal pore architecture that consisted of both macropores and mesopores within sheet-like two-dimensional microstructures. The average pore sizes for C_{60} crystals obtained from the CCl_4/benzene (80:20) system becomes 15–25 nm. This mesopore size can be increased with a gradual increase in the mixing ratio of CCl_4. In addition, specific gravimetric capacitance increased with increasing proportion of CCl_4 during the preparation process, and reached one order of magnitude greater than that of pristine C_{60}.

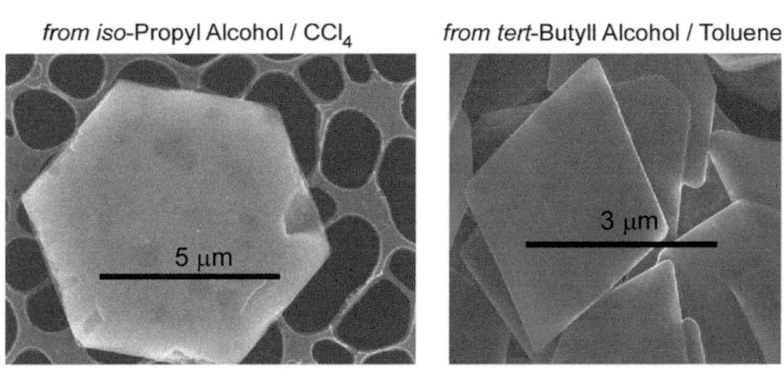

Fig. 3.3 Hexagonal and rhombic morphology crystals of C_{60} fullerene

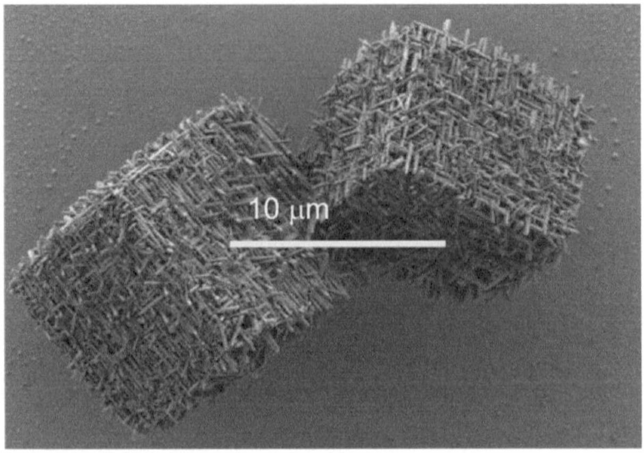

Fig. 3.4 Bucky cubes prepared from a C_{60}-$AgNO_3$ organometallic heteronanostructure

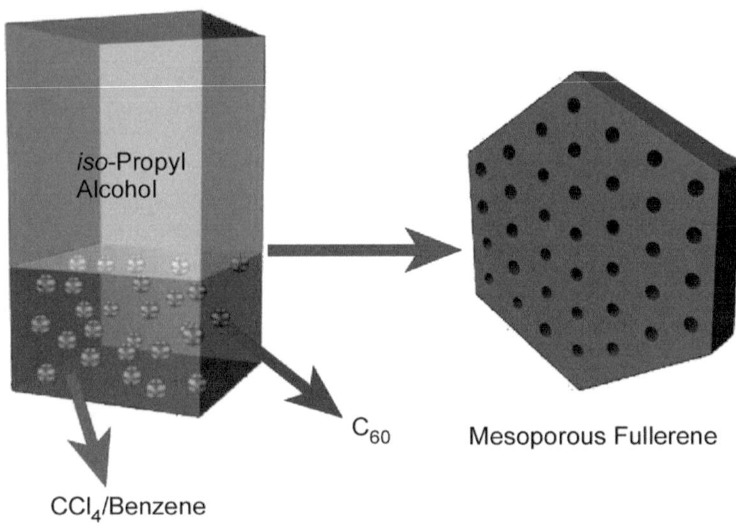

Fig. 3.5 Formation of mesoporous fullerene crystals

Self-assembly of copolymers having blocks with different polarities often provides microscopically phase-separated structures. Charvet et al. synthesized block copolymers with both photoactive electron donor porphyrins and electron acceptor fullerenes (C_{60}) using living ring-opening metathesis polymerization (ROMP) [21]. Wire-like one-dimensional microstructures were formed by simple spin-coating or drop-casting of its chloroform solutions onto solid supports (Fig. 3.6). The wire structures consist of regularly alternating domains arranged perpendicular to their long axes as observed by transmission electron microscopy (TEM). The diameter of the wire structures appears to be determined by the length of the copolymer chain

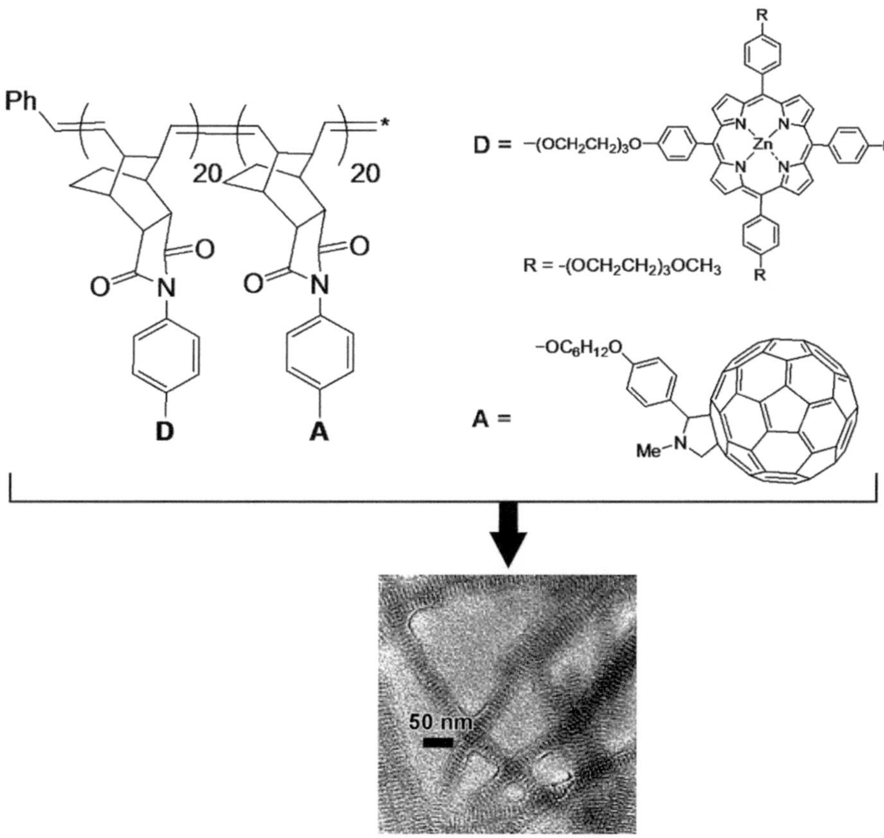

Fig. 3.6 Wire-like self-assembled structures of donor/acceptor block-copolymer

while the width of the polymer building block defines the internal domain size. The prepared films assembled on gold gap electrodes displayed photocurrent switching with sharp on/off responses upon alternating white light irradiation and darkness. Charge carriers seem to be generated at the microscopic interface between porphyrin and fullerene domains through photoinduced electron transfer from porphyrins to fullerenes.

Using appropriate techniques such as Langmuir-Blodgett (LB) method [22, 23] and layer-by-layer (LbL) assembly [24–26], functional materials can be assembled into predesigned structures. For example, sensing materials were prepared through LbL assembly of mesoporous carbon CMK-3 and polyelectrolytes on an electrode of a quartz crystal microbalance (QCM) [27]. Sensing performance of the LbL film of CMK-3 was investigated in aqueous solution from frequency shifts upon adsorption of tannic acid and related substances. The frequency shifts upon adsorption of tannic acid greatly exceeded those for catechin and caffeine. The superior adsorption capacity for tannic acid likely results from interaction with the carbon surface through π–π interactions and hydrophobic effects as well as size fitting of tannic acid to the CMK-3. Interestingly, the quantity of tannic acid adsorbed onto the LbL film of CMK-3 materials at equilibrium showed a sigmoidal profile at low concentrations, suggesting highly cooperative adsorption behaviour from confinement effects during adsorption.

Mesoporous carbon capsules synthesized using zeolite crystals as templates were also assembled on QCM sensors using the LbL technique (Fig. 3.7) [28]. Detection of various volatile substances adsorbed onto the carbon capsule LbL films in air phase was compared by in situ QCM frequency decrease. Aromatic hydrocarbons such as benzene and toluene could be more easily detected than aliphatic hydrocarbons, such as cyclohexane. For example, sensitivity for benzene is almost five times larger than that for cyclohexane, despite their very similar vapour pressures, molecular weights, and structures. The results obtained indicate that π–π interactions are crucial for the adsorption of volatiles to the carbon capsule film. In addition, impregnation with additional recognition components can alter sensor selectivity. The carbon capsule film impregnated with lauric acid exhibited large affinities for non-aromatic amines. In contrast, impregnation of dodecylamine into the carbon capsule films resulted in a strong preference for acetic acid.

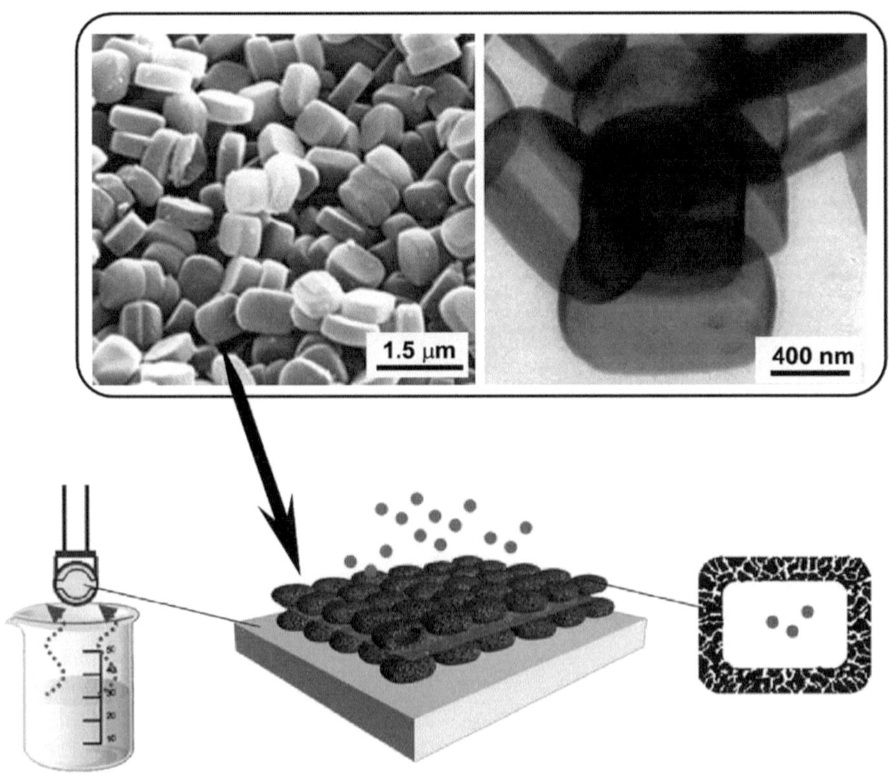

Fig. 3.7 Sensing system with LbL films of mesoporous carbon capsules on a QCM plate

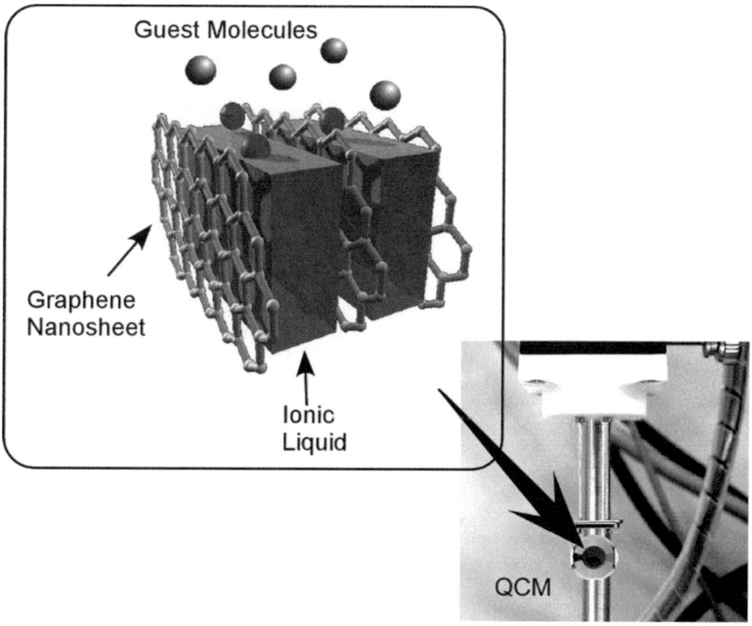

Fig. 3.8 Sensing system with LbL films of graphene nanosheets and ionic liquid on a QCM plate

Similarly, LbL films fabricated from graphene and ionic liquids could be used as a gas sensor (Fig. 3.8) [29]. Graphene oxide sheets were first prepared by oxidation of graphite under acidic conditions followed by their reduction to graphene sheets in the presence of ionic liquids in water. Because the resulting composites of graphene sheet/ionic liquid are charge-decorated nanosheets, they can be assembled alternately with anions such as poly(sodium styrenesulfonate) by LbL adsorption. Exposure of the LbL films (immobilized on QCM plates) to various saturated vapours caused an in situ decrease in frequency of the QCM due to gas adsorption. The composite LbL films showed significantly higher selectivity (more than ten times) for benzene vapour over cyclohexane. Control of the electronic properties of the graphene/ionic liquid LbL films upon gas adsorption was also demonstrated.

LbL assembled films of mesoporous capsules can also be applied for drug delivery system (DDS) applications [30, 31]. Hollow silica capsules, containing hierarchical micro- and nanospaces with an internal void microspace ($1{,}000 \times 700 \times 300\ \text{nm}^3$) at their interiors, and mesopores (average diameter 2.2 nm) in the silica walls, were assembled on a QCM sensor surface using LbL technique with the aid of silica nanoparticles and polyelectrolytes (Fig. 3.9). After immersing the capsule LbL film (on a QCM resonator) in water and drying under nitrogen flow, evaporation of the trapped water from the capsule interior was measured in air. Interestingly, the frequency shifts upon water evaporation from the capsule LbL films showed a stepwise release profile even in the absence of an external stimulus. The observed stepwise release is assumed to originate from a combination

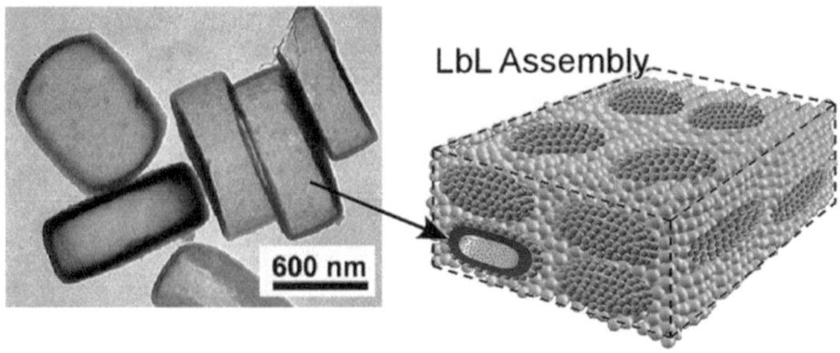

Fig. 3.9 LbL films of mesoporous silica capsule and silica nanoparticles containing hierarchical micro- and nanospaces

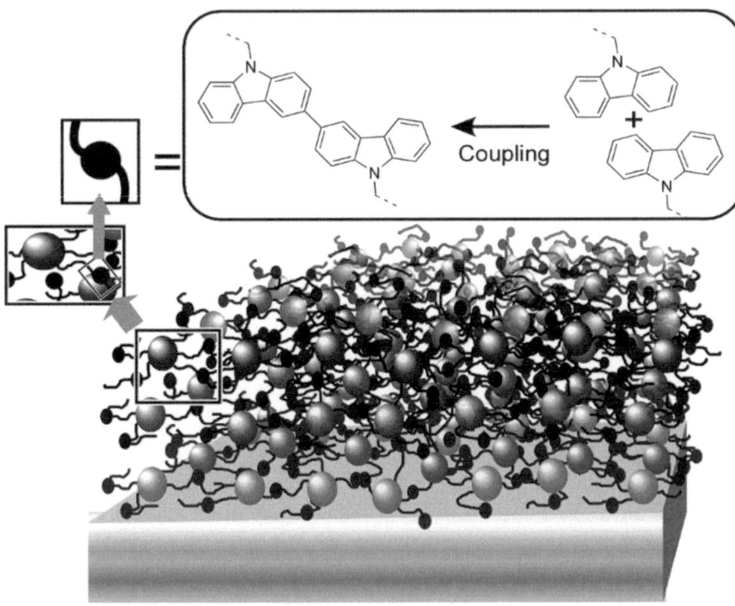

Fig. 3.10 Electrochemical coupling layer-by-layer (*ECC-LbL*) assembly

of two processes: water evaporation from the pores and capillary penetration into the pores. This release profile was also demonstrated for the controlled release of various fluid drugs including fragrance molecules. It is a rare example of a stimulus-free controlled release system for drug delivery, which has features of controlled loading/release with great utility for the development of energy-less and clean stimulus-free DDS applications.

Apart from electrostatic interactions, the LbL process can be operated involving various driving forces. Li et al. has developed a new LbL process, so called electrochemical coupling layer-by-layer (ECC-LbL) (Fig. 3.10) [32, 33]. This LbL assembly is driven using electrochemical reactions (*N*-alkylcarbazole dimerization) based on electrical control signals from electrodes. For this ECC-LbL assembly, various carbazole derivatives carrying distinctive donor (porphyrin and fluorene) and acceptor (fullerene) moieties were synthesized. For example, alternating layers of porphyrin and fullerene were prepared by cyclic voltammetric (CV) sweeps. Because assembly was conducted by electrical commands from the electrodes, films at the required location with the required timing could be prepared. As a potential application of the ECC-LbL films, photoelectronic performance was demonstrated using a prototype thin-film p-n heterojunction device with the structure ITO/porphyrin (donor)/fullerene (acceptor)/Al. In addition, ECC-LbL in area-selective patterned mode was also demonstrated using a patterned ITO electrode with a shaped mask. Area-selective assembly in patterning mode could be performed by varying the spatial application of voltage. Therefore, this ECC-LbL assembly can be coupled with the products of top-down nanotechnology such as miniaturized integrated circuits.

3.4 Macroscopic Materials with Internal Nanostructures

Transcription of assembled structures into other materials represents a powerful method to synthesize macroscopic-scale and/or macroscopic quantities of materials having precise internal nanostructures. A typical strategy for structural transcription is template synthesis as can be seen in the synthesis of mesoporous materials [34]. Mesoporous silica materials can be prepared by silica formation around template micelle assemblies followed by template removal by appropriate methods such as calcination and solvent extraction. Mesoporous carbon, which is a carbon material with regular mesopores, can be synthesized using mesoporous silica as a hard template. In the latter case, the carbon source was introduced into the mesopore channels and then carbonized, followed by selective removal of the silica framework. These mesoporous materials can be synthesized in bulk quantity and have precise internal nanostructures.

Because of the varieties of template assembly and materials available, numerous mesoporous materials have been prepared. Vinu et al. demonstrated synthesis of mesoporous carbon nitride using both carbon and nitrogen sources during template synthesis similar to that for mesoporous carbon materials [35]. The novel strategy of elemental substitution was also proposed for preparation of mesoporous boron nitride and mesoporous boron carbon nitride [36]. In addition, such mesoporous materials can be easily fuctionalized through modification at pore surfaces and/or immobilization of functional units within mesopores. For example, confinement of biomimetic systems in mesoporous media is an important subject as can be seen in the incorporation of proteins [37], amino acids [38], and vitamins [39] in mesopore channels.

Figure 3.11 shows preparation of biomimetic hybrid structures of peptide assemblies and mesoporous silica frameworks using an organosilane compound having amino acid with a condensable head and a cleavable alkyl tail [40]. This

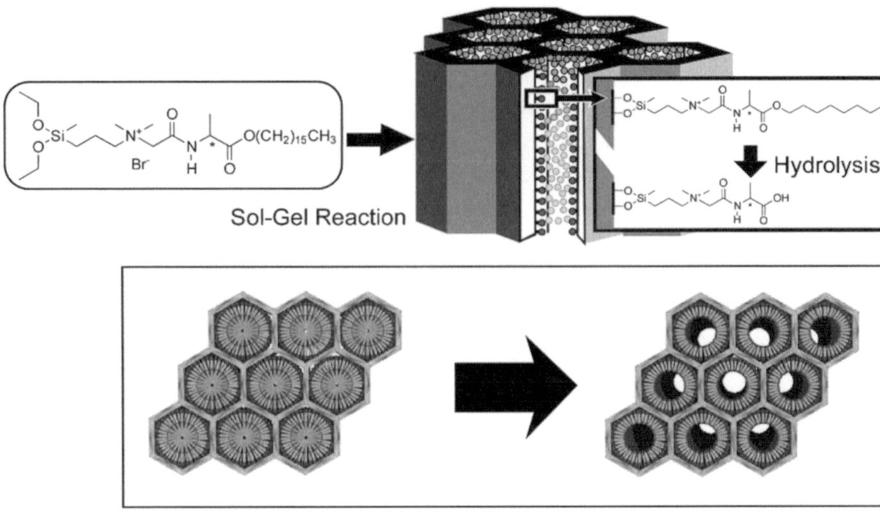

Fig. 3.11 Modification of mesoporous silica channels with amino acid residues

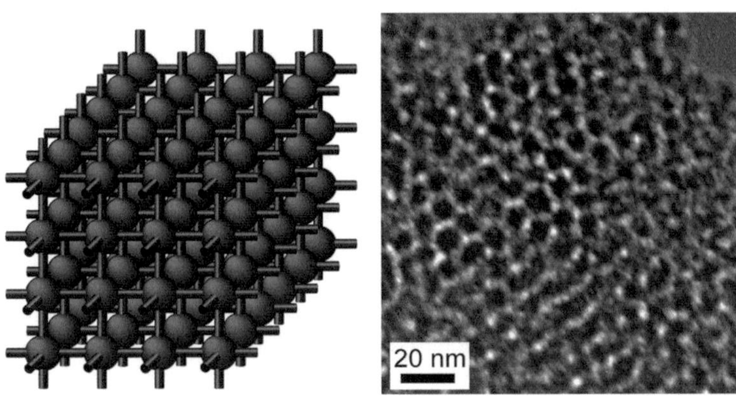

Fig. 3.12 Mesoporous nanocarbon material (carbon nanocage)

approach enables both dense modification of the pore interior with biofunctionality and excellent accessibility for incoming guests. Functional units such as amino acid residues can be covalently attached to the silica framework and cleavage and removal of the alkyl tail by selective hydrolysis of the ester at the C-terminal leaves open pores. Aida and coworkers used the hybrid structure (uncleaved material) in reactor applications for the acetalization of a ketone, such as cyclohexanone, in ethanol under mild conditions [41].

Structural control of mesoporous materials can be atained through selection of the template structure. For example, Vinu et al. developed a novel mesoporous nanocarbon material, so called carbon nanocage (Fig. 3.12), using three-dimensional, large cage-type, face-centered-cubic mesoporous silica materials

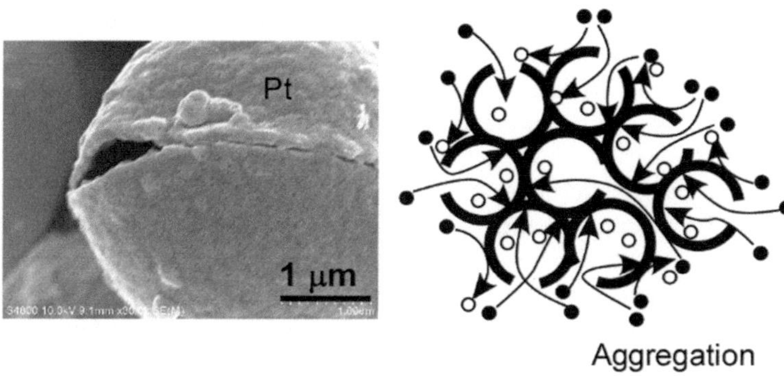

Fig. 3.13 An open-mouthed metallic microcapsule (metallic cell)

(KIT-5) as hard templates [42]. Images of high-resolution transmission electron microscopy (TEM) confirmed highly regular structures and nitrogen adsorption/desorption isotherms revealed its specific surface area (1,600 m^2/g) and specific pore volume (2.1 cm^3/g) that are much larger than those reported for conventional mesoporous carbon materials. Because of its large pore volume, carbon nanocage exhibited superior properties in separation applications such as complete removal of dyestuffs from water [43]. Furthermore, one-pot separation of tea components, such as catechin and tannic acids, was also demonstrated where competitive adsorption of catechin and tannic acid resulted in highly selective adsorption of tannic acid (about 95 %).

Mandal et al. reported fabrication of metallic (platinum) microcapsules with sufficient accessibility to their interiors through template synthesis using polystyrene spheres (Fig. 3.13) [44]. Each capsule has one or two 'mouths' that allow facile diffusion of materials from exterior to interior, a feature that suggested the label metallic cell. More importantly, the interior surfaces of these hollow structures of metallic cells are free from surface area loss associated with particle agglomeration. In particular, metallic cells exhibited enhanced activity for carbon monoxide oxidation at a rather low reaction temperature. Time courses of carbon monoxide conversion by oxygen confirmed that Pt metallic cells exhibited significantly superior catalytic activities over Pt on polymer spheres and Pt powder. While the activity ratio of the Pt metallic cell to Pt-coated spheres remains ca. 2.5-fold at room temperature, it reached up to ca. fivefold for functions at elevated temperature. The proposed concept is also applicable to alloys of Pt, Pd, and/or Rh and may provide ideal solutions to demands in practical applications such as exhaust catalysts for automobile technologies.

3.5 Future Approaches for Access from Macro to Molecule

As mentioned in the sections above, various kinds of materials with controlled structures in the range from molecular-level dimensions to visible size can be prepared. These materials have potential in practical uses. If we can easily manipulate these nanostructured materials using simple processes such as hand operation, such materials might become much more useful in our everyday lives. Methodologies for facile access from the macroscopic world to nano-sized or micro-sized structures should be established for generalization of forefront nanotechnology in a wide range of human activities.

Placing materials at interfacial media may provide a possible means to access this future target. For example, compression and expansion of macroscopic dimensions while maintaining molecular-level thickness of a film is possible for Langmuir films. These interfacial structures can couple macroscopic motions and molecular-level functions [45]. An example is shown in Fig. 3.14. In this example, molecular machines containing a central ring (1,6,20,25-tetraaza[6.1.6.1]-paracyclophan

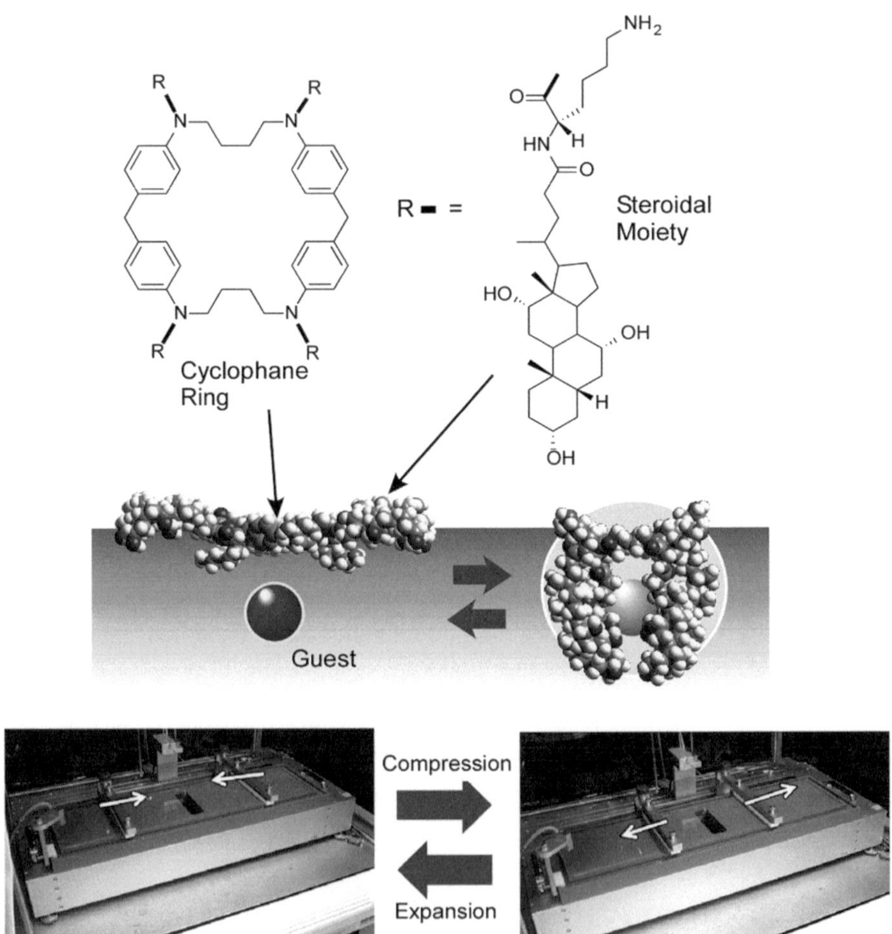

Fig. 3.14 Capture and release of a guest molecule with molecular machine driven by macroscopic mechanical motions

cyclic core) and four rigid cholic acid walls are operated by mechanical macroscopic motions at the air-water interface [46, 47]. The molecular machine can catch and release guest molecules based on hand-motion-like microscopic motions. Similarly, control of molecular phenomena including chiral recognition [48, 49], nucleic acid base discrimination [50], glucose detection by indicator displacement assay [51], and controllable drug delivery [52] by macroscopic mechanical motions have been demonstrated using interfacial media and related structures. These are realistic examples of accessing molecular-level functions from the macroscopic world. Beyond these examples, various nanostructured materials contained by interfacial media will be an important key in the near-future of technology and nanotechnology.

Acknowledgement This work was partly supported by World Premier International Research Center Initiative (WPI Initiative), MEXT, Japan and Core Research for Evolutional Science and Technology (CREST) program of Japan Science and Technology Agency (JST), Japan.

References

1. Ariga K, Hill JP, Lee MV, Vinu A, Charvet R, Acharya S (2008) Challenges and breakthroughs in recent research on self-assembly. Sci Technol Adv Mater 9:014109
2. Li M, Ishihara S, Ji Q, Akada M, Hill JP, Ariga K (2012) Paradigm shift from self-assembly to commanded assembly of functional materials: recent examples in porphyrin/fullerene supramolecular systems. Sci Technol Adv Mater 13:053001
3. Ariga K, Li M, Richards GJ, Hill JP (2011) Nanoarchitectonics: a conceptual paradigm for design and synthesis of dimension-controlled functional nanomaterials. J Nanosci Nanotechnol 11:1–13
4. Ariga K, Ishihara S, Abe H, Li M, Hill JP (2012) Materials nanoarchitectonics for environmental remediation and sensing. J Mater Chem 22:2369–2377
5. Ariga K, Ji Q, McShane MJ, Lvov YM, Vinu A, Hill JP (2014) Inorganic nanoarchitectonics for biological applications. Chem Mater 24:728–737
6. Ramanathan M, Shrestha LK, Mori T, Ji Q, Hill JP, Ariga K (2013) Amphiphile nanoarchitectonics: from basic physical chemistry to advanced applications. Phys Chem Chem Phys 15:10580–10611
7. Mori T, Sakakibara K, Endo H, Akada M, Okamoto K, Shundo A, Lee MV, Ji Q, Fujisawa T, Oka K, Matsumoto M, Sakai H, Abe M, Hill JP, Ariga K (2013) Langmuir nanoarchitectonics: one-touch fabrication of regular-size nanodisks at the air-water interface. Langmuir 29:7239–7248
8. Ariga K, Ji Q, Mori T, Naito M, Yamauchi Y, Abe H, Hill JP (2013) Enzyme nanoarchitectonics: organization and device application. Chem Soc Rev 42:6322–6345
9. Ariga K, Mori T, Hill JP (2013) Interfacial nanoarchitectonics: lateral and vertical, static and dynamic. Langmuir 29:8459–8471
10. Shrestha LK, Ji Q, Mori T, Miyazawa K, Yamauchi Y, Hill JP, Ariga K (2013) Fullerene nanoarchitectonics: from zero to higher dimensions. Chem Asian J 8:1662–1679
11. Hill JP, Wakayama Y, Schmitt W, Tsuruoka T, Nakanishi T, Zandler ML, McCarty AL, D'Souza F, Milgrom LR, Ariga K (2006) Regulating the stability of 2D crystal structures using an oxidation state-dependent molecular conformation. Chem Commun 2320–2322
12. Hill JP, Wakayama Y, Akada M, Ariga K (2007) Self-assembly structures of a phenol-substituted porphyrin in the solid state: hydrogen bonding, Kagomé lattice, and defect tolerance. J Phys Chem C 111:16174–16180

13. Hill JP, Wakayama Y, Akada M, Ariga K (2009) Two-dimensional molecular array of porphyrin derivatives with bright and dark spots as a model of two-digit molecular-dot memory. Synth Met 159:765–768
14. Ariga K, Lee MV, Mori T, Yu X-Y, Hill JP (2010) Two-dimensional nanoarchitectonics based on self-assembly. Adv Colloid Interface Sci 154:20–29
15. Ariga K, Ito H, Hill JP, Tsukube H (2012) Molecular recognition: from solution science to nano/materials technology. Chem Soc Rev 41:5800–5835
16. Oishi Y, Torii Y, Kato T, Kuramori M, Suehiro K, Ariga K, Taguchi K, Kamino A, Koyano H, Kunitake T (1997) Molecular patterning of a guanidinium/orotate mixed monolayer through molecular recognition with flavin adenine dinucleotide. Langmuir 13:519–524
17. Oishi Y, Kato T, Narita T, Ariga K, Kunitake T (2008) Formation of nano-domains confined in two-dimensional molecular plane. Langmuir 24:1682–1685
18. Sathish M, Miyazawa K, Hill JP, Ariga K (2009) Solvent-engineering for shape-shifter *pure* fullerene (C_{60}). J Am Chem Soc 131:6372–6373
19. Shrestha LK, Sathish M, Hill JP, Miyazawa K, Tsuruoka T, Sanchez-Ballester NM, Honma I, Ji Q, Ariga K (2013) Alcohol-induced decomposition of olmstead's crystalline Ag(I)-fullerene heteronanostructure yields 'Bucky Cubes'. J Mater Chem C 1:1174–1181
20. Shrestha LK, Yamauchi Y, Hill JP, Miyazawa K, Ariga K (2013) Fullerene crystals with bimodal pore architectures consisting of macropores and mesopores. J Am Chem Soc 135:586–589
21. Charvet R, Acharya S, Hill JP, Akada M, Liao M, Seki S, Honsho Y, Saeki A, Ariga K (2009) Block-copolymer-nanowires with nanosized domain segregation and high charge mobilities as stacked p/n heterojunction arrays for repeatable photocurrent switching. J Am Chem Soc 131:18030–18031
22. Acharya S, Hill JP, Ariga K (2009) *Soft* Langmuir-Blodgett technique for *hard* nanomaterials. Adv Mater 21:2959–2981
23. Ariga K, Yamauchi Y, Mori T, Hill JP (2013) What can be done with the Langmuir-Blodgett method? Recent developments and its critical role in materials science. Adv Mater 25:6477–6511
24. Ariga K, Hill JP, Ji Q (2007) Layer-by-layer assembly as a versatile bottom-up nanofabrication technique for exploratory research and realistic application. Phys Chem Chem Phys 9:2319–2340
25. Ariga K, Ji Q, Hill JP, Bando Y, Aono M (2012) Forming nanomaterials as layered functional structures towards materials nanoarchitectonics. NPG Asia Mater 4:e17
26. Ariga K, Yamauchi Y, Rydzek G, Ji Q, Yonamine Y, Wu KC-W, Hill JP (2014) Layer-by-layer nanoarchitectonics: invention, innovation, and evolution. Chem Lett 43:36–68
27. Ariga K, Vinu A, Ji Q, Ohmori O, Hill JP, Acharya S, Koike J, Shiratori S (2008) A layered mesoporous carbon sensor based on nanopore-filling cooperative adsorption in the liquid phase. Angew Chem Int Ed 47:7254–7257
28. Ji Q, Yoon SB, Hill JP, Vinu A, Yu J-S, Ariga K (2009) Layer-by-layer films of dual-pore carbon capsules with designable selectivity of gas adsorption. J Am Chem Soc 131:4220–4221
29. Ji Q, Honma I, Paek S-M, Akada M, Hill JP, Vinu A, Ariga K (2010) Layer-by-layer films of graphene and ionic liquid for highly selective gas sensing. Angew Chem Int Ed 49:9737–9739
30. Ji Q, Miyahara M, Hill JP, Acharya S, Vinu A, Yoon SB, Yu J-S, Sakamoto K, Ariga K (2008) Stimuli-free auto-modulated material release from mesoporous nanocompartment films. J Am Chem Soc 130:2376–2377
31. Ji Q, Acharya S, Hill JP, Vinu A, Yoon SB, Yu J-S, Sakamoto K, Ariga K (2009) Hierarchic nanostructure for auto-modulation of material release: mesoporous nanocompartment films. Adv Funct Mater 19:1792–1799
32. Li M, Ishihara S, Akada M, Liao M, Sang L, Hill JP, Krishnan V, Ma Y, Ariga K (2011) Electrochemical coupling layer-by-layer (ECC-LbL) assembly. J Am Chem Soc 133:7348–7351
33. Li M, Ishihara S, Ji Q, Ma Y, Hill JP, Ariga K (2012) Electrochemical coupling layer-by-layer (ECC-LbL) assembly in patterning mode. Chem Lett 41:383–385

34. Ariga K, Vinu A, Yamauchi Y, Ji Q, Hill JP (2012) Nanoarchitectonics for mesoporous materials. Bull Chem Soc Jpn 85:1–32
35. Vinu A, Ariga K, Mori T, Nakanishi T, Hishita S, Golberg D, Bando Y (2005) Preparation and characterization of well ordered hexagonal mesoporous carbon nitride. Adv Mater 17:1648–1652
36. Vinu A, Terrones M, Golberg D, Hishita S, Ariga K, Mori T (2005) Synthesis of mesoporous BN and BCN exhibiting large surface areas via templating methods. Chem Mater 17:5887–5890
37. Vinu A, Miyahara M, Ariga K (2005) Biomaterial immobilization in nanoporous carbon molecular sieves: influence of solution pH, pore volume and pore diameter. J Phys Chem B 109:6436–6441
38. Vinu A, Hossain KZ, Satish Kumar G, Ariga K (2006) Adsorption of L-histidine over mesoporous carbon molecular sieves. Carbon 44:530–536
39. Hartmann M, Vinu A, Chandrasekar G (2005) Adsorption of vitamin E on mesoporous carbon molecular sieves. Chem Mater 17:829–833
40. Zhang Q, Ariga K, Okabe A, Aida T (2004) Condensable amphiphile with cleavable tail as 'lizard' template for sol-gel synthesis of functionalized mesoporous silica. J Am Chem Soc 126:988–989
41. Otani W, Kinbara K, Zhang Q, Ariga K, Aida T (2007) Catalysis of a peptidic micellar assembly covalently immobilized within mesoporous silica channels: importance of amphiphilic spatial design. Chem Eur J 13:1731–1736
42. Vinu A, Miyahara M, Sivamurugan V, Mori T, Ariga K (2005) Large pore cage type mesoporous carbon, carbon nanocage: a superior adsorbent for biomaterials. J Mater Chem 15:5122–5127
43. Ariga K, Vinu A, Miyahara M, Hill JP, Mori T (2007) One-pot separation of tea components through selective adsorption on pore-engineered nanocarbon, carbon nanocage. J Am Chem Soc 129:11022–11023
44. Mandal S, Sathish M, Saravanan G, Datta KKR, Ji Q, Hill JP, Abe H, Honma I, Ariga K (2010) Open-mouthed metallic microcapsules: exploring performance improvements at agglomeration-free interiors. J Am Chem Soc 132:14415–14417
45. Ariga K, Mori T, Hill JP (2012) Mechanical control of nanomaterials and nanosystems. Adv Mater 24:158–176
46. Ariga K, Terasaka Y, Sakai D, Tsuji H, Kikuchi J (2000) Piezoluminescence based on molecular recognition by dynamic cavity array of steroid cyclophanes at the air-water interface. J Am Chem Soc 122:7835–7836
47. Ariga K, Nakanishi T, Terasaka Y, Tsuji H, Sakai D, Kikuchi J (2005) Piezoluminescence at the air-water interface through dynamic molecular recognition driven by lateral pressure application. Langmuir 21:976–981
48. Michinobu T, Shinoda S, Nakanishi T, Hill JP, Fujii K, Player TN, Tsukube H, Ariga K (2006) Mechanical control of enantioselectivity of amino acid recognition by cholesterol-armed cyclen monolayer at the air-water interface. J Am Chem Soc 128:14478–14479
49. Michinobu T, Shinoda S, Nakanishi T, Hill JP, Fujii K, Player TN, Tsukube H, Ariga K (2011) Langmuir monolayer of cholesterol-armed cyclen complex that can control enantioselectivity of amino acid recognition by surface pressure. Phys Chem Chem Phys 13:4895–4900
50. Mori T, Okamoto K, Endo H, Hill JP, Shinoda S, Matsukura M, Tsukube H, Suzuki Y, Kanekiyo Y, Ariga K (2010) Mechanical tuning of molecular recognition to discriminate the single-methyl-group difference between thymine and uracil. J Am Chem Soc 132:12868–12870
51. Sakakibara K, Joyce LA, Mori T, Fujisawa T, Shabbir SH, Hill JP, Anslyn EV, Ariga K (2012) A mechanically-controlled indicator displacement assay. Angew Chem Int Ed 51:9643–9646
52. Izawa H, Kawakami K, Sumita M, Tateyama Y, Hill JP, Ariga K (2013) β-Cyclodextrin-crosslinked alginate gel for patient-controlled drug delivery systems: regulation of host-guest interactions with mechanical stimuli. J Mater Chem B 1:2155–2161

Chapter 4
A Brief History of Nanoscience and Foresight in Nanotechnology

Štefan Luby, Martina Lubyová, Peter Šiffalovič, Matej Jergel, and Eva Majková

Abstract Nanotechnology as a natural continuation of microtechnology introduced a new bottom-up approach in the building of structures. In this paper we summarize a brief history of nanoscience and nanotechnology by documenting the main milestones on the roadmap of this branch since the beginning of the twentieth century. We discuss the new properties of materials and structures appearing in the nanoworld that originate from both classical and quantum phenomena. We provide a critical analysis of inflated versus realistic expectations of the new technology. Attention is also paid to risks and regulations in the field, as well as codes of conduct of responsible nanoscientists and specific aspects of nanoethics that open a new chapter in ethical studies. The study elaborates on five foresight topics covering the building of structures atom-by-atom, the possibilities to close the appearing nano-divide, the future of silicon, single-particle devices, and sustainability in the field and single-particle devices. Among single particle devices the focus is on the transistors and sensors. We also highlight the role of social sciences and humanities in nanoscience and nanotechnology in the fields such as philosophy, psychology, security, the protection of privacy and intellectual property rights. Ethics is the main area of these activities.

Abbreviations

AMD	Advanced Micro Devices
AFM	Atomic force microscope
CNT	Carbon nanotube

Š. Luby (✉) • P. Šiffalovič • M. Jergel • E. Majková
Institute of Physics, Slovak Academy of Sciences, Dúbravska 9, Bratislava 84511, Slovakia
e-mail: stefan.luby@savba.sk; peter.siffalovic@savba.sk; matej.jergel@savba.sk; eva.majkova@savba.sk

M. Lubyová
Institute for Forecasting, Slovak Academy of Sciences, Šancova 56, Bratislava 81105, Slovakia
e-mail: progluby@savba.sk

M. Bardosova, T. Wagner (eds.), *Nanomaterials and Nanoarchitectures*,
NATO Science for Peace and Security Series C: Environmental Security,
DOI 10.1007/978-94-017-9921-8_4

CNTFET	Field effect transistor with channel from nanotube
D	Dimension
EC	European Commission
ERC	European Research Council
FET	Field effect transistor
FP	Framework program of EC
GMR	Giant magnetoresistance
IBM	International Business Machines
IC	Integrated circuit
ICT	Information and communication technology
MFM	Magnetic force microscope
MBE	Molecular beam epitaxy
NMP	Nanoscience, nanotechnologies, materials and producing technologies, thematic area of FP7
N&N	Nanoscience and nanotechnology
NNI	National Nanotechnology Initiative, USA
NP	Nobel Prize
RAM	Random access memory
ROM	Read only memory
SPM	Scanning probe microscope, into this category belong AFM, MFM and STM
STM	Scanning tunneling microscope

4.1 Introduction

Nanosience and nanotechnology (N&N) represent a rapidly expanding research area. The main aims of the discipline cover design, preparation, manipulation and application of materials and structures having at least one dimension within the interval from 1 to 100 nanometers (nm). At the left margin this interval touches the quantum world and the emerging picotechnology. At the right margin the interval adjoins the submicrometer region defined by the range of 100–1,000 nm. It was introduced into the terminology of microelectronics when the 1 μm boundary has been penetrated. Typical nano-objects are molecules, atomic clusters, nanocrystallites, nanoparticles, nanowires, nanolayers, etc.[1] In order to illustrate the fast progress and main principles of nanoscience and nanotechnologies, in the Box 4.1 below we provide selected statements by well-known scientists and Nobel Prize laureates in the field.

[1]For comparison the length of carbon bond and diameter of DNA are 0.2 and 2 nm, respectively, the size of bacteria equals to approx. 200 nm.

Box 4.1. Statements About N&N

Atoms cannot be perceived by senses. They exist only in our imagination. Ernst Mach, 1900

The principles of physics do not speak against the possibility of manipulating things atom-by-atom. Richard Feynman, 1959

Nanotechnology is the art of building devices at the ultimate level of finesse, atom-by-atom. Richard Smalley, 2000

Nanotechnology is …a way of precise, controlled building, with incidentally environmental benignity built in by design. Roald Hoffmann

Nanotechnology has given us the tools…to play with the ultimate toy box of nature – atoms and molecules….The possibilities to create new things appear limitless. Horst Stromer

Those who control nanotechnology will lead the industry. M. Nakamura, Hitachi

Biggest breakthroughs in nanotechnology are going to be in the new materials. T. Kirkpatrick, General Electric

Our ability to reap the long-term benefits of nanotechnology will depend on how well industry and governments manage the safety and performance of the first generation of nanotechnology products. A. Maynard, University of Michigan

Sources: [R6, R10]

4.2 Main Milestones in the Development of Nanoscience and Nanotechnology

In Box 4.2 below we enumerate some of the main milestones as summarized by Stix [1] and they were completed by us. We use a broader concept including zero dimension (0D) molecules, clusters, nanoparticles, one-dimension (1D) nanowires, nanotubes and two-dimensional (2D) layers. This means that also spintronics is covered here as a branch of nanoscience.[2] The milestones summarized in Box 4.2 document that the fundaments of nanoscience have been laid since the early twentieth century.

[2]In spintronics information is carried by both charge and spin of electron. Information is lost at the length l_{sf} of the spin-flip. l_{sf} is about 10–40 nm in cobalt and 40–140 nm in copper, which are two common metals used in spin valves. Therefore spintronics belongs to N&N.

Box 4.2. Summary of the Main Milestones

[M1] A. Einstein estimated the diameter of sugar molecule at about 1 nm.
[M2] M. Knoll and E. Ruska invented electron microscope, NP.
[M3] R. Feynman presented his talk "Plenty of room at the bottom".
[M4] A. Y. Cho and J. Arthur demonstrated MBE deposition of monolayers.
[M5] N. Taniguchi conceived the word *nanotechnology.*
[M6] A. J. Heeger, A. G. MacDiarmid and H. Shirakawa discovered conducting (conjugated) polymers.
[M7] G. Binnig and H. Rohrer invented STM, NP.
[M8] R. F. Curl, H. W. Kroto and R. E. Smalley discovered fullerenes, NP.
[M9] G. Binnig, C. F. Quate and C. Berger invented AFM.
[M10] K. E. Drexler published his book *Engines of creation.*
[M11] T. A. Fulton and G. J. Dolan following the idea by D. Averin and K. Likharev from 1985 fabricated single electron transistor.
[M12] A. Fert and P. Grünberg discovered GMR, NP.
[M13] B. Dieny et al. invented spin valve.
[M14] S. Iijima described carbon nanotubes.
[M15] C. Dekker et al. fabricated FET from the carbon nanotube (CNTFET).
[M16] N. Garcia et al. discovered ballistic magnetoresistance.
[M17] S. Lloyd published ultimate physical limits to computation.
[M18] W. Clinton announced NNI of USA.
[M19] V. Derycke et al. integrated CNTFETs to form logic gates.
[M20] A. Geim and K. Novoselov isolated graphene, NP.
[M21] N&N Action plan for Europe was declared.
[M22] G. Meyer made out single molecule switch.
[M23] F. Schedin et al. detected individual gas molecule on the surface of graphene
[M24] M. C. Lemme et al. fabricated graphene FET.
[M25] The force to move an atom over surface was measured at Almaden Res. Ctr.
[M26] K. I. Bolotin et al. determined the mobility of electrons (2×10^5 cm^2/Vs) in free hanging graphene.
[M27] A. Bérut et al. experimentally verified the minimum amount of energy kTln2 (Landauer principle, 1961) required to change one bit of information

4.3 Extrapolation of Microtechnology into the Nanoworld

N&N is a continuation of five decades of miniaturization and growing level of integration described by the Moore's law. Its original formulation was born in 1965 when G. Moore noticed that the number of components per integrated circuit for minimum component cost increased roughly by a factor of 2 annually between 1962 and 1965 [2, 3]. He predicted that the rate would remain constant for at least 10 years and by 1975 the number of components would reach 65,000.

His prediction proved to be accurate: the number of transistors on integrated circuits has been doubling approximately every 2 years since 1965. In this regard, Intel executive D. House introduced a notion of 18 months as the period needed for doubling the chip performance (i.e. the combination of density of transistors and their speed of operation) [3, 4]. However, nowadays the Moore's law approaches a physical barrier [5, R10]. The future of electronics is therefore discussed more vividly than before and the role of N&N grows.

N&N currently represent a common denominator for the recent developments in chemistry/nanochemistry, biology/nanobiology and medicine/nanomedicine, whenever these disciplines touch upon the dimension of a single molecule. The interactions obey the laws of physics which are valid at the level of elementary particles (therefore we do not speak about nanophysics). N&N are the point of confluence of physics, chemistry, biology, material research and engineering. Interdisciplinary education involving the above mentioned branches is unavoidable for the further development in this field [R6].

4.4 Scanning Tunneling Microscope and Manipulation of Materials at the Nano-Level

The scanning tunneling microscope (STM) was a breakthrough in the investigation of material with atomic resolution and in atom-by-atom manipulation. STM has been invented at the IBM Lab in Zürich [6] (G. Binnig and H. Rohrer were awarded the Nobel Prize). Afterwards also other types of scanning probe microscopes (SPM) were created – atomic force microscope (AFM) and magnetic force microscope (MFM). In these devices the information is not obtained by using radiation but by mechanical sharp tip scanning over the surface at a very small distance and imaging its topography [7]. All the devices have the resolution in sub-nm area, which is not limited by diffraction. They are sometimes compared to Braille-reading because the epistemological basis of STM is not analogous to visual perception but it is in fact tactile. STM does not "see" the atoms – it "feels" them and provides the outcomes of measurement in a digital representation [8]. These devices are also used for manipulation of atoms or molecules. In this way the Feynman vision [9] of maneuvering matter atom-by-atom has been fulfilled at least at the laboratory scale, because the process is too slow for practical applications. Anyway, the SPMs heralded the emergence of N&N [10].

Nowadays the structures are not synthesized atom-by-atom but from bigger blocks, e.g. from nanoparticles composed of thousands of atoms. This additive process is called *bottom – up* approach. It is complementary to the *top – down* formation of structures, broadly used in micro- and nanoelectronics, where the substrate (mostly silicon) covered by various layers is patterned by removing certain parts with the help of optical or electron lithography and chemical or physical etching. Both approaches have their place in the history of technology starting from architecture.[3]

There are also hybrid technologies that take the advantage of both above mentioned approaches. An example is given by Kunze [11]. First a top-down approach is used when furrows into polymer layer are dynamically ploughed by vibrating tip of AFM followed by the subsequent chemical etching leading to grooves. The second step is bottom-up filling of grooves by nanoparticles.

4.5 "There Is Plenty of Room at the Bottom"

The famous lecture by Feynman [9] was inspired by the continual progress in reaching lower and lower temperatures and achieving of higher and higher vacuum. Both fields seemed to be "bottomless". Feynman generalized this observation towards the manipulating and controlling things at ever smaller scale. This type of activity is also prolific for new inventions. Examples are represented by small bearings running dry because heat escapes away from the small device rapidly, or amorphous materials which are more appropriate for miniaturization purposes than metals where the grain structure is a limiting factor.

However, the most visionary Feynman's ideas were those of the miniaturization of recording media linked to almost bottomless growth of the density of information. He assumed that in order to write one bit of information, at least 100 atoms would be needed in order to secure that none of the information is lost through diffusion. Then he estimated how the information from 24 millions of volumes of the Library of Congress, British museum and National Library in France (about 10^{15} bits) could be recorded: under the above mentioned assumption one would need 10^{17} atoms. The density of atoms in silicon is $5.1 \times 10^{19}/\text{mm}^3$, so the required 10^{17} atoms would fill a cube of material with the edge of slightly more than 0.1 mm.

Fifty years after the lecture [9] the information density of semiconductor media produced by AMD, IBM or Intel is about $5 \times 10^6/\text{mm}^2$, i.e. one billion of bits per chip with one-transistor memory cells. For 10^{15} bits we need 10^6 chips and the equivalent volume of Si is about 0.5×10^6 cm^3. This is 10^{12} times more than the cube with the edge of 0.1 mm. At the same time the miniaturization approaches the saturation regime and the Moore's law is expected to hit its margins within a decade

[3]Old amphitheatres and dwellings digged into rocks, like in Italian town Matera, are examples of the first and second approach, respectively.

[5]. Thus the Feynman's vision was too optimistic, however, it was farseeing and mobilizing for the development on N&N.

4.6 New Phenomena and Qualities in the Nanoworld

New properties and qualities of materials and structures that appear when we enter the nanoworld are discussed in [9]. These properties and qualities originate from classical and quantum phenomena.

In the realm of classical physics we must consider:

(a) large surface to bulk ratio;
(b) small size of nano-entities;
(c) self-assembly of nanoparticles.

Ad (a), 10-nm particle has 10^5 of atoms on its surface, while in a 1-nm particle almost all atoms are on the surface [12]. This manifests itself in the decrease of the melting temperature. Gold nanoparticles with the diameter of 2 nm melt already at 500 °C because of lower cohesion of the crystalline lattice due to smaller number of bonds to the neighboring atoms. *Per analogiam,* in the nano-world many properties are changing dramatically: insoluble substances become soluble, non-inflammable things are burning, chemically inert material like Au particles show catalytic properties. From the same point of view we must consider the toxicity. Chemically toxic materials like As, Cd, Co show toxicity in bulk [13]. In the nano-world we also speak about physical toxicity. Large effective surface of structures generates free radicals with threatening consequences, such as, for example, inflammations of lungs, fibrosis, or tumors [14]. In this way a conception that all nanoparticles are toxic was born. Nevertheless, in medical applications dominate iron-oxide nanoparticles that are safe, as far as we know; moreover, iron is metabolized [15, 16].

Ad (b), small nanoparticles penetrate through membranes, opaque films become transparent, nanoparticle solutions are colored depending on their size. The properties are used, for example, in the glass manufacture where this effect has been in fact applied since ancient times – an example is the colloid suspension of gold called *Cassius purpur*. An example of medical application is given by the *molecular Trojan horse* capsules [17] designed for patients affected by genetic diseases. For patients suffering from the lack of specific enzyme and whose his immune system does not allow introducing the enzymes directly into their bodies, hollow capsules (50–500 nm) are used. The capsules protect the substance inside and enable its exchange with the environment via protein channels. Another example of utilizing the small dimensions are small nanoclusters in the matrix of material (e.g. ceramics) that increase its strength (Fig. 4.1) [18].

Ad (c), an important phenomenon is the self-assembly of small nanoparticles due to the so-called van der Waals attraction, stearic repulsion and magnetic interaction between them (if particles are magnetic). Through the self-assembly mechanism, regular arrays are formed (Fig. 4.2) that have various applications in devices, e.g. in gas sensors [19].

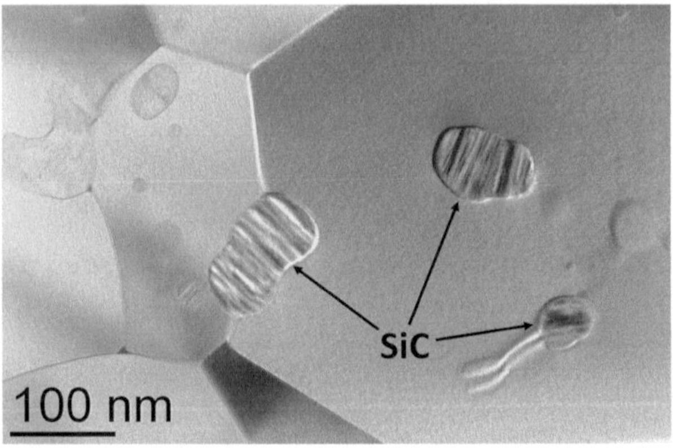

Fig. 4.1 Nano-inclusions of silicon carbide (*SiC*) within the matrix improve the room temperature strength of Si_3N_4 ceramics from 870 MPa to 1.2 GPa. The deformation at temperature over 1,200 °C is less than 1 % a year at the load 100 MPa (With the courtesy of P. Šajgalík, Institute of Inorganic Chemistry SAS, Bratislava [18])

The most interesting phenomena are introduced through entering the territory of quantum physics. Some of them could be exploited in the future quantum computing and quantum encryption [20]. On the other side, it is also important to keep in mind the quantum limits of classical devices: when the mean-free-path of electrons is larger then the device dimensions, the wave nature of electron manifests itself, thus changing the dielectric constant, energy gap and other properties [21]. The quantum behavior is observed not only in electrical properties, but also in thermal conductivity and heat flow [22]. The quantum phenomena limit also the magnetic recording density. For example, in media based on high anisotropy FePt nanoparticles the recording density is 43 Tb/in^2 due to the thermal fluctuations and spontaneous magnetization reversal even at low temperature of 1.3 K [23].

The onset of the quantum size effects due to the discreteness of the electronic level structure is explained by Yacamán et al. [24]: according to the Kubo criterion, quantum size effect requires the spacing between levels $\Delta >> kT$ (where k is the Boltzmann constant and T is temperature). In Au nanoparticles this criterion is fulfilled if number of atoms is less than 400, i.e. the size is about 2 nm. With small semiconducting nanoparticles (3–4 nm) we observe the enlargement of the energy gap [25], which should be borne in mind in the case of semiconductor nanoparticle sensors.

4.7 N&N in the Post-Hype Era

New technologies typically pass through the cycle of hyperbolic expectations. Five phases of this cycle can be characterized as follows:

(i) launching the technology;
(ii) over – optimistic expectations and predictions;
(iii) des-illusion;
(iv) enlightenment and realistic views;
(v) expectations matched at the level of productivity.

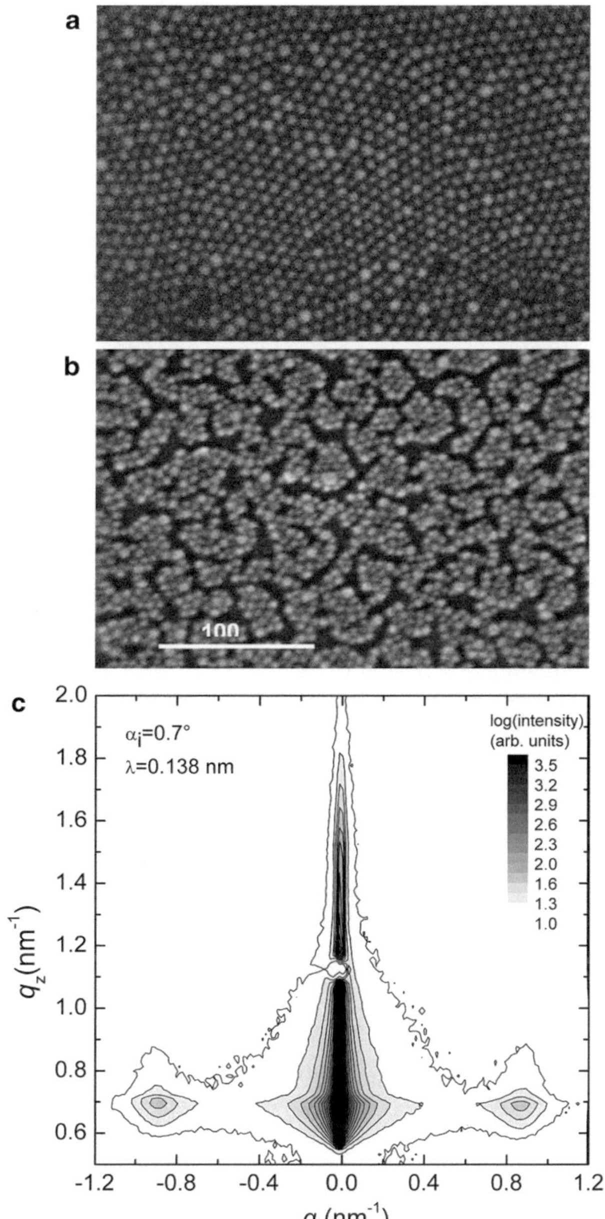

Fig. 4.2 Fe_2O_3 nanoparticle arrays for gas sensors before (**a**) and after (**b**) removal of protecting insulating surfactant in order to achieve the electrical conductivity. In (**c**) GISAXS spectrum in the reciprocal space with well developed side wings typical for high regularity of array (Institute of Physics SAS, GISAXS measurement at the synchrotron beam)

At present N&N enter the phase of productivity and therefore safety and security become the imperative of the day [26].

Unrealistic expectations and plans were formulated also at the launch of the NNI [R1]. It was expected that before 2020 we would have materials ten-times stronger than steel having only a fraction of its weight, that tumors could be diagnosed at the stage of having a few cells, or that the whole information from the Library of Congress will be recorded in the volume of one sugar cube. However, from Sect. 4.5 it follows that nowadays the number of "sugar cubes" needed would be about 10^5 and that before 2020 an improvement of at most one order of magnitude could be reached.

Nevertheless, the progress is tremendous and it manifests itself in many fields, such as for example [R6, R10]:

- high strength materials;
- durable composites;
- new surface coatings;
- improved medicaments;
- artificial skin, bones and cardiac tissues;
- high efficiency plasmonic solar cells;
- CNT transistors and low loss cables;
- sensors and nanosensors;
- filters for decontamination of water;
- plastic items from plant-derived raw materials (instead of petrochemicals);
- thermochromic self-cleaning and anti-fog windows, etc.

The world investments into N&N between 2000 and 2015 are estimated as 250 bil. US$. In FP7 over 3 bil. € was spent and ERC supported 300 N&N related projects [R10].

4.8 Risks and Regulations, Code of Conduct

N&N generate many concerns related to health risks and environmental threats. One may ask to what extent could be N&N harmful. The question is related to the issues of use or misuse. On the one hand, knife, dynamite or alcohol can be potentially harmful if used in an improper way. On the other hand, there is one significant difference – while knife is visible to the naked eye, nanoscale entities are invisible to unmediated senses.

Motivated by these concerns, a new journal *Nanoethics* was launched in 2007. A year later the European Commission adopted a voluntary Code of Conduct [R7]. For the purpose of the latter the nanoobjects are defined as nanoparticles and their aggregations at nanoscale, nanosystems, nanomaterials, nanostructured materials and nanoproducts. N&N research is defined to encompass all research activities dealing with matter at the nanometric scale from 1 to 100 nm. The Code is based on a set of general principles, such as the following:

(i). N&N research should be comprehensible to the public;
(ii). N&N research activities should be safe, ethical and contributing to sustainable development;
(iii). N&N research should be conducted in accordance with the precautionary principle, anticipating potential environmental health and safety impacts.

Moreover, research authorities and standardization bodies should endeavor to adopt N&N standard terminology to facilitate the communication of the scientific evidence. Researchers should guarantee that scientific data and results are duly peer-reviewed before being widely disseminated outside the scientific community in order to ensure their clarity and balanced presentations. These recommendations are elaborated into 13 specific responsibilities which are of general validity, not being unique only to N&N [27].

Similarly like in other research areas, a crucial problem in N&N is whether the researcher would be able to stop his or her work to avoid possible dangerous consequences after recognizing them. This seems to be problematic because the inventions are too appealing to inventors, their institutions, or industries. Moreover, inventions and discoveries appear often simultaneously at two or more places when the threshold level of knowledge is reached. The remedy seems be brought by even more technology [28].

The publicly shared sentiments originate from the fear of self-replication of nanosystems, uncontrolled robots, trans-human creatures, biological weapons, toxicity of nanomaterials, latent fingerprints, "molecularly naked" patients, etc. Nanoelectronics with ubiquitous sensors and cameras interconnected with "thirsty" computers represent a typical "Big Brother" technology [29, 30]. While some of the fears are real, others – such as the self-replication of systems or trans-human creatures, belong to the world of science fiction. They are propagated also due to Drexler's publications [31, 32] about assemblers, i.e. devices having a submicroscopic robotic arm under computer control, which should allow for the construction of large atomically precise objects and for their replication.[4]

On the other hand, the worries related to the loss of privacy were formulated first of all as a consequence of the fast development of microelectronics. The toxicity of nanomaterials as a potential cause of serious illnesses deserves much attention. Nevertheless, at present in Europe there are no lethal cases registered.

For the field of N&N incremental and cooperative risk management approach was recommended that consists of four consecutive steps [34]:

1. information gathering/dissemination;
2. multi stakeholders norms, self-regulation;
3. enforced self-regulation;
4. hard law/legislation.

[4]A self-replicating device was described earlier by von Neumann [33].

Table 4.1 Basic nanoparticles used in N&N

Metals	Oxides	Compounds	Semiconductors
Ag, Al, Au, Co, Cu,	Al_2O_3, Fe_2O_3, SiO_2,	AgCu, $BaTiO_3$,	CdS, CdTe, Si,
Ni, Pd, Pt, Zn	TiO_2, WO_3, ZnO	CuNi, MoS_2	GaAs, InP

4.9 Ubiquitous Nanoparticles

N&N work with many types of nanoparticles (Table 4.1). Some of them are chemically toxic. Threats like cancerogenity, mutagenity, genotoxicity are related to their composition and size. Respiration and inhalation cause troubles if particles are smaller than 5 μm. The essential equation reads: $hazard = risk \times exposition$ [13, 14]. Therefore, the first measure is to avoid exposition. The dangerous channels are related to inhalation and skin contact. Particles stored in liquids are less dangerous than aerosols. Safety precautions involve activities such as toxicological screening, measuring of particle concentration in the air, water and food, controlling of explosive and flammable mixtures with air or oxygen and pyrophoric matters [35].

On the other hand, nanoparticles can provide remedy to various diseases or health-related problems. Well-known are the antimicrobial properties of silver nanoparticles. However, the overuse of nanosilver products could pose a danger to the needed microorganisms and enable the flourishing of resistant strains [36].

4.10 Foresight in Nanotechnology: The Main Issues

4.10.1 Building of Structures Atom-by-Atom: Yes or No?

Feynman's idea of manipulation things atom-by-atom permeated through N&N as a "romantic variant". Process of manipulation is time consuming and it could be calculated that composing macroscopic structures even with the speed of light needs approximately as much time as the period elapsed since the Big Bang. An outspoken critic of molecular assembly as advocated by Drexler [31] was Smalley, the "father of nanotechnology" [37]. His two main objections were termed "sticky fingers problem" and "fat fingers problem". In other words, the former problem expresses that it is difficult to position atom on another atom of comparable size and stickiness (a similar problem arises when manipulating a scotch tape which sticks to the fingers). One can immobilize one of the atoms, but this complicates the process. Similarly, the latter problem expresses that it is difficult to localize in the nanometer area of molecular assembly all the necessary manipulators with their tips ("fingers") [38] (a similar problem arises when we touch two icons on the display of the phone). Here we should mention also the diffusion of atoms due to lattice vibrations that destabilize the composed structures. Overcoming the diffusion by cooling is another source of complication.

These arguments are not accepted unanimously and the search for new approaches will by no means continue. The above mentioned limitations do not exclude the synthesis of materials and/or devices with new properties by controlled manipulation of their microstructure on the atomic level. The procedures comprise isolated, substrate-supported or embedded nanometer-sized particles, wires, films and surface regions of bulk materials [39]. The low speed of the molecular assembly could be circumvented by simultaneous manipulation of many atoms by multi-tip SPM instrument. This idea was formulated by STM inventors even in eighties of the previous century. Production of features with line-width of 15 nm and areal densities of 100–200 Gbin^{-1} using this approach was discussed by Moriarty [10]. Nevertheless, there the goal was not a bottom-up construction of a structure, but the development of SPM as a patterning tool that can surpass the resolution of lithographic techniques without sacrificing speed. A parallel cantilever probe platform for nanoscale applications is in Fig. 4.3 [40].

At present nanotechnology is based on chemical technologies – e.g. colloid chemistry which cannot guarantee the reproducibility at the level of a single atom. This is because it is not single atoms but groups of 5–15 atoms that enter into the chemical reactions simultaneously.

We may conclude that the atom-by-atom manipulation of matter is not in contradiction with physics, but at present the formation of structures is practically not conceivable and has no clear applications. This fact does not downgrade the Feynman's brilliant vision.

Fig. 4.3 Smart parallel cantilever probe platform for nanoscale analysis and synthesis [40] (With the courtesy of I. Kostič, Institute of Informatics SAS, Bratislava)

4.10.2 Closing the Nano-Divide?

The division of the world by information and communication technology into rich and poor parts (the digital divide) seems to be a reality and the belief that the progress of technology will remove the disparity of economically divided world seems to be less and less plausible. A. Penzias, Nobel Prize winner and a qualified expert in the field of information technology formulated the following conclusions to this end [41]:

- Electron equipment helps to make informed decisions. A healthy flow of information separates winners from losers.
- Information technologies widen the gap between the rich and poor people.
- Information technology helps to overcome the shortage of human and material resources.

It can hardly be assumed that in even more complex and broader field of nanotechnology the situation will be different and that any nano-divide could be bridged easily. Apparently, the spread of technologies from developed world is fostered [42] and nanotechnology products do reach the developing world. Nevertheless, the developed countries will continue to hold the cutting edge (in the absence of this gradient the driving force would cease to exist).

Some advocates of closing the nano-divide argued that nanotechnology facilities and laboratory equipments are relatively cheap. This is far from reality. Nowadays the nanotechnology research requires complex infrastructure including synchrotrons and top level equipment in clean laboratories. The research and production of smaller and smaller entities demands ever more complicated and expensive instruments (Fig. 4.4). The shift in microelectronics from diffusion furnace to ion

Fig. 4.4 Laboratory equipment for GISAXS analysis of nanoparticle arrays substituting to some extend the synchrotron beam facility (Institute of Physics SAS, Bratislava)

implanter, from light lithography to its electron counterpart, from Petri dish with acid to ion beam etching confirms this statement. This shift has been discussed in the context of microelectronics as Moore's second law [4].

4.10.3 The Future of Silicon

Since 60 years silicon has been a dominant semiconductor in microelectronics and recently in nanoelectronics. During this period various competitors entered the scene with the ambition to compete with silicon or to replace it altogether.

- During the 1980-ties the conception of magnetic bubbles in permalloys (A. Bobeck, Bell Labs.) for the construction of non-volatile memories (maintaining information also after switching off the current supply) appeared. However, the concept was not competitive enough to withstand the fast developing non-volatile silicon flash RAM memories.
- Another promising concept in the course of 1980-ties were the superconducting logic circuits and memories based on Josephson tunnel junctions that were elaborated at the IBM T. J. Watson Center in USA at the level of prototypes. The switching time of these devices was as small as 1 ps and the noise levels were low due to operating at low temperatures. However, the latter was also perceived as a big disadvantage and thus silicon withstood this challenge, too.
- GaAs has been since a long time considered as a successor of silicon within the semiconductor family. It is reasonable to assume that it will keep this position also in the foreseeable future and even beyond. GaAs and other $A^{III}B^{V}$ semiconductors have special applications in fast circuits and in optoelectronics but they do not threaten the strategic position of silicon.

This position relies on unlimited raw material reserves for the production of Si, high quality oxide used as passivation and insulation layer and patterning mask. The sufficient working temperature interval of Si is also strategically convenient. Another contributing factor to be borne in mind is the economic inertia of producers who tend to adhere to the existing expensive infrastructure (like e.g. Fig. 4.5). The cost of new production facilities grow exponentially – along with the number of components per integrated circuit. In the course of four decades the costs increased from the order of magnitude of ten million to ten billion US$ [4, 43].

As mentioned above, the Moore's law is nowadays approaching its physical barrier [5], which raises again the question of the future development of electronics. The increasing demand for higher quantity and faster information processing could be solved by the way of multi-parallelism, similarly like in the supercomputers with thousands of cores. At the same time, nanotechnology offers many particular improvements of the silicon technology, such as, for example, carbon nanotubes in the interconnections of ICs, better cooling of circuits with high density of components, miniature lithium batteries, etc. [R10].

Fig. 4.5 Electron lithography apparatus in the clean room: air is maintained at well-controlled temperature and humidity level and it is filtered and circulated (With the courtesy of I. Kostič, Institute of Informatics SAS, Bratislava)

Along with the above-mentioned improvements due to nanotechnology we should mention also the new principles in nanoelectronics, such as spintronics, especially in magnetic MRAM memories, or non-volatile PC RAM (phase change RAM) memories employing the Ovshinsky switching in materials like $Ge_2Sb_2Te_5$ or Sb_2Te_3. These chalcogenide compounds change their resistance by switching between amorphous and crystalline states. Such devices have been studied already since the 1970-ties and today they matured: 8 Gb memory with design rules of 20 nm was presented in 2012 [44].

A great amount of research efforts has been dedicated to magnetic memories composed of self-assembled magnetic, e.g. FePt nanoparticles [45]. (An example of assembling of Co nanoparticles is in Fig. 4.6). For recording media they must be converted into the high anisotropy LI_0 phase and it is desirable that they have a common magnetic easy axis. The LI_0 phase can be formed by annealing. Unfortunately, annealing in excess of 600 °C is required and this leads to the particle agglomeration and sintering.

Some old concepts can be seen returning at a higher level. For example, a gate-insulated vacuum channel transistor with nano-gap formed by standard silicon process was published by Han et al. [46]. A cut-off frequency 0.46 THz was obtained and the operation voltage can be decreased to the level of semiconductor devices. Still a massive research effort will be needed to develop this concept further.

Despite the fact that nanotechnology offers a wide array of improvements in microelectronics, we cannot reasonably expect that any remarkably new *disruptive technology* would appear (c.f. [R10]). This expectation is of primary importance notably for Europe, where semiconductor sector employs more than 100,000 persons directly and many other in an indirect way.

Fig. 4.6 Self-assembled artificial crystal composed from Co nanoparticles with the diameter of 11 nm (Institute of Physics SAS, Bratislava)

4.10.4 N&N and Sustainability

According to Smalley [47] the top ten problems of humanity for next 50 years in the order of priority are as follows: (1). Energy, (2). Water, (3). Food, (4). Environment, (5). Poverty, (6). Terrorism and war, (7). Disease, (8). Education, (9). Democracy, (10). Population. N&N help to cope with at least seven of them. Therefore they are assisting in strive for sustainability of the twenty-first century. The main contributions are expected due to the variety of new materials, such as:

- Green chemistry and high performance materials: ceramics, composites, solvents, paints, coatings, catalysts, membranes,
- Energy: solar and fuel cells, storage of hydrogen, batteries, lightening,
- Information: spintronic and photonic components, flat displays,
- Engineering: sensors, biosensors, analytic equipments,
- Health: medical treatments, gene therapy, implants, reparation of tissues, antimicrobial coatings, nanoparticle drug delivery, bioimaging, bone regeneration.

An evidence that main activities in N&N cover the area of new materials is given by [48] and especially by [49]: the frequency of N&N publication is decreasing in the following sequence: material research, chemistry, biomedicine, physics, informatics, environment, health, cognitive sciences, clinical medicine, geosciences and agriculture. A representative overview of N&N research was presented at the

19th International Vacuum Congress, Paris, 2013. Amongst 1,500 contributions the following topics were presented in the descending order of number of papers:

- Single entities with the prefix nano: particles, tubes (carbon, boron), wires, islands, fibers, clusters, corals, dots, horns, pillars, ribbons, rods, sheets, troughs, whiskers.
- Hybrid and composite structures: nanoalloys, nanocomposites, bimetallic nanoparticles, core-shell nanoparticles.
- New materials: graphene, silicene, carbyne, nanodiamond.
- Components: CNT cathodes, nanocontacts, nanoelectrodes.
- Devices: solar cells, nanosensors, nanobiosensors, graphene sensors, plasmonic sensors, DNA sensors, photonic sensors, FET, nanostructured capacitors, interference mirrors, nanoswitches, spin transistors, single molecule junctions.
- Inorganic applications: catalysis, coatings (hard, anti-corrosion, plasma facing), art conservation and protection, energy conversion, lightening, stiffness tuning.
- Organic/life applications: nanoparticles in medicine, antibacterial coatings, bioimaging, bone regeneration.

4.10.5 Single-Particle Devices

In Sect. 4.10.1 we called into question the building of nanostructures by manipulation of single atoms. Recently there is another field whereas the single atom or, more generally, single-particle (molecule, atom, electron, photon) approach is feasible: single particle devices (transistors, sensors) open a new era of top nano-applications. While they embrace different structures and principles, the common denominator is the operation or reaction involving only one particle. We characterize some of these devices in the Box 4.3 below.

Box 4.3. Selected Examples of Single-Particle Devices

A metallic single atom transistor structure from two silver crystals within the gap of 50 nm forms an atomic scale contact. It is submerged in electrolyte [50]. Switching between on and off states is achieved by the gate voltage controlled relocation of single silver atom at the apex of one metallic tip.

Another type of *single atom silicon transistor* with standard configuration has one STM positioned phosphorus atom between atomic scale electrodes [51]. The operation of the device is controlled by electrostatic gate.

Broadly studied is the *single electron transistor of MOSFET type*: a channel composed from metallic or semiconducting island of about 10 nm is isolated from source and drain by two tunnel junctions [52–54]. Transistor turns on and off each time when one electron is added to the island. Switching

(continued)

Box 4.3 (continued)
is controlled by the gate voltage. The behavior of the device is entirely quantum-mechanical.

Superconducting nanowire detector cooled below its critical temperature with a cross-section of $5 \times 100\ \mathrm{nm}^2$ is able to detect even a single photon [55, 56]. Incident photon breaks Cooper pairs and reduces the local critical current below that of the bias current.

Graphene sensor of the micrometer size is able to recognize individual events when a gas molecule attaches or detaches to or from graphene surface, respectively [57]. The high sensitivity is due to the extremely low electronic noise of the device.

Chemically localized surface plasmon resonance sensing with single Ag nanoparticle (nanotriangle with the height of 50 nm and in-plane width of 100 nm) was described by Yonzon et al. [58]. The limit of detection was estimated to be well below 1,000 molecules for small molecule adsorbents.

4.11 The Role of Social Sciences and Humanities in N&N

N&N has so far benefited mainly from the results of natural and technical sciences. Nevertheless, social sciences and humanities can increasingly contribute to the development of N&N. The potential fields of intervention include scientometry [48, 49], specifics of intellectual property rights [59], the economic context of N&N, sociology and psychology of consumers, security and protection of privacy. A common denominator of the enumerated activities is ethics [29, 30].

Ethical evaluation of nanotechnology usually follows the aspects of safety, in particular in relation to sanitary, food and environmental risks. However, the nanotechnology affects also fundamental rights related to freedom, equality and autonomy [60]. The discussion of these aspects is often shifted from social and political field to metaphysical [61] and theological [62] problems.

Technology was traditionally considered as merely neutral and only its use was subject to moral evaluation. Nowadays technology becomes a direct object of ethics [60, 63]. N&N appears to be one of the first areas of scientific activity in which a long-standing belief is being challenged: the belief that society is solely responsible for what happens when a researcher's work viewed as neutral and enabling is applied in a particular social context.

Social and economic consequences and ethical aspects of nanotechnologies represent an unknown and underestimated area that may need urgent attention very soon. For example, a practical question might arise as to what would happen if nanomedicine would allow for prevent ageing by making repairs at the cellular level? If everybody could live hundreds of years what would happen to our economy and society? Would only elite get such treatment? [64]

So far there are no good reasons to claim that N&N will give rise to fundamentally new category of ethical problems. Bacchini concludes that there are merely new occurrences of the well-known problems [65]. He admits that a new ethical problem could occur "if – and only if – every possible means of paraphrasing it as an old ethical problem is such that at least one group of ethically relevant arguments supporting certain ethical conclusion about it has no corresponding group of paraphrased arguments supporting the corresponding ethical conclusion about the old ethical problem".

Religious reactions to nanotechnology are analyzed by Toumey under two main topics [66]. According to the former, religious persons worry that nanotechnology would contribute to re-defining human nature in a way that diminishes its religious understanding. According to the latter, the worries are related to the fear that the control of science would lead to adverse consequences for equality and justice. Other actual problems include queries such as whether the perfectionism inherent in nanotechnology questions our relation to nature and god [60], or whether nanotechnology represents a step towards understanding humans as technological beings [67].

The large extent of new qualities offered by N&N is challenging also philosophy. In Sect. 4.10.2 we mentioned that the small size of the studied structures and large dimensions of the used facilities are complementary to one another. Complementarity is one of basic physical principles manifesting itself also in the fact that penetrating into the nano-world corresponds to increasing progress in the understanding of macro-world. New materials produced by nanotechnology open the door into the space, depth of oceans and interior of the Earth. The role of nanotechnology in unraveling the mysteries of cosmos is discussed by Schattenburg [68].

It is noteworthy that complementary and symmetric to each other are also the manipulation with matter at about 10^{-9} m at the one side and landing on the Moon at a distance from the Earth of about 4×10^{8} m on the other side. Both achievements are thanks to the humans who typically dwell in the middle of the metric scale at the length of about 1 m.

4.12 Closing Remarks

While the limited extent of this study did not allow us to cover many research topics that pave the way for nanotechnology progress towards new practical applications, we attempted to summarize the main milestones in the development of N&N and to shed light on the potential of the discipline as well as the threats that it can pose to the mankind. Nanotechnology is currently moving beyond the discovery of basic building blocks, i.e. nanotubes, to their applications in future products [69]. New inspirations are provided by biomimetics that studies the structure and function of biological systems as models for the design and engineering of materials and devices

[70] – an example is the spider's fiber with the strength exceeding that of steel by the factor of 6. A rapid development is observed in the field of nano-measurement. Therefore, standardizations and regulations related to the commercialization of nanotechnology, its safety and the related environmental threats play an important role, especially in Europe. It seems that currently the N&N field moves from the development of new materials and structures to their systemization, standardization and applications. N&N products are conquering the world markets. Under these circumstances the role of researchers and their responsibility continue to grow.

Acknowledgements This work was supported by the project VEGA, Bratislava, contract 2/0162/12, 2/0010/15 and Center of Excellence of SAS "CESTA", contract III/2/2011

References

Research Papers

1. Stix G (2001) Little big science. Sci Am 285(3):32
2. Moore GE (1998) Cramming more components onto integrated circuit. Proc IEEE 86:82, Reprinted from Electron Mag 38(8):114 (1965)
3. Moore's law (2013) http://en.wikipedia.org/wiki/Moore\T1\textquoterights-law. Accessed 4 Nov 2013
4. Schaller B (2013) The origin, nature and implication of "Moore's law". http://reserach.microsoft.com/en-us/um/people/gray/Moore_Law/html. Accessed 14 Nov 2013
5. The end of Moore's law is on the horizon, says AMD. http://www.pcworld.com/article/2032913. Accessed 3 Apr 2013
6. Binnig G, Rohrer H, Gerber C, Weibel E (1982) Surface studies by scanning tunneling microscopy. Phys Rev Lett 49:57
7. Luby S, Majkova E (1997) Semiconductors. In: Hajko V (ed) Physics in experiments. VEDA, Publication House of SAS, Bratislava, p 163
8. Bontems VK (2011) How to accommodate of the invisible? The 'halo' of 'nano'. Nanoethics 5:175
9. Feynman RP There's plenty of room at the bottom. (Caltech Eng Sci J, 1960), http://www.zyvex.com/nanotech/feynman.html. Accessed 26 Feb 1999
10. Moriarty P (2001) Nanostructured materials. Rep Prog Phys 64:297
11. Kunze U (2002) Nanoscale devices fabricated by dynamic ploughing with an atomic force microscope, review. Superlattice Microst 31:3
12. Ramachandra Rao CN, Kulkarni GU, Thomas PJ, Edwards PP (2000) Metal nanoparticles and their assemblies. Chem Soc Rev 29:27
13. Gould P (2004) Nanoparticles probe biosystems. Mater Today 7(2):36
14. Warheit DB (2004) Nanoparticles: health impact? Mater Today 7(2):32
15. Tartaj P, Morales MP, Gonzáles-Carreño T, Veintemillas-Verdaguer S, Serna CJ (2005) Advances in magnetic nanoparticles for biotechnology applications. J Magn Magn Mater 290–291:28
16. Gupta AK, Gupta M (2005) Synthesis and surface engineering of iron oxide nanoparticles for biomedical applications. Biomaterials 26:3395
17. Meier W (2002) Vision – science and innovation made in Switzerland, Special issue. Science Com Ltd, Bern, p 16
18. Sajgalik P, Hnatko M, Lences Z, Dusza Z, Kasiarova M (2006) In situ preparation of Si_3N_4/SiC nanocomposites for cutting tools application. Int J Appl Ceram Technol 3:41, Special Issue

19. Luby S, Chitu L, Jergel M, Majkova E, Siffalovic P, Caricato AP, Luches A, Martino M, Rella R, Manera MG (2012) Oxide nanoparticle arrays for sensors of CO nad NO_2 gases. Vacuum 86:590
20. Kaye P, Laflamme R, Mosca M (2007) An introduction to quantum computing. Oxford University Press, New York, p 288. ISBN 9780198570493
21. Tsu R (2001) Challenges in nanoelectronics. Nanotechnology 12:625
22. Kouwenhoven LP, Venema LC (2000) Heat flow through nanobridges. Nature 404:943
23. Held GA, Grinstein G (2001) Quantum limit of magnetic recording density. Appl Phys Lett 79:1501
24. Yacamán MJ, Ascencio JA, Liu HB, Gardea-Torresdey J (2001) Structure shape and stability of nanometric sized particles. J Vac Sci Technol B 19:1091
25. Li M, Li JC (2006) Size effects on the band-gap of semiconductor compounds. Mater Lett 60:2526
26. Hersam M (2011) Nanoscience and nanotechnology in the posthype era. ACS Nano 5:1
27. McGinn R (2010) Ethical responsibilities of nanotechnology researchers: a short guide. Nanoethics 4:1
28. Grinbaum A (2010) The nanotechnological golem. Nanoethics 4:191
29. Ebbesen M (2008) The role of the humanities and social sciences in nanotechnology research and development. Nanoethics 2:1
30. Toumey C (2007) Privacy in the shadow of nanotechnology. Nanoethics 1:211
31. Drexler KE (1986) Engines of creation. Anchor, New York
32. Drexler KE (1981) Molecular engineering: an approach to the development of general capabilities for molecular manipulation. Proc Natl Acad Sci U S A 78:5275
33. von Neumann J, Burks AW (eds) (1966) Theory of self-reproducing automata. University of Illinois Press, Urbana/London, pp 64–87
34. Marchant GE, Sylvester DJ, Abbott KW (2008) Risk management principles for nanotechnology. Nanoethics 2:43
35. Huber DL (2005) Synthesis, properties, and applications of iron nanoparticles. Small 1:482
36. Eggelson K http://www.sciencedaily.com/releases/2012/04/120428000220.htm. Accessed 9 Oct 2013
37. Smalley R http://en.wikipedia.org/wiki/Richard_Smalley. Accessed 10 June 2013
38. Nanotechnology http://en.wikipedia.org/wiki/Nanotechnology
39. Gleiter H (2000) Nanostructured materials: basic concepts and microstructure. Acta Mater 48:1
40. Rangelow IW, Kostic I et al (2007) Piezoresistive and self-actuated cantilever arrays for nanotechnology applications. Microelectron Eng 84:1260
41. Penzias A (1989) Ideas and information. Penguin, Canada
42. International Workshop on Nanotechnology (2013) Serpon Indonesia, Oct 2013
43. Roukes M (2001) Plenty of room, indeed. Sci Am 285:42
44. Fujisaki Y (2013) Review of emerging new solid-state non-volatile memories. Jpn J Appl Phys 52:040001
45. Terris BD, Thomson T (2005) Nanofabricated and self-assembled magnetic structures as data storage media. J Phys D Appl Phys 38:R199
46. Han J-W, Oh JS, Meyyappan M (2012) Vacuum nanoelectronics: back to the future? – gate insulated nanoscale vacuum channel transistors. Appl Phys Lett 100:213505
47. Smalley RE (2003) Top ten problems of humanity for next 50 years. Paper presented at the conference on energy and nanotechnology, Rice University, Houston, 3 May 2003
48. Kozhukharov V, Machkova M (2013) Nanomaterials and nanotechnology: European initiatives, status and strategy. J Chem Tech Metall 48:3
49. Leydesdorff L, Rafols I (2009) A global map of science based on the ISI subject cathegories. J Am Soc Inf Sci Technol 60:348
50. Xie F-Q, Maul R, Augenstein A, Obermair C, Starikov EB, Schön G, Schimmel T, Wenzel W (2008) Independently switchable atomic quantum transistors by reversible contact reconstruction. Nano Lett 8:4493

51. Rossier JF (2013) Single-atom devices: quantum engineering. Nat Mater 12:480–481
52. Fulton TA, Dolan GJ (1987) Observation of single-electron charging effects in small tunnel junctions. Phys Rev Lett 59:109
53. Tilke A, Pescini L, Blick RH, Lorenz H, Kotthaus JP (2000) Single-electron tunneling in silicon nanostructures. Appl Phys A 71:357
54. Mandal S (2013) Single electron transistor. Int J Innov Eng Technol 2:408
55. Semenov AD, Gol'tsman GN, Korneev AA (2001) Quantum detection by current carrying superconducting film. Phys C 351:349
56. Natarajan CM, Tanner MG, Hadfield RH (2012) Supraconducting nanowire single-photon detector: physics and applications. Supercond Sci Technol 25:063001
57. Schedin F, Geim AK, Morozov SV, Hill EW, Blake P, Katsnelson MI, Novoselov KS (2007) Detection of individual gas molecules adsorbed on graphene. Nat Mater 6:652–655
58. Yonzon CR, Stuart DA, Zhang X, McFarland AD, Haynes CL, van Duyne RP (2005) Towards advanced chemical and biological nanosensors – an overview. Talanta 67:438
59. Bastani B, Fernandez D (2002) Intellectual property rights in nanotechnology. Thin Solid Films 420–421:472
60. Larrére C (2010) Ethics and nanotechnology: the issue of perfectionism. Int J Philos Chem 16:19
61. Dupuy JP (2004) Pour une evaluation normative du programme nanotechnologique. Ann Mines 27
62. Sandel MJ (2007) The case against perfectionism: ethics in the age of genetic engineering. Belknap, Cambridge, MA
63. McGinn R (2008) Ethics and nanotechnology: views of nanotechnology research. Nanoethics 2:101
64. Ethics and Nanotechnology http://www.understandingnano.com/nanotechnology-ethics.html. Accessed 9 Dec 2013
65. Bacchini F (2013) Is nanotechnology giving rise to new ethical problems? Nanoethics 7:107
66. Toumey C (2011) Seven religious reactions to nanotechnology. Nanoethics 5:251
67. Grunwald A, Julliard Y (2007) Nanoethics 1:77
68. Schattenburg ML (2001) J Vac Sci Technol B 19:219
69. Rashba E, Gamota D (2003) J Nanopart Res 5:401.
70. http://en.wikpedia.org/wiki/Biomimetics. Accessed 6 Dec 2013

Thousands of pages of reports and plans are produced every year in the field of N&N by governments, institutions, scientific societies and companies. Here ten of them with the supranational importance are summarized. Comprehensive national reports are indexed in [R5]

[R1] National Nanotechnology Initiative, president Clinton WJ, 2000 (think tank of the White house, 1998)

[R2] Nanostructure Science and Technology, National Science and Technology Council, R. W. Siegel, chair, Loyola Coll. Maryland, 1999

[R3] Bhushan B (ed) (2004) Springer handbook of nanotechnology. Springer, Berlin/Heidelberg/New York, 1221 pp

[R4] N&N Action plan for Europe 2005–2009, COM (2005) 243, 9. 6. 2005

[R5] UNESCO (2006) The ethics and politics of nanotechnology. UNESCO, Paris. http://www.unesco.org/shs/ethics

[R6] Dosch H, Van de Voorde MH (eds) (2008) Gennesys white paper. Max-Planck Inst. für Metallforschung, Stuttgart. ISBN 978-3-00-027338-4

[R7] Commission Recommendation on a Code of Conduct for Responsible Nanosciences and Nanotechnology Research, C (2008) 424

[R8] N&N Opportunities and Uncertainties, Royal Society, Royal Acad Eng, 2004, retrieved 2011

[R9] Graphene Flagship, EC, January 2013, grant of 54 M€/30 months, 75 participants 17 countries, ICT & Mater & Energy & Health

[R10] Nanotechnology: the invisible giant tackling Europe's future challenges, DG Res Inov Ind Techn EUR 13325 EN, 2013. ISBN 978-92-79-28892-0

[M1] Einstein A (1905) Eine neue Bestimmung der Moleküldimensionen. Ann der Physik 19:289
[M2] Knoll M, Ruska E (1932) Das Elektronenmikroskop. Z Physik 78:318
[M3] see ref. [9]
[M4] Arthur JR (1968) Interaction of Ga and As2 molecular beams with GaAs surfaces. J Appl Phys 39:4032
[M5] Taniguchi N (1974) On the basic concept of 'nano-technology. Proc Int Conf on Production Eng, Tokyo, 1974, Part II (Japan Society of Precision Engineering)
[M6] Shirakawa H, Louis EJ, MacDiarmid AG, Chiang CK, Heeger A (1977) Synthesis of electrically conducting organic polymers: halogen derivatives of polyacetylene, (CH)x. J Chem Soc, Chem Comm 16:578
[M7] Binnig G, Rohrer H, Gerber Ch, Weibel E(1982) Surface studies by scanning tunneling microscopy. Phys Rev Lett 49:57
[M8] Kroto HW, Heath JR, O'Brien SC, Curl RF, Smalley RE (1985) Buckminsterfullerene. Nature 162–163:318
[M9] Binnig G, Quate CF, Gerber C (1986) Atomic force microscope. Phys Rev Lett 56:93
[M10] see ref. [31]
[M11] see ref. [52]
[M12] Fert A, Nguyen Van Dau F, Petroff F, Etienne P, Creuzet G, Friedrich A, Chazelas J (1988) Giant magnetoresistance of (001)Fe/(001)Cr magnetic superlattices. Phys Rev Lett 61:2472
Binasch G, Grünberg P, Saurenbach F, Zinn W (1989) Enhanced magnetoresistance in layered magnetic structures with antiferromagnetic interlayer exchange. Phys Rev B 39:4828
[M13] Dieny B, Speriosu VS, Gurney BA, Parkin SSP, Wilhoit DR, Roche KP, Metin S, Peterson DT, Nadini S (1991) Spin-valve effect in soft ferromagnetic sandwiches. J Mag Mag Mater 93:101
[M14] Iijima S (1991) Helical microtubules of graphitic carbon. Nature 354:56
[M15] Tans SJ, Verschueren ARM, Dekker C (1998) Room-temperature transistor basen on a single carbon nanotube. Nature 393:49
[M16] Garcia N, Munoz M, Zhao YW (1999) Magnetoresistance in excess of 200 % in ballistic Ni nanocontacts at room temperature and 100 Oe. Phys Rev Lett 82:2923
[M17] Lloyd S (2000) Ultimate physical limits to computation. Nature 406:1047
[M18] see ref. [R1]
[M19] Derycke V, Martel R, Appenzeller J, Avouris P (2001) Carbon nanotube inter- and intramolecular logic gates. Nano Lett 1:453
[M20] Novoselov KS, Geim AK, Morozov SV, Jiang D, Zhang Y, Dubonos SV, Grigorieva IV, Firsov AA (2004) Electric field-effect in atomically thin carbon films. Science 306:666
[M21] see ref. [R4]
[M22] Liljeroth P, Repp J, Meyer G (2007) Current-induced hydrogen tautomerization and conductance switching of naphthalocyanine molecules. Science 317:1203
[M23] see ref. [57]
[M24] Lemme MC, Echtermeyer TJ, Baus M, Kurz H (2007) A graphene field effect device. IEEE El Dev Lett 28:1
[M25] Ternes M, Lutz CP, Hirjibehedin F, Giessibl FJ, Heinrich AJ (2008) The force needed to move an atom on a surface. Science 319:1066
[M26] Bolotin KI, Sikes KJ, Jiang Z, Klima M, Fudenberg G, Hone J, Kim P, Stromer HL (2008) Ultrahigh electron mobility in suspended graphene. Sol St Comm 146:351
[M27] Bérut A, Arakelyan A, Petrosyan A, Ciliberto S, Dillenschneider R, Lutz E (2012), Experimental verification of Landauer's principle linking information and thermodynamics. Nature 483:187

Subjectcode: Z14000 Material Science, Nanotechnology

Chapter 5
Halloysite Clay Nanotube Composites with Sustained Release of Chemicals

Joshua Tully, Rawil Fakhrullin, and Yuri Lvov

Abstract Halloysite is a naturally occurring nanometer scale tube that is capable of both enhancing the physical properties of a material and functionalizing the material. The addition of halloysite into polymeric materials increases the composite physical strength because of the shape and stability of these 50-nm diameter and ca. 1,500 nm length tubes. Whereas the unique chemical and physical characteristics of halloysite allow for loading drugs, biomacromolecules, anti-corrosion agents, flame-retardant agents, and metal nanoparticles followed by their controlled release. Therefore, by loading a chemical of interest inside of the tubes and then mixing the modified halloysite with various materials one will not only be able to make stronger materials but make them smarter and provide sustained functionality that would otherwise not be possible.

5.1 Introduction

Halloysite is naturally occurring aluminosilicate clay that is chemically similar to kaolinite [1–4]. In fact, halloysite is essentially a rolled plate of kaolin. This family of clay minerals includes halloysite, kaolinite, montmorillonite, talc, vermiculite, and others. Indeed, if one researched halloysite there will inevitably be references and comparisons to other clays. This is because their chemical structure contains an abundance of hydroxyl groups on the outer surface. These groups allow the clays to participate in electrostatic or hydrogen bonding in polar solvents and interact favorably with common polymers like nylon, epoxy resins, and polyvinyl chloride. As mentioned earlier halloysite is a rolled plate of kaolinite and the implications of this define the usefulness of halloysite. Kaolinite is essentially a rectangular plate of

J. Tully • Y. Lvov (✉)
Institute for Micromanufacturing, Louisiana Tech University, Ruston, LA 71272, USA
e-mail: jtu006@latech.edu; ylvov@coes.latech.edu

R. Fakhrullin
Institute of Fundamental Medicine and Biology, Kazan Federal University, Kazan, Tatarstan, Russian Federation
e-mail: kazanbio@gmail.com

M. Bardosova, T. Wagner (eds.), *Nanomaterials and Nanoarchitectures*, NATO Science for Peace and Security Series C: Environmental Security,
DOI 10.1007/978-94-017-9921-8_5

$Al_2Si_2O_5(OH)_4$ layers stack on top of each other. This creates a sandwich of silicate molecules hydrogen bonded to the hydroxyl groups of the aluminum hydroxide molecules (see Figs. 5.1 and 5.2 for a visual representation). Halloysite differs from kaolinite in that it is a single sheet of kaolinite rolled along y-axis creating 15–20 layers ordered in a spiral manner with periodicity of 0.72 nm in the dehydrated state.

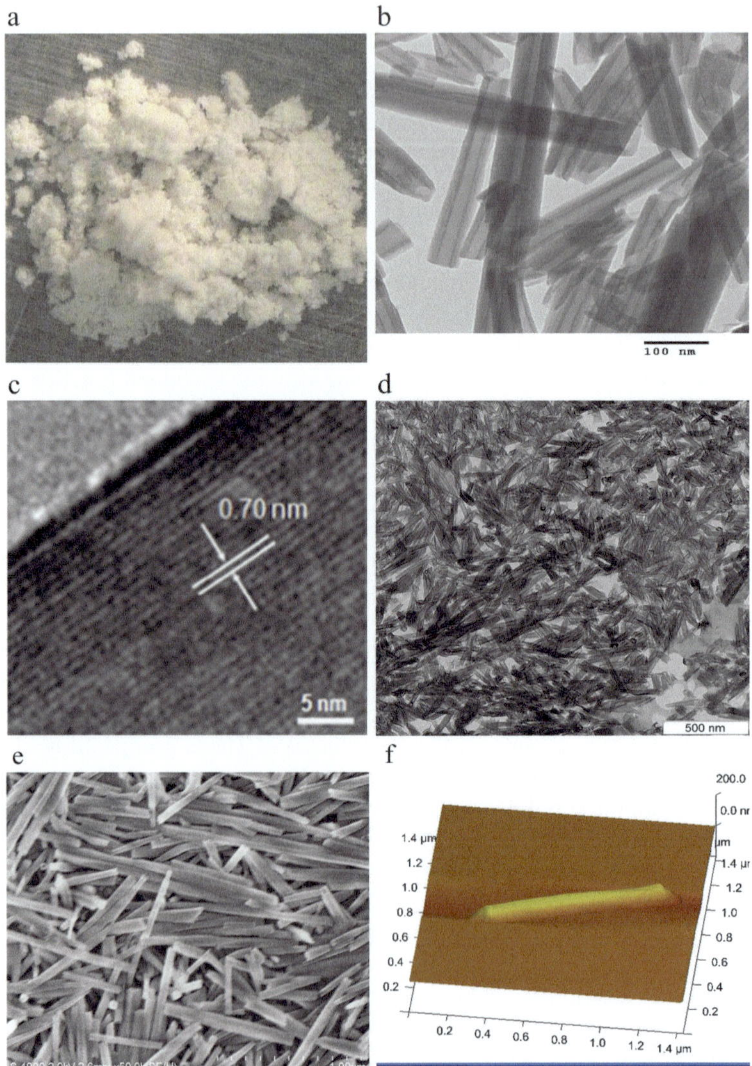

Fig. 5.1 Halloysite images; (**a**) powdered halloysite, (**b–e**) TEM of halloysite tubes including tube cross-section, (**e, f**) SEM and AFM image of the clay nanotubes from Applied Minerals Inc.

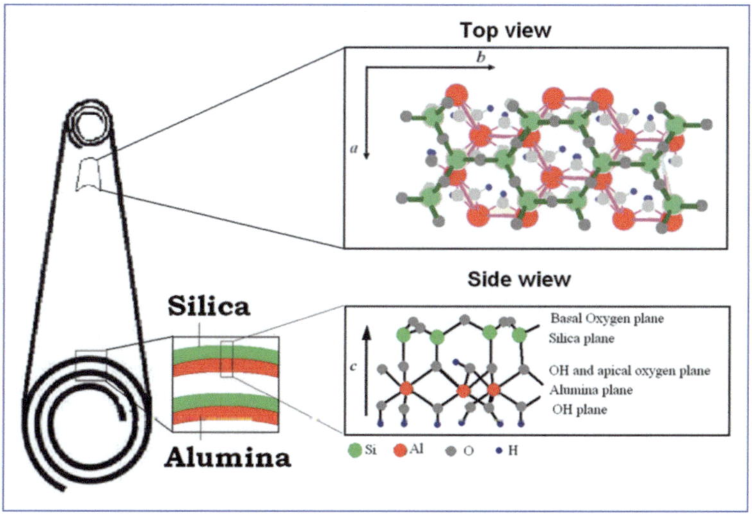

Fig. 5.2 Halloysite crystal structure; tubular morphology, *side view* and *top view* of the halloysite layer; *a*, *b*, *c* are crystalline directions

This creates a tube-like structure with an average outerdiameter of 50–60 nm and average lumen diameter of 10–20 nm with a length of 1 ± 0.5 μm.

It is the unique elongated shape of halloysite with an axis ratio of ca. 30:1 that makes it such an amazing dopant in many materials. The nanotubes act similarly to carbon nanotubes when added to a material. However, they are more miscible with polar polymers, especially with biopolymers. They behave like a skeleton or the frame of a house. When added into polymeric compound, for example, the nanotubes are able to pierce several pieces of polymer and hold them together like a skewer of meat. This obviously is going to increase the tensile strength of the material since in order to separate the polymers one must break the tubes, which are many times stronger than the polymer itself. There is a limit to the benefits of adding halloysite to a material. It is often found that the addition of 5–10 % halloysite nanotubes gives a 50–100 % increase in composite strength. Going over this amount will typically make the material weaker than the unmodified material and adding less than this amount often provides no benefit.

Actually, carbon nanotubes are much stronger than halloysite nanotubes. So there must be some benefit to using halloysite over carbon nanotubes. Indeed, there are many advantages to halloysite despite being weaker than carbon nanotubes. These advantages include cost, biocompatibility, and the ability to load chemicals inside of the lumen of the tube. Carbon nanotubes are extremely expensive, good quality single wall nanotubes are in the range of 500–1,000 USD per gram, whereas halloysite nanotubes cost 2–3 USD per kilogram. Not only is halloysite much cheaper but it is also completely natural and biocompatible. Carbon nanotubes are carcinogenic to organism, whereas it has been shown that halloysite is less toxic to cells than salt [3, 4]. Finally, halloysite nanotubes can be loaded with a variety of molecules ranging from proteins and polymers to drugs and anti-corrosion agents. In contrast single wall carbon nanotubes are very thin (lumen of ca. 1 nm diameter) and are terrible containers capable of loading only the smallest of molecules like water and some ions.

It is the ability of halloysite to contain chemicals that really makes it unique. This means that not only can one add halloysite to a material to improve its physical properties but the material can also be functionalized or “smart”. This is because once a chemical is loaded inside of the tubes it can be stored almost indefinitely when the tubes are dried then released in wet environments. One can imagine this being similar to a pill. The rate of drug release depends on the rate of decomposition of the capsule which influences the rate of diffusion of the drug out of the capsule. With halloysite, the rate of diffusion is controlled by the 10 nm diameter of the lumen [2]. By applying this property of halloysite nanotubes to polymeric paint, it was possible to create latex composite that was able to fix cracks and scratches by releasing self-healing compounds [2]. Another example is loading halloysite nanotubes with antiseptics (e.g. brilliant green or gentamicin) and then monitoring the enhanced inhibition of bacteria growth due to prolonged release of the agents. It was found that halloysite was able to provide a steady supply of antibiotics and inhibit bacteria growth, including multidrug resistant strains, for many more hours than by applying the medicine traditionally [4–7].

Another interesting property of halloysite that is being investigated now is its use as an additive for corrosion protection and flame retardance. For corrosion protection, clay nanotubes are loaded with anticorrosion agents, such as benzotriazole or hydroxyquinoline, and added into paints at 5 % w/w. The loaded halloysite nanotubes stay in the dried paint for years until becoming opened when a scratch or crack occurs and then they slowly release protective agents to cure or prevent corrosion underneath the trauma site [7–9]. It is not entirely clear how halloysite interacts with the combustion process to inhibit burning, but it works. From ongoing research in our laboratory it is possible to added 5–10 % w/w halloysite and significantly (2–3 times) reduced the duration of the fire and smoke development. One theory is that a combination of the endothermic decomposition of aluminum hydroxides found in the clay and the extra water that exists helps quench the fire. This is a promising line of research because not only are the halloysite tubes effective on their own, but they can also be loaded with flame retardants for a dual approach to enhance the fire safety of a material.

5.2 Halloysite Structure

Halloysite forms in volcanic zones as rolled aluminosilicate sheets with an ideal formula of $Al_2Si_2O_5(OH)_4 \times nH_2O$, which is similar to kaolinite with additional water between the adjacent layers [1–10]. When halloysite is mined from deposits, it has hydrated form with $n = 2$, and after drying at elevated temperature it irreversibly loses water transforming into more common product called halloysite – 7Å (indicating its interlayer spacing) or just commercial halloysite. Pure halloysite is white mineral (Fig. 5.1). The discoloration (slightly brown) is often due to the replacement of silicon or aluminum molecules with other metals such as iron or titanium.

Figure 5.2 gives scheme of halloysite tube structure. The average hydrodynamic size of halloysite nanotubes, as determined by dynamic light scattering, is 150–200 nm; however, the tubes often exist as larger aggregates and larger sizes are not unusual. Halloysite tubes have porosity distribution extending up to 30 nm, with most pores less than 10 nm. Its specific surface area is in the range of 50–150 m^2/g (depending on natural deposit and processing) and cation exchange capability ranges from 0.1 to 0.7 M/kg [2–18]. Hydrated halloysite tubes can intercalate a variety of salts (potassium, ammonium and cesium acetates) in the layers between their walls [19–21] and small organic compounds such as urea, formaldehyde, glycerol, dimethylsulfoxide [1].

Halloysite has different surface chemistry at the inner and outer sides of the tubes: silica layer is relevant to the outer surface of tube, while the alumina is in the inner (lumen) surface. Aluminum and silicon oxides have different ionization and surface charge which is evident from zeta-potentials of their colloids in water (Fig. 5.3). Alumina has positive charge up to pH 8.5, while silica is negative above pH 1.5 allowing for the selective loading of negatively charged molecules inside the halloysite lumen [22].

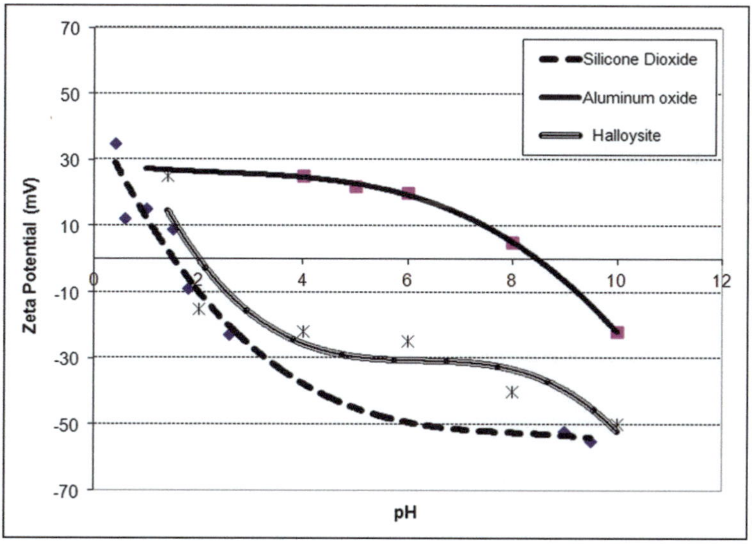

Fig. 5.3 Zeta-potential for halloysite nanotubes, silica and alumina nanoparticles (compounds of similar size and surface chemistry) (Reprinted with permission from Vergaro et al. [17], 820–826. Copyright 2010 American Chemical Society)

Halloysite Biocompatibility Halloysite is natural material and it does not pose any risk to the environment. The first biocompatibility experiments were performed on a halloysite by coating a plastic or glass surface by layer-by-layer (LbL) assembly with the cationic polymers poly(ethyleneimine) and polylysine. The addition of a thin halloysite layer, which is negatively charged on the surface, on top of the cationic polymer layer improved the adhesion of human dermal fibroblasts and they maintained cellular phenotype [21–27]. In the case of a polymethylmethacrylate bone cement implant doped with 5 % of halloysite, it was shown that the number of nanotubes is ca. 10^{10} within 10 μm of the surface. Eventually, as the implant degrades, these clay nanoparticles may be released to the surrounding tissue. Exposure poses little to no risk because the assumed concentration will be hundreds of times less than the data given in Fig. 5.8c for cellular safe halloysite concentrations of 0.1 mg/mL. This is because of the nanotubes sub-micrometer size, which allows for removal by macrophages.

The toxicity of halloysite was tested after 48-h incubation with fibroblast and human breast cells [17]. It was observed that halloysite is non-toxic to the cells and is even much less harmful than ordinary table salt (NaCl). The percentage of live cells was determined by the Celltiter-96™ reagent and was measured at various concentration of additives (halloysite was added at concentrations 0.1–100 mg/mL of culture and NaCl was a negative control). The data showed that the addition of even as much as 0.1 wt% of halloysite to the cell culture was not lethal. Therefore, one can conclude that halloysite nanotubes are safe for biomedical application; however, cannot be injected intravenously because they are not biodegradable (Fig. 5.4).

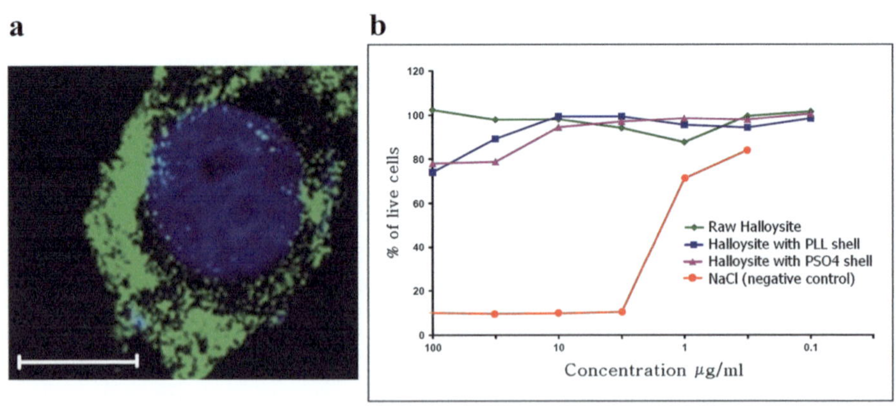

Fig. 5.4 Localisation of HNTs (*green*, FITC labelled) outside HeLa nuclei (*blue*, Hoechst staining) (**a**). Scale bar is 10 μm. Viability assay of MCF-7 cells on halloysite nanotubes and halloysite coated with cationic polylysine (*PLL*) and protamine sulfate (after 48 h) (**b**) (Reprinted with permission from Vergaro et al. [17], 820–826. Copyright 2010 American Chemical Society)

Encapsulation and Sustained Release Halloysite nanotubes can be loaded with broad variety of substances from simple organic and inorganic molecules to high molecular weight polymers and biologically active substances including drugs, enzymes and DNA [3]. Halloysite entraps molecules in different ways including adsorption to the external and internal walls of the tubes [28, 29], intercalation [1, 29] and, most essential, loading into lumen accompanied by condensation/crystallization of the loaded materials [3, 6, 7, 9, 14, 22, 27]. The possibility of pulling-in (sucking) chemicals from high concentrated solutions (and from melt) into its 15 nm diameter lumen seems to be the most remarkable for applications [5, 6].

Empty lumen of the clay tubes is a good nanocontainer for loading active chemical agents allowing smart functions of plastic composites doped with halloysite. A general procedure for the halloysite loading is as follows (Fig. 5.5): First, halloysite is mixed as a dry powder with saturated (highly concentrated) solution of the compound (melted compounds could be also used). Solvents have to be of low viscosity, provide high solubility for the agents and wet halloysite walls (water, acetone and ethanol). Once halloysite is added to the solution, the suspension is stirred and transferred to a jar evacuated with vacuum pump. Suspension should be kept under vacuum for 10–30 min and allowed to atmospheric pressure. Slight fizzling of the solution under vacuum indicates that air being removed from the tubes. Once the vacuum is broken, solution enters into lumen and loaded compound condensates within the tube. This has to be repeated 3–4 times to increase the loading efficiency. Typically, we reached 10–15 wt% loading which is close to theoretical estimation for 15-nm diameter halloysite lumens. After loading, tubes are washed to remove loosely attached substance from the external walls, dried and embedded into polymers [2, 5, 6, 28–36].

Concentration gradient plays a major role in loading the tubes; therefore, fast drying solvents with low viscosities such as acetone or ethanol are preferable for organic substances. High dielectric constant of the solvent is also desirable since it provides ionization of the hydroxyl groups of the halloysite walls, stabilizing halloysite suspension and enhancing loading of the negatively charged molecules in the lumens [22]. Water is also efficient for loading [6] which is the standard solvent

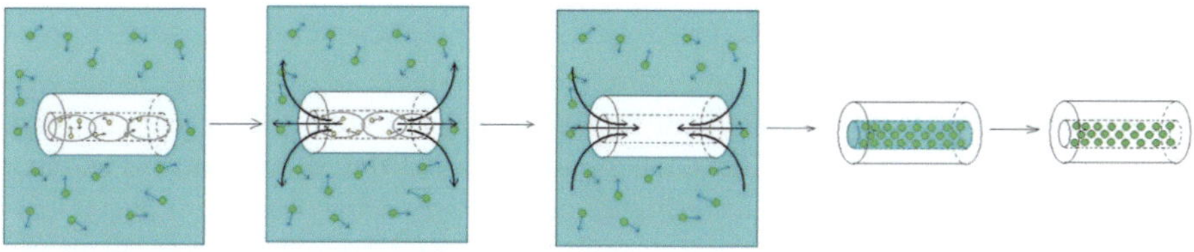

Fig. 5.5 Loading halloysite nanotubes with benzotriazole (Reprinted with permission from Abdullayev et al. [63]. Copyright 2009 American Chemical Society)

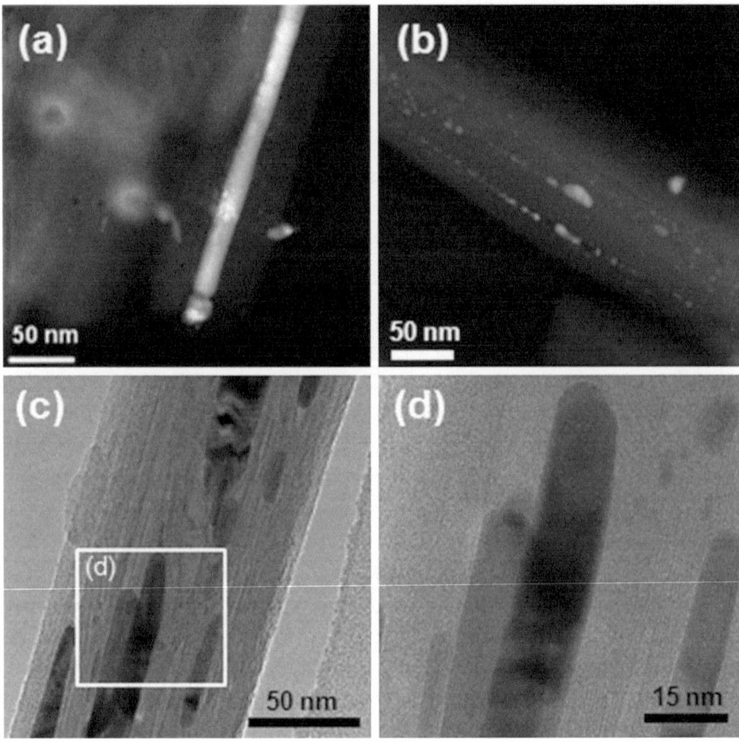

Fig. 5.6 HAADF-STEM (**a**, **b**) and HR-TEM (**c**, **d**) images of silver nanorods within halloysite tubular templates (Reprinted with permission from Abdullayev et al. [34], 4040–4046. Copyright 2011 American Chemical Society)

for biomacromolecules. For example, a method for loading silver metal inside of halloysite is presented by utilizing silver acetate (Fig. 5.6). First, the tubes were loaded in a solution saturated with silver acetate, dried, and heated at 300 °C to decompose the silver acetate into silver metal. After drying, the lumen of the tubes was filled with silver, which indicates the ability to loaded molecules efficiently inside of the tubes [34].

Enhanced Lumen Capacity One current disadvantage of halloysite is the 10–20 % loading efficiency. This is well below the typical 50–60 % active chemical content in polymeric microcapsules. By exploiting the different interior and exterior chemistry (inner: alumina; outer: silicon oxide), we were able to enhance halloysite's loading efficiency to 30–40 % by finding conditions, heating and sulfuric acid, that were able to etch the inside of the tubes from the average diameter of 15 nm to 25–30 nm (Fig. 5.7) [35].

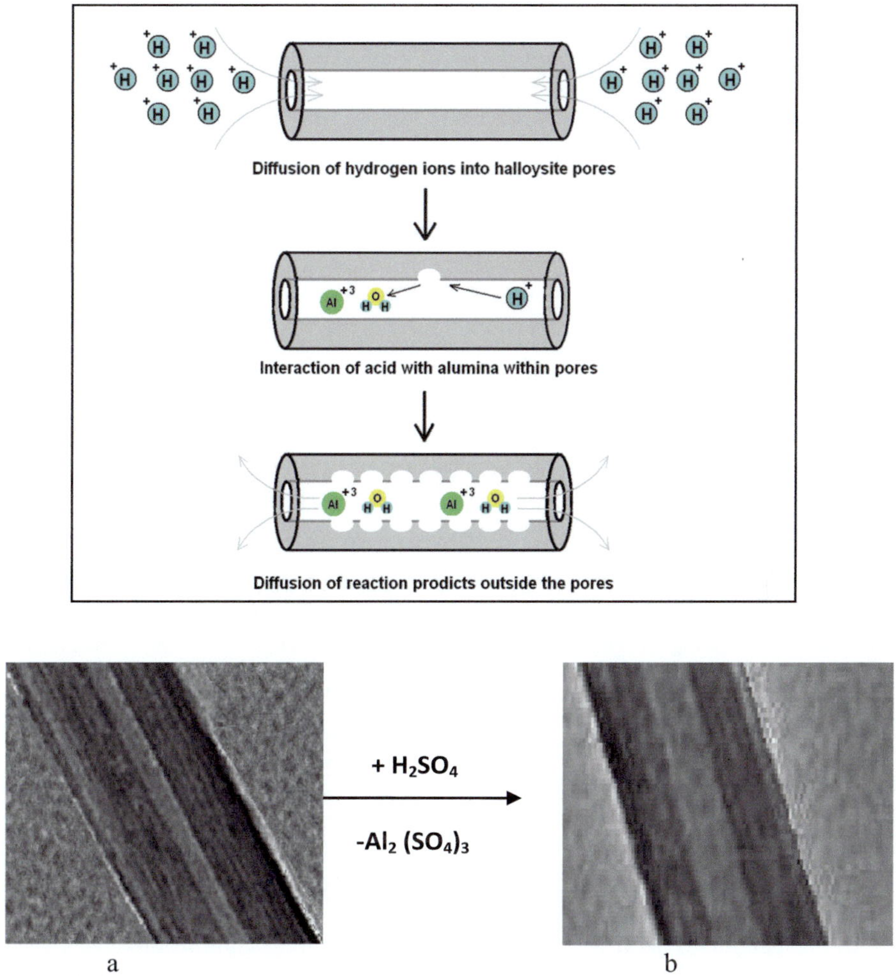

Fig. 5.7 The reaction and TEM images of halloysite tubes before (**a**) and after (**b**) treatment with H_2SO_4. Average lumen diameter increases from 10 to 20 nm, allowing the increase of loading up to 30 % (Reprinted with permission from Abdullayev et al. [35], 7216–7226. Copyright 2012 American Chemical Society)

The sequential treatment of halloysite with organosilane coupling agents offers a simple way to adsorb another compound on the silica outermost surface (Fig. 5.8). The tube inside was coated with octadecyl phosphonic acid and external part with octadecyl silane [30, 36].

Metal-Imidazole and Metal Triazole Shells It is also desirable to control the duration over which molecules are released from the nanotubes back into the environment. Through the interactions of triazoles and imidazoles with transition

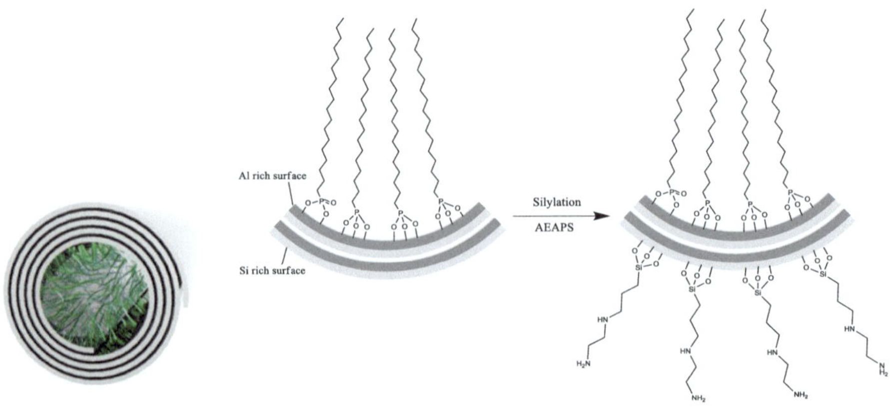

Fig. 5.8 Schematic illustration of bifunctionalization of the silica-alumina oxide surface of halloysite by octadecyl phosphonic acid and subsequent silylation (Reproduced with permission from Yah et al. [30], 1853–1859. Copyright 2012 American Chemical Society)

metal ions it is possible to create a thin layer coating the halloysite tubes that also covers the ends of the tubes. The composite layer creates a plug or stopper that is capable of prolonging the release of the loaded molecules back into the environment.

A general approach is to form a thin coating on the halloysite external surface through sequential adsorption of the film-forming components A and B. An insoluble film, formed by complexation of A and B, covers the entire surface of the halloysites and clogs the tube openings and any defects. (Scheme 5.1). Alternatively, component A may be loaded into the halloysite lumen and then tubes may be rinsed with component B.

Formation of the stoppers requires only quick rinsing of the loaded tubes with the solutions of triazoles and transition metal ions. Figure 5.9a shows that encapsulating halloysite nanotubes with a copper-benzotriazole coating significantly decreased the brilliant green release rate. One can control the release rate by varying the concentration of the metal ions in the external bulk solution [37]. Having a high concentration of metal ions in the solution causes a thicker coating to form which makes it more difficult for molecules to move out of the tubes thereby reducing the release rate and vice-versa.

In Fig. 5.9b an overlapping image of the nitrogen and oxygen maps taken on halloysites coated with copper-mercaptobenzimidazole film is given. Oxygen arises from the halloysite nanotubes and nitrogen from the coating films. As one can see, the film covered the entire halloysite surface. Such coated halloysite may contain loaded chemical with much longer release (e.g. corrosion inhibitors and antifouling agents) and may be doped into polymers to provide additional functions.

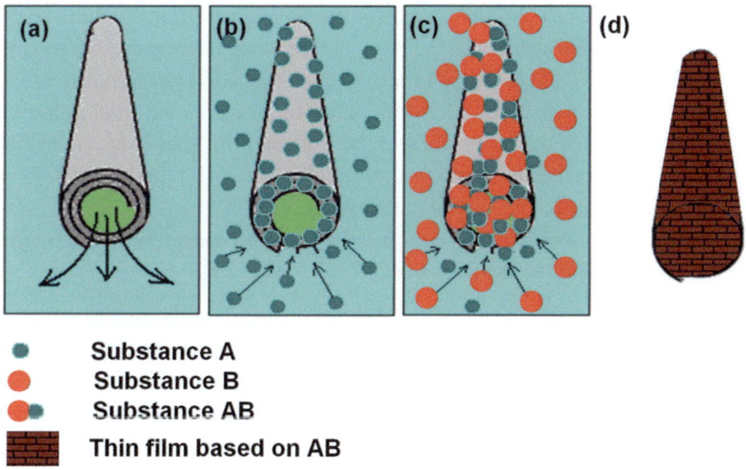

Scheme 5.1 Coating formation through consecutive rinsing with the solutions that contain film forming substances A and B; (**a**) leakage of the bioactive substances from halloysite, (**b**) adsorption of the component A, (**c**) rinsing with component B resulting in A/B complexation on the halloysite surface, (**d**) halloysite tightly sealed A/B surface coating

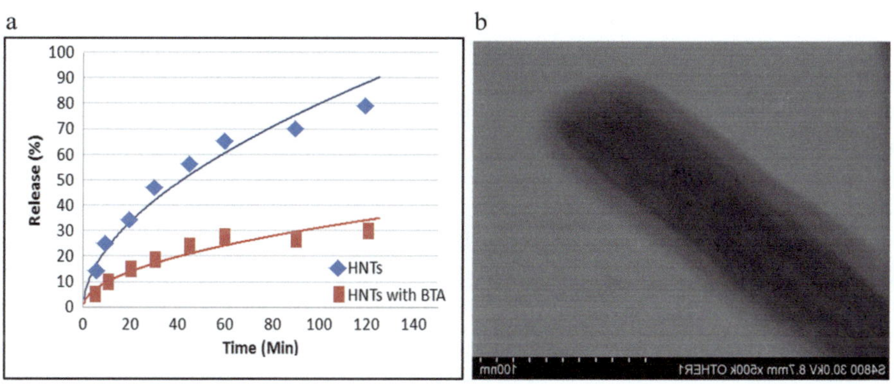

Fig. 5.9 (**a**) Release profiles of brilliant green from halloysite nanotubes (*HNTs*) and from HNTs encapsulated with copper-benzotriazole complexes (*HNTs with BTA*). (**b**) TEM image of the halloysites covered with copper-mercaptobenzimidazole

5.3 Polymer-Halloysite Nanocomposites

Ways to Prepare Halloysite-Polymer Composites (1) Dispersing fine halloysite tubes in monomer and conducting in situ polymerization. In this method, monomers interact with the halloysite surface and form uniform suspension. Sonication may be required to achieve a uniform dispersion of tubes. Then an initiator is added to the system to start polymerization (it also can be initiated by light, heat, etc.). Uniformity of the composite depends on the diffusion of the monomers between nanotubes [6, 20, 27]. (2) Solution casting involves dispersion of halloysite in the polymer solution (often sonicated). Then, the solvent is evaporated to obtain the composites [26, 28, 38]. (3) The third method involves addition of halloysite directly into the melted polymer and blending with an extruder or mechanical mixer to obtain a uniform distribution of the nanotubes in the polymer matrix [4, 6, 24, 31–34] (e.g., polyamides [24] and polybutadiene terephtalate [4]). (4) Layer-by-layer deposition of halloysite in alternation with polycations for coatings was made by sequential substrate dipping in clay and polycation solutions (an easier method would be by spraying each solution onto the substrate) (Fig. 5.21) [24]

Polymer-clay nanocomposites allow for increased tensile and flexural strength [36], improved thermal resistance and flame retardance [39, 40], decreased gas and liquid permeability [36] and increased biodegradability [6, 36]. The two most commonly used fillers for polymer nanocomposites are platy clays such as montmorillonite, hectorite and saponite, and carbon based fillers such as carbon fibers, nanotubes and graphene. Halloysite nanotubes have certain advantages over platy clay minerals. Platy clay sheets are strongly stacked to each other which requires difficult and costly exfoliation process to obtain good dispersion within polymer matrix [2, 40–43]. Halloysite readily dispersed in many polymers without any surface treatments (e.g., polyamides, polyethylene tereftalates, epoxy based polymers, and polysaccharides) [2, 3]. The advantages of utilizing halloysite over carbon based alternatives was discussed in the introduction.

For hydrophobic polymers like polyolefins, halloysite requires minimal treatment to make its surface compatible with the polymer. Aggregated halloysite tubes require less shear forces to be dispersed in polymers as compared with the exfoliation of the platy clays [4, 39]. In addition, halloysite has empty lumen which can be loaded with chemically and biologically active agents to provide extra functions for composite self-healing [7], anticorrosion [6, 9, 14] and antioxidant properties, enzymatic [23] and antibiotic [34] activity. Empty tubular lumen of halloysite is effective in capturing reaction products upon inflammation of polymers which increases flame retardance. For example, 10 % halloysite doping increased the temperature stability of polypropylene from 351 °C to 425 °C [40, 41].

Halloysite tubes usually are well dispersed within polar and medium-polar polymer matrixes but in some formulation aggregates of about 5–10 μm were detected (Fig. 5.10). In order to further improve halloysite dispersion within polymers, specific additives that interact with halloysite external surface via hydrogen bonding like melamine, diphenyl guanine, β-cyclodextrin (β-CD) were proposed [40]. These additives allow formation of hydrogen bonding bridged networks within polymer matrix causing restricted movement of polymer chains under applied strain and increasing composite toughness and tensile strength. Additives capable of forming electron transfer complexes like benzothiazoles or benzoxazoles were also shown to remarkably improve halloysite dispersion within polypropylene.

Current polymeric paints contain many additives on the basis of nanoparticles such as titanium dioxide (rutile), silica, clay, mica, etc. [43]. Some of these nanoparticles are added to improve the properties of the paint while others are added

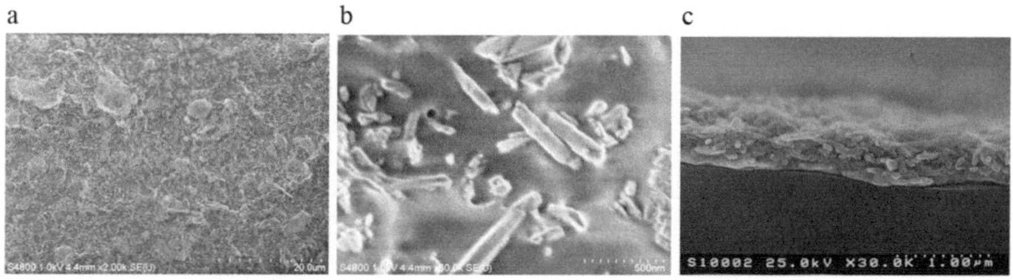

Fig. 5.10 (a, b) SEM image of halloysite-polypropylene nanocomposite with 56 % halloysite. (c) 300-nm thick porous multilayer coating of alternating halloysite and polyethyleneimine (halloysite/polymer ratio 20:1)

just for the reduction of the product price. Halloysite nanotubes offer great benefits for the paint industry because these particles are readily miscible with a large variety of the products, significantly improve their mechanical strength and adhesion, and may provide slow release of protective agents (anticorrosion, antimicrobial, antifouling, antioxidants). Halloysite-based composite metal coating suppressed *E. Coli* and *S. Aureus* bacterial contamination on paint surface. Halloysite tubes were loaded with silver before mixing with polyurethane paint. Ag^{+} ions slowly leaked from the tubes and killed bacteria. Paint surface and additionally a clear zone of inhibition at the paint edges were observed in both the cases. Contrary to the addition of bare silver nanoparticles, halloysite core-shell structures provided better color stability for coating [34].

Layer-by-Layer (LbL) Assembly of Halloysite-Polycation Multilayers Sequential adsorption of negatively charged halloysite and positively charged biocompatible polyelectrolytes and proteins offers a simple method for organized thin film fabrication [25, 44]. Halloysite can be adsorbed on any solid surface in alternation with cationic polymers such as polyethyleneimine, polylysine, or chitosan. Also, the films can be made to any thickness with a precision of 1 nm. This technique may be useful for microelectronic and biomedical devices, such as biosensors and implants. A LbL-halloysite film was prepared by alternatively dipping a quartz substrate into an aqueous PEI (3 mg/mL) and a dispersion (2 mg/mL) of halloysite. The thickness of seven halloysite/PEI bilayers was 360 nm which corresponds to the average bilayer thickness of 52 nm. Recall that the diameter of halloysite is approximately 50 nm, which is indicative that the main component of the coating is halloysite glued with a thin polycation interlayer. Figure 5.11 shows that for the multilayer assembly of halloysite nanotubes requires alternation of negatively charged halloysite with positively charged PEI. It is shown that repeated layering of halloysite only or in alternation with negative PSS did not allow the film to grow. SEM images show that the halloysite film is highly porous having many voids between nanotubes. With LbL methods more complicated “architectural” multilayers with alternation of different components are also possible where halloysite, polycation, and silica are repeated four times in the multilayer (Fig. 5.11c) [25].

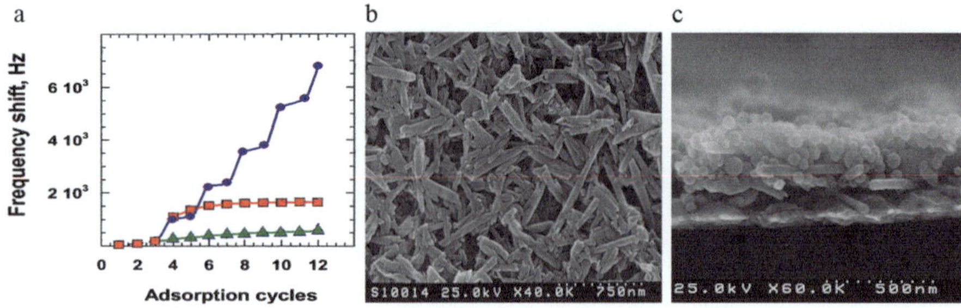

Fig. 5.11 (**a**) LbL assembly of halloysite/PEI (*circles*), halloysite/halloysite (*squares*) and halloysite/PSS (*triangles*) monitored by quartz crystal microbalance: frequency shift is proportional to the film thickness, (**b**) SEM image of halloysite/PEI film (*top view*); (**c**) Multilayer of (PEI/halloysite/PEI/50-nm silica)$_4$ on silver electrode (*side-view*)

An assembly of halloysite nanotubes loaded with an enzyme alcohol dehydrogenase (ADH) was also accomplished. ADH catalyzes oxidation of alcohols into aldehydes and ketones in the presence of cofactor nicotinamide adenine dinucleotide (NAD). By LbL-assembling ADH and NAD loaded halloysites in alternation with PEI, a functional films capable of oxidizing alcohols to aldehydes and ketones was prepared [25]. An advantage of layer-by-layer method is that any substrates may be coated. Metals, polymers, silicon wafer, glass, wood fibers and living cells were selectively LbL-coated with polymeric multilayers [44]. This allowed deposition of biocomposite thin films for surface modification of medical implants or cell encapsulation. Halloysite coating enhances better growth of human dermal fibroblasts as compared with coating with silica or montmorillonite. Thicknesses of nanoparticle/polyelectrolyte composite films containing five bilayers were 30, 35, 60 and 265 nm for montmorillonite, silica, titania and halloysite, respectively. Halloysite and titania based LbL-films were rougher than those of silica and montmorillonite with the roughness being 80 ± 30 and 50 ± 20 nm, respectively [39].

Mechanical properties of the composites are affected by halloysite dispersion through the polymer matrix. Lazzaro et al. [38, 45] explained halloysite distribution within composites with pectin by percolation theory, which states that for a system of independent particles of cylindrical shape dispersed in continuous medium (like polymer), a transition in particle-particle interaction take place when volume fraction of nanoparticles varies in the range from 10 to 20 wt%. Therefore, at concentrations of halloysite below 11 % interaction between halloysites and polymer predominates favoring their uniform distribution within composite while at above 20 % halloysite starts forming interconnected network resulting in increase of composite tensile strength and brittleness. Figure 5.12 shows that halloysite is uniformly distributed within pectin matrix below 30 % and forms clusters above this concentration. Composites based on pectin can be used as water soluble pouches for detergents, softeners and for medical delivery. Surprisingly, this composite was found efficient for sculpture cleaning in architecture restoration process [38].

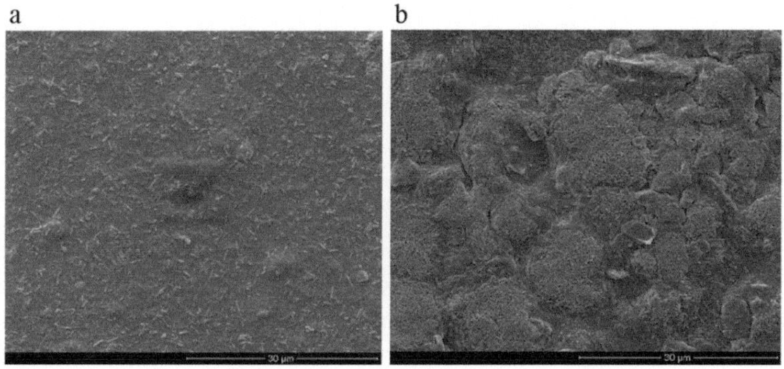

Fig. 5.12 SEM images of halloysite-pectin nanocomposites with (**a**) 10 % and (**b**) 30 % of halloysites (Reproduced with permission from Cavallaro et al. [38], 1158–1163. Copyright 2011 American Chemical Society)

Thermal degradation temperature increased from 238 °C to 250 °C when halloysite content ranged from 50 to 80 wt%. The control experiment with an addition of platy laponite and kaolinite caused much less increase in thermal stability. This effect was attributed to partial entrapment of pectin within halloysite lumen which was also supported by the measurements on halloysite molar volume in biocomposite matrix [38].

Composites of halloysite with cationic chitosan are quite common and more advanced preparation with acrylic acid cross-linking was also made [46]. Addition of halloysite rendered formation of calibrated pores in the polymeric matrix which was useful for the waste water treatment (e.g. adsorption capacity of ammonium ions increased to 41 mg/g for the composite with 50 % halloysite).

5.4 Polymeric Protective Coating Doped with Halloysite

The hollow tubular structure of the halloysite nanotubes makes it very perspective for using as nanocontainers for encapsulating corrosion inhibitors and other active agents for active and passive corrosion protection [7, 47]. These tubes were utilized for producing self healing coatings. Self healing effect is demonstrated in the Scheme 5.2 below. First, halloysite nanotubes, loaded with desired substances for self healing effect, is added to the liquid paint and applied to the metal surface. Loaded substances stay in the halloysite lumen as long as no crack is occured on the paint. Once paint is cracked loaded agents leak to the corrosive environment and stop the corrosion process, either by healing the coating or by inhibiting the corrosion process. Both of these approaches were analyzed in the current work. First we demonstrate self healing process by using corrosion inhibitors.

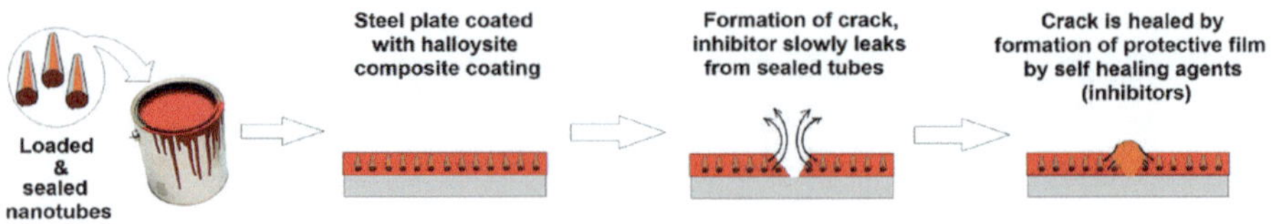

Scheme 5.2 Schematic demonstration of the self healing effect (Reproduced with permission from Joshi et al. [64]. Copyright 2013 American Chemical Society)

Inhibitors for Corrosion Protection Benzotriazole, 2-mercaptobenzothiazole and 2-mercaptobenzimidazole was used as corrosion inhibitor for the protection of the metals (Scheme 5.3). These inhibitors form two dimensional thin film with the metal ions on the metal surface, which insulates metal from the corrosive environment and protects it from corrosion (Fig. 5.13).

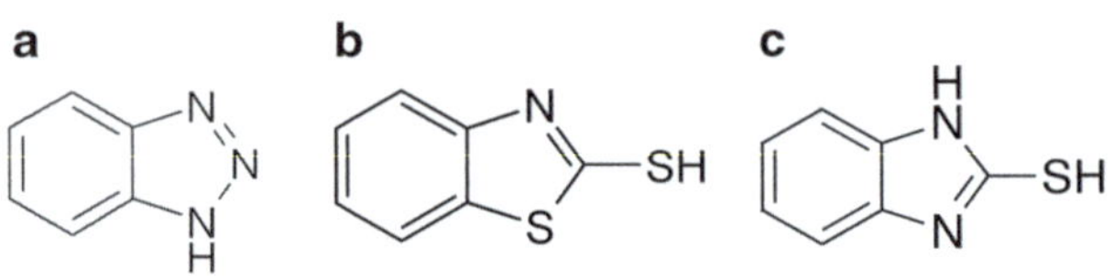

Scheme 5.3 Chemical structures of benzotriazole (**a**), 2-mercaptobenzothiazole (**b**) and 2-mercaptobenzimidazole (**c**)

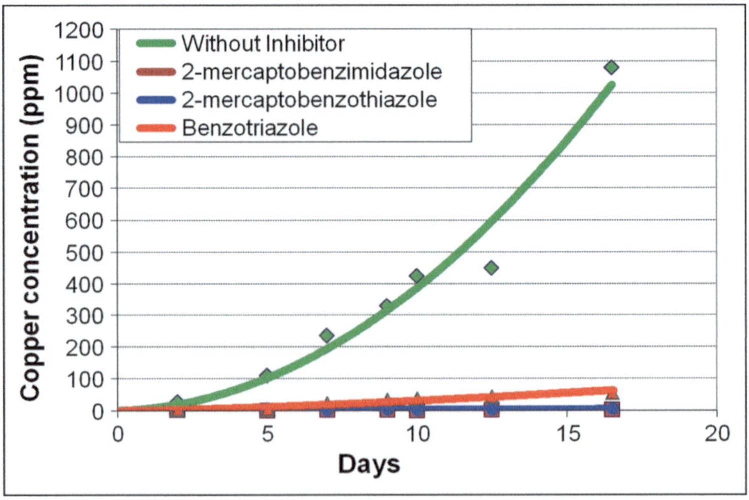

Fig. 5.13 Copper concentration change during corrosion test performed by dipping copper strips in salty water with 30 g/L NaCl and 0.01 M corrosion inhibitor

Corrosion inhibition efficiency of these inhibitors was evaluated on copper strips in corrosive media containing 30 g/L NaCl. Concentrations of the inhibitors were 0.1 M. Corrosion process was evaluated by measuring the concentration of the copper ions in the corrosive media. Concentration of the copper ions is constantly increasing in the absence of corrosion inhibitor, while corrosion process was significantly suppressed in the presence of corrosion inhibitors. Corrosion inhibition efficiency of the corrosion inhibitors were calculated by the formula:

$$I.E.\,(\%) = \frac{v_{uninhibited} - v_{inhibited}}{v_{uninhibited}} * 100 = \frac{\left[Me^{+n}\right]_{uninhibited} - \left[Me^{+n}\right]_{inhibited}}{\left[Me^{+n}\right]_{uninhibited}} * 100$$

where v is the corrosion reaction rate and $[Me^{+n}]$ is the concentration of the metal ions in the corrosive environment due to the corrosion process (in this case metal is copper). Corrosion inhibition efficiencies of the benzotriazole, 2-mercaptobenzothiazole and 2-mercaptobenzimidazole were 95.0 %, 99.6 % and 99.7 % respectively.

The process of corrosion inhibition by abovementioned corrosion inhibitors is based on the formation of the 2-D thin film on the basis of copper-inhibitor complex that covers the entire metal surface and insulates it from the corrosive environment. Once the surface of the metal is completely covered with the insulate film reaction between inhibitor and copper stops. Kinetics of this process follows second order reaction kinetics, i.e. reaction rate depends on both inhibitor concentration and the concentration of the unreacted copper atoms at the surface of the metal.

$$\mathrm{Cu}^{+2} + \mathrm{Inh} \rightarrow \ [\mathrm{Cu}\,(\mathrm{Inh})]_{\mathrm{n}}$$

Rate of the reaction can be calculated by following differential equation:

$$Rate = -\frac{d\,[Inh]}{dt} = k * [Cu] * [Inh]$$

If we assume that only monolayer of the corrosion inhibitor deposits on copper surface and insulating film covers the metal surface uniformly then following equations can be obtained.

$$a(t) = \frac{a_0 a_\infty}{a_0 + (a_0 - a_\infty) * \exp\left(-a_\infty kt/\,(a_0 - a_\infty)\right)}$$

where *a0* and $a\infty$ are the initial and final concentrations of the benzotriazole, *a(t)* is the concentration of the benzotriazole at time *t* and *k* is the reaction rate constant. Derivation of this equation is given in supporting information.

In Fig. 5.14 kinetics of the corrosion inhibitor interaction with copper wafers are shown. As one can see from the images theoretical calculations fit with the experimental data points reasonably well. Reaction rate constant were determined from the best fit of the model with the experimental points and values of *k*

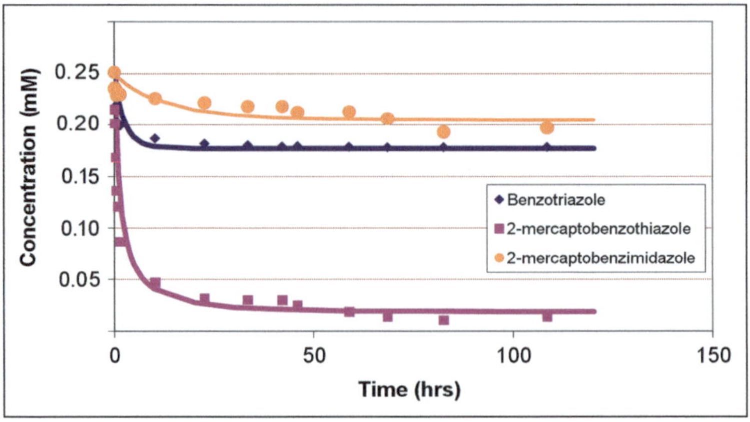

Fig. 5.14 Adsorption of corrosion inhibitors; benzotriazole, 2-mercaptobenzothiazole and 2-mercaptobenzimidazole on copper metal surface in water

were equal to 0.13, 0.65 and 0.015 for benzotriazole, 2-mercaptobenzothiazole and 2-mercaptobenzimidazole respectively. Initial concentrations of the corrosion inhibitors were 0.25 mM in all of the cases and final concentrations of the unreacted corrosion inhibitors were 0.17, 0.21 and 0.02 mM for benzotriazole, 2-mercaptobenzimidazole and 2-mercaptobenzothiazole respectively.

In Fig. 5.15 XRD spectrums of the halloysite samples loaded with corrosion inhibitors are described Sect. 5.2. A basal reflection of the halloysite nanotubes remained unchanged after loading process indicating that corrosion inhibitors were loaded mainly due to the precipitation inside the pores. No peaks of benzotriazole and 2-mercaptobenzothiazole were found in the XRD spectra showing that loaded corrosion inhibitors are mainly in amorphous form. On the other hand basal reflection of 2-mercaptobenzimidazole at 22.5° corresponding to 4.0 nm inter-planar distance is clearly visible from the XRD spectrum (Fig. 5.15c).

Analysis of Corrosion Inhibitor Release Characteristics In Fig. 5.16 release profiles of three corrosion inhibitors; benzotriazole, 2-mercaptobenzothiazole and 2-mercaptobenzothiazole from halloysite are shown. Release rates of benzotriazole and 2-mercaptobenzothiazole are close to each other more than 80 % being within 40 h while release rate of 2-mercaptobenzimidazole is a bit slower (only 50 % in 50 h).

The Peppas model [22, 48] was the best model describing the release profile of the corrosion inhibitors from the halloysite samples ($M_t/M_\alpha = kt^n$, where M_t is the amount of substance released at time t, M_α is the amount released at infinite time, n is the exponential constant related to the release mechanism, and k is a constant). Values for k are 64.0, 60.0 and 12.5 for benzotriazole, 2-mercaptobenzothiazole and 2-mercaptobenzimidazole respectively, while the values of n are 0.10, 0.08 and 0.37. Lower value of k and higher value of n for 2-mercaptobenzothiazole

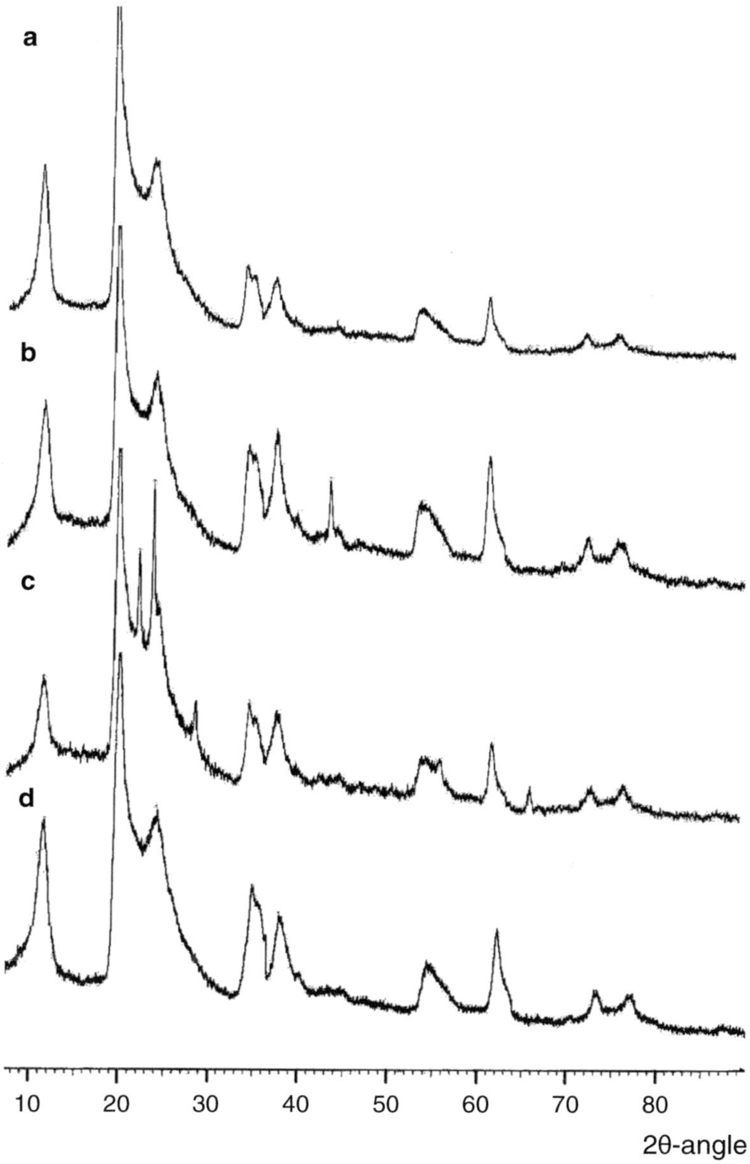

Fig. 5.15 X-Ray Powder Diffraction Spectrums of the empty halloysite nanotubes (*D*) and halloysite loaded with benzotriazole (*A*), 2-mercaptobenzothiazole (*B*), and 2-mercaptobenzimidazole (*C*). Cu-Kα radiation source was used

is associated with smaller initial burst for this substance. Loading efficiencies of the inhibitors were 4.0 %, 8.0 % and 20.0 % by weight for benzotriazole, 2-mercaptobenzothiazole and 2-mercaptobenzimidazole respectively. High loading

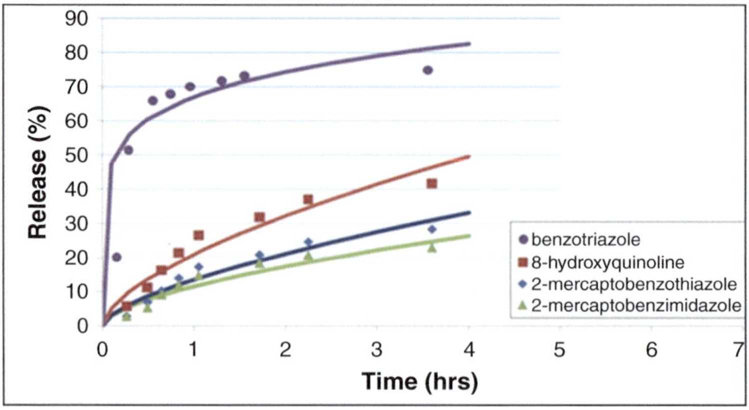

Fig. 5.16 Release profiles of three different organic corrosion inhibitors; benzotriazole, 8-hydroxyquinoline, 2-mercaptobenzothiazole and 2-mercaptobenzimidazole from halloysite nanotubes in water

Fig. 5.17 (**a**) Clay nanotube image, location in paint and (**b–e**) corrosion development on scratched painted metal samples after 9 months in simulated sea water. Steel: (**b**) – usual paint coating, and (**c**) – paint/halloysite composite loaded with dodecylamine. Aluminum 2024 alloy coated with transparent polyurethane paint: (**d**) – usual paint, and (**e**) – paint/halloysite composite loaded with hydroxyquinoline

efficiency of the 2-mercaptobenzimidazole indicates that significant amount of the inhibitor was loaded into the tube.

Figure 5.17 shows a corrosion development on iron strips coated by usual paint and paint with inhibitor loaded halloysite. A cross-like scratch was made on both samples before immersing them into a corrosive water solution. One can see an improvement of corrosion protection for coating with halloysite loaded paint.

Therefore, we can develop formulations for loading halloysite clay nanotubes with anticorrosion inhibitors specific to steel, copper and aluminum and can produce these materials as powder additive to existing paints.

Flame Retardance Triphenyl phosphate (on which we have preliminary results, below) is selected for encapsulation. Figure 5.18 is demonstrating flame torch experiments at StormWall facilities which revealed two-time slower flame spreading speed for halloysite doped paint as compared to the standard flame retardant paint (black in the picture below is burnt pant on the boards, and white is preserved latex

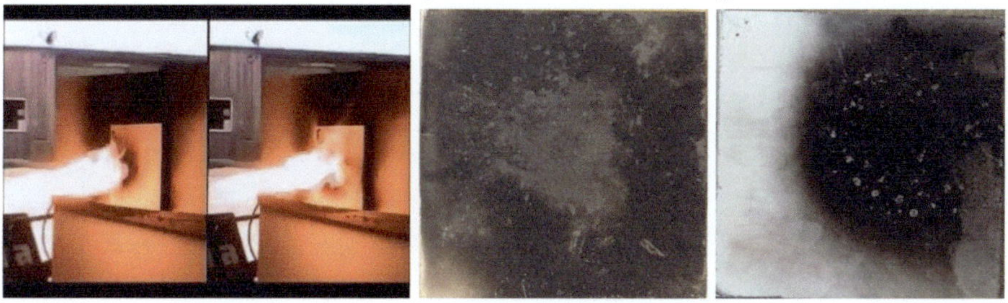

Fig. 5.18 Comparison of 30 s flame tests (**a**) for usual white latex paint coating (**b**) and halloysite-triphenyl phosphate doped paint (**c**) has shown an essential improvement in preventing spreading of the paint burning for halloysite formulation (*right image*) (Image is provided by A. Joshi and Y. Lvov, Louisiana Tech University)

paint) Even larger flame-retardance effect we are expecting after loading halloysite with flame retardance agents.

5.5 Bioactive Composites with Sustained Controlled Release of Drugs and Proteins

Halloysite nanotubes were studied as containers for loading and sustained release of cosmetic additives, antiseptics, drugs, proteins [3, 6, 21, 48–52] and DNA [53]. Halloysite loading is simpler than other conventional encapsulation methods that require additional chemical treatments, having various side effects associated with it, and provides higher loading efficiency. For example, sol–gel silica shells used for entrapment of drugs and proteins release alcohol during hydrolysis procedure which has detrimental effects on bioactivity of the loaded substances. Besides, the cost of the organosilanes is much higher than halloysite. Encapsulation with glycerol loaded halloysite matrix may provide more effective environment for bacterial cell protection [21].

The first work on halloysite loading with drugs were demonstrated by using khellin, a lypophilic vasodialator, used for promoting hair growth, oxytetracycline HCl, a water soluble antibiotic, and NAD [5]. Oxytetracycline-HCl was loaded from a saturated water solution, while khellin was loaded from melt and nicotinamide adenine dinucleotide was loaded from 5 % aqueous polyvinylpyrrolidone to increase viscosity. Release of oxytetracycline was completed in 30 h while khellin release extended over 500 h. The hypertension drug anti-angina drug diltiazem-HCl was also reported as having a 3 h release cycle [5, 49–51]. All these drug release rates obtained in sink conditions are much longer than direct dissolution of these compounds in water (usually few minutes) indicating nanopore controlled diffusion [22]. This approach was extended for tetracycline HCl encapsulation with halloysite

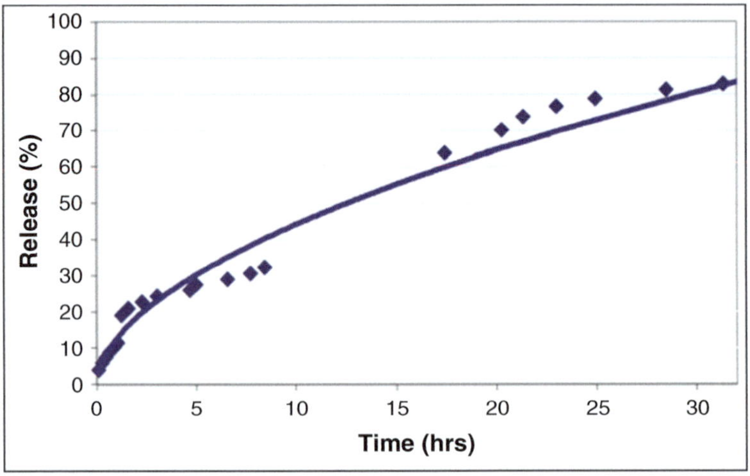

Fig. 5.19 Glycerol release profiles from halloysite

through complexation with chitosan and proposed for dog teeth fillings [50]. Halloysite tubes doped into polymers can contain drugs for longer time with very low leakage and with triggered release when tube ends are open to environment (e.g., in defects and/or microcracks) [52–55].

One of the cosmetic applications of halloysite is encapsulation of the skin moisturizing agent glycerol (Fig. 5.19). The loading efficiency of glycerol was 20 wt%. Solvation of glycerol loaded inside of halloysite was much slower than its bulk dissolution: only 30 % of the glycerol was released within 10 h [21].

Antiseptic, brilliant green microcrystals dissolved in water within 5 min while only 80 % was released from halloysite within 2 h, Brilliant green released from halloysite in two steps: 60 % released from halloysite main lumen and 40 % much slower released from smaller pores (like interlayer pockets). The release follows the first order kinetics with exponential function; $M_t = M_\infty * [1 - e^{-\alpha t}]$, where M_∞ is the amount of brilliant green released over infinitively long time and α is the release rate constant. 95 % release of brilliant green is achieved within 30 h which is 200 times slower than its dissolution rate without encapsulation [37, 52].

Release characteristics of three drugs, furosemide, nifedipine and dexamethasone are given in Fig. 5.20. All these drugs dissolve in water within 10–20 min but loading into halloysite extends their release to 10–50 h [22]. Encapsulation in halloysite reduced drug release time more than 25 times: 70 % of dexamethasone, 50 % of furosemide and 40 % nifedipine were released in the first 4 h.

Halloysite was loaded with an antibiotic, tetracycline, mixed with 0.2 % solution of chitosan in acetate buffer supplemented with 20 % poloxamer, 0.5 % polyethylene glycol and 1 % octylcyanoacrylate. This composite was capable of delivering antibiotics for up to 2.5 months and retained its syringeability for over 9 months. An in vivo study was performed by injecting this composite to the surgical pockets

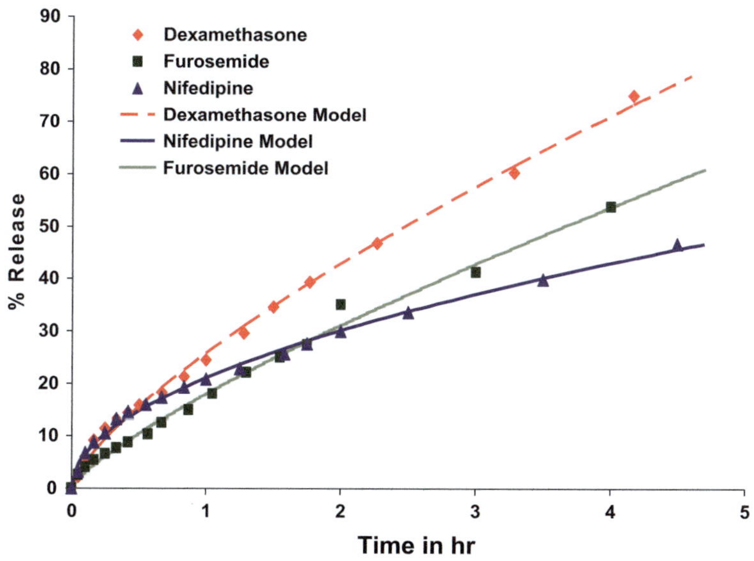

Fig. 5.20 Release of drugs from clay nanotubes (*dashed lines*) and dissolution of naked drug crystals (*solid lines*) in water. Drugs were loaded into halloysite using 50 % ethanol-water

created in the mouth of dogs. Bacterial growth was inhibited by release of antibiotic [50].

Halloysite-PVA composite for delivery of tetracycline HCl was prepared by drying aqueous suspension of the drug-loaded halloysite and PVA [54]. Halloysite-cellulose composite for the sustained delivery of anticancer opioid Fentanyl was made in [55]. However, drug release kinetics was moderate: 90 % of the drug was released in 7–8 h. Formation of the tube end stopperzss allows for much longer (100–200 h) drug release [56, 57].

Protein Loading and Slow Release Proteins are a large class of biomacromolecules that can serve many functions including transporting molecules, catalysis (proteins which catalyze a reaction are called enzymes), and cellular signaling. Many companies and researchers are particularly interested in the catalytic activity of many enzymes. Some of the fastest enzymes, such as carbonic anhydrase, increase the rate of the reaction by a factor of 10^6 or greater. Traditional metal catalysts simply cannot compete with enzymes, but enzymes also have another advantage. Enzymes can be very selective of the substrate or not at all, and enzymes are both stereo selective and stereospecific. However, enzymes do have some disadvantages that can be difficult to overcome. The biggest problem with enzymes is their chemical structure and that their three dimensional conformation affects their ability to catalyze a reaction. This means that the environment of the reaction must be relatively mild since proteins often lose their activity above 60 °C and at pH 2 units above or below their optimum. Not only are proteins sensitive to their environment but they can also be difficult to obtain, often making them somewhat expensive. There is a way to alleviate some of these issues through a technique known as enzyme immobilization which is applied in our approach of loading proteins and enzymes into halloysite.

Enzyme immobilization is an area of research with the goal of finding ways to attach enzymes to solid substrates so that they may be reused or their stability may be increased. It is theorized that immobilizing a protein onto a substrate increases its stability because it decreases the number of conformations the protein can observe. If an enzyme has *n* total conformations and *a* conformations in which it has catalytic activity then the probability that it is in its active form is *a*/*n*. Assuming that immobilizing the protein does not reduce *a*, which sometimes is the case, then by reducing *n* the probability that a protein is in an enzymatically active conformation must increase. A variety of approaches have been developed over many years and include chemical fixation, adsorption to silica or clay [58], immobilization in polymer matrices via in-situ polymerization reactions, and various other techniques [59]. The method we have developed at our university utilizes halloysite nanotubes and takes advantage of its interesting surface chemistry and shape. By exploiting the unique loading mechanism of halloysite nanotubes we are able to load different enzymes and proteins inside of the tubes, observe their release, then determine how much, if any, protein was left inside of the tube.

In general, the standard procedure for loading any chemical inside of halloysite nanotubes can also be utilized for proteins. However, it is recommended to use ice during sonication to prevent overheating the sample and denaturing the proteins. Another change that should be accounted for is that proteins should not be dried in the oven after the sample is washed. Typically after preparing a sample it is advantageous to dry the sample to prevent release. This is typically done in a 60 °C oven; however, due to the thermal sensitivity of proteins, this is not an option. Therefore if the samples must be dried then it should be done in a desiccator or by freeze-drying. Freeze-drying is performed by freezing the sample then subliming the water. There are some concerns that this could adversely affect the protein but, many proteins are often preserved in the fashion and come prepared in this fashion when bought from a chemical supplier.

The goal of our research was to determine if proteins could be loaded inside of halloysite nanotubes, if pH had any effect on loading efficiency, whether we could control the location of adsorption (on the surface of the tube or in the lumen), what are the release kinetics, and finally if there any protein left inside of the tubes after release. We have developed strong evidence we can load proteins inside of halloysite or on the surface by adjusting the pH in the loading solution. We also have evidence that we have proteins that release from the tube and proteins that reside inside of the tube permanently. Table 5.1 shows the proteins that were tested, their respective isoelectric points (pI), and their subunit weight. When the pH of the solution is equal to the pI of a protein then the protein will have a net charge of zero. One implication of this state is that normally a protein's solubility is extremely low at the isoelectric point.

Table 5.1 A concise summary of the proteins related to the loading and release studies. Where possible an EC number (Enzyme Commission number) was used to provide an identifier so information about a particular enzyme can be easily found. Two proteins included in the study, myoglobin and bovine serum albumin, are not enzymes from a protein database (Data from the table was compiled from BRENDA [65])

Enzyme	EC number	Molecular mass (kDa)	pI (isoelectric point)
Bovine serum albumin	N/A	70 (monomer)	5.0
Glucose oxidase	1.1.3.4	65 (dimer) = 130	4.2
Lipase	3.1.1.3	50 (dimer) = 100	5.0
Myoglobin	N/A	17 (dimer) = 34	7.0
Pepsin	3.4.23.1	35 (monomer)	1.0
Urease	3.5.1.5	90 (hexmer) = 540	5.0

The isoelectric point allows one to manipulate the net charge of the molecule by adjusting pH. This means that a pH higher than the isoelectric point the molecular will have a net negative charge while at a lower pH it will have a net positive. Loading typically needs to be in a saturated solution (5–10 % w/v). It should be noted that the listed weights corresponds to summarized subunits which means that urease, for example, weighs approximate 540 kDa because it is a homohexamer. There does not seem to be any correlation between weight and the loading efficiency of a protein. However, there does seem to be a correlation between the secondary structure of the protein and loading efficiency. Generally, speaking proteins with a secondary structure that is mostly α-helixes such as glucose oxidase and bovine serum albumin tend to have very low loading. Protein composed more equally of both α-helixes and β-sheets tend to have better loading.

In Fig. 5.21, the release curves of a number of proteins are shown. In these experiments the proteins were both loaded and released in water at pH 6.5 and studied over a period of 6 h. To determine the amount of protein release we utilized UV-spectrophotometry and read the absorbance of each aliquot at 280 nm. In this graph it is shown that close to 100 % release was achieved in 5–6 h. However, this is not entirely true because it is difficult to know exactly how much was loaded inside of the tube (may be some of the proteins are bound tightly and were not released).

In this set of experiments the loading pH was above the isoelectric point and our proteins had a net negative charge. This means that adsorption onto the outer negative surface of the tubes is unfavorable whereas the positively charge inner lumen would provide electrostatic interaction endorsing tube inner loading.

In the opposite experiment (positive proteins) we saw similar release patterns with one important difference. When loaded at low pH (below pI), the loading efficiency, calculated as the total mass of protein released divided the mass of halloysite nanotubes used, was significantly higher. Obviously a lot of positive proteins are adsorbed on the tube negative outer-surface, though external adsorption is more common (and less interesting for halloysite).

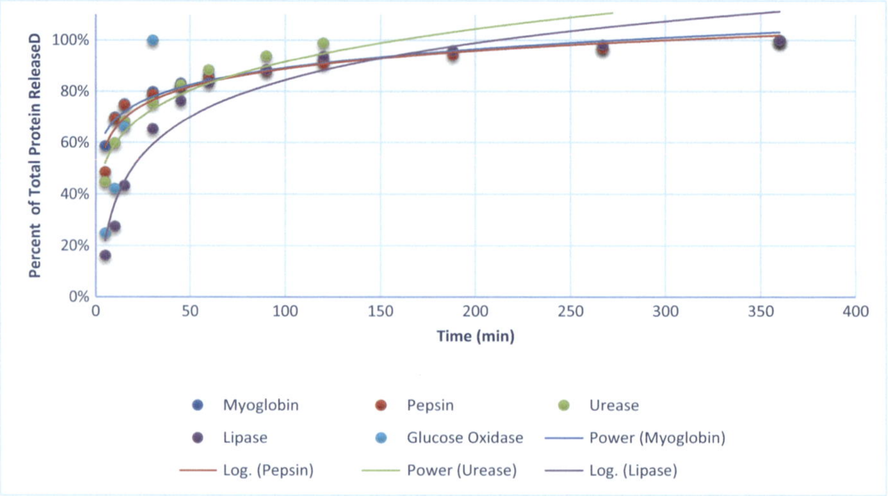

Fig. 5.21 Six proteins release studies performed on loaded halloysite tubes. All of the studies were done with proteins loaded in deionized water and released in deionized water (typical loading is 4–9 wt%)

In attempt to find whether some proteins are remaining in the tube lumen unreleased, we retrieved protein release data both though release curves (Fig. 5.21) and thermo gravimetric analysis (Fig. 5.22). For proteins loaded at a pH above their isoelectric point (negative) we observed about 3–5 % loading efficiency (probably mostly inside the tubes) and at a pH below their isoelectric point (positive) we found loading efficiencies between 10 % and 15 % (probably, mostly on outside surface).

An important result of our experiments was that there are some amounts of protein still residing inside of the halloysite nanotubes even after our release studies are complete (Fig. 5.23). We performed an experiment with lipase in an attempt to quantify how much protein is left inside of the tubes after 24 h release. Following the standard procedure as outlined in the chapter, the sample the sample to sat overnight in water after being dispersed. From the TGA analysis, 76 ± 7 % of initially loaded protein remained after being allowed to release overnight.

This immobilized protein has the potential to be more stable than in its free form and reusable. Preliminary experiments in our laboratory have shown that the glucose oxidase and urease do indeed survive the loading procedure. However, we have not explored enough to quantify if stability is increased nor if the enzymatic activity has been affected. We may only state that it is possible to load proteins inside of halloysite nanotubes at 3–5 % (w/w) efficiency, place them into an aqueous solution in which they will release ca. 1/3 of the proteins from inside of the tubes, then the residual 2/3 of protein immobilized in the lumen will retain their catalytic activity for weeks or months. Time-extended biocatalysis was also obtained for horse radish peroxidase and laccase immobilized in halloysite [60].

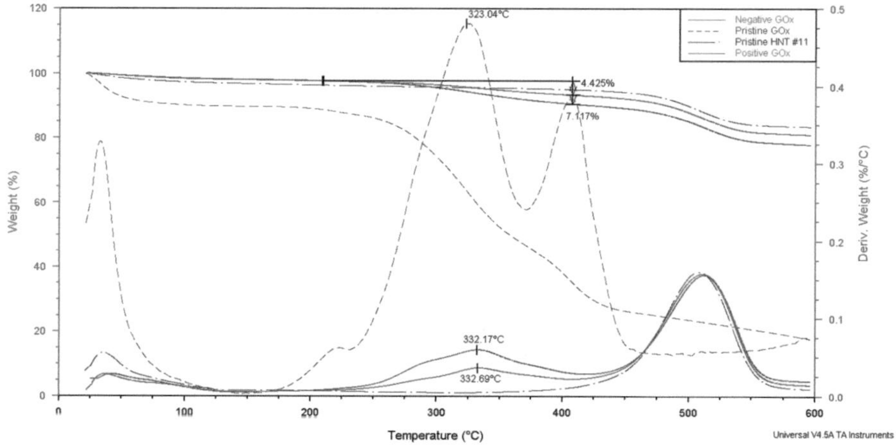

Fig. 5.22 TGA analysis of glucose oxidase loaded inside of halloysite and adsorbed to the surface. The *curves* at the *top* of the graph show the loss weight (shown as the weight percent which is the current weight divided by the original weight of the sample) with respect to temperature. Whereas the *curves* at the *bottom* show $\frac{dw}{dT}$ where w is equal to the weight percent. The maximum at 500 °C is due to structural changes in the halloysite

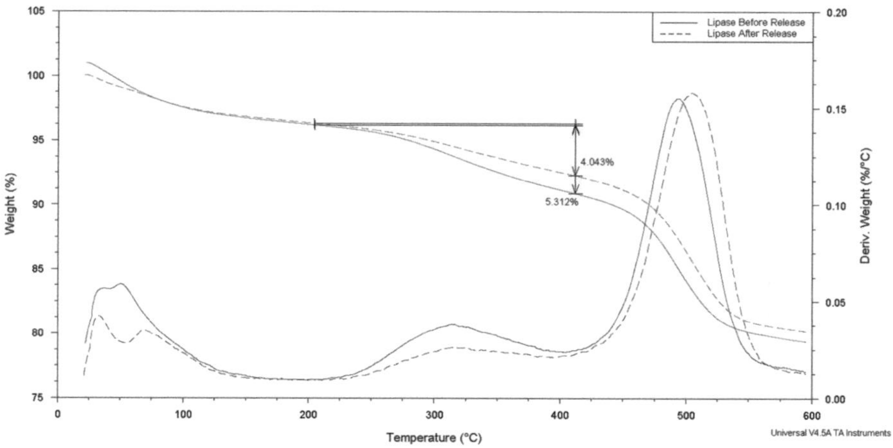

Fig. 5.23 TGA results after loading lipase (*solid line*) and after releasing the loaded protein (*dashed line*). The *line* representing the protein before release was offset one unit to account for the extra water that was in the sample (this is the weight percent lost in the first 100–150 °C)

Figure 5.24 shows the release kinetics of the ubiquitous protein insulin over a period of 200 h. In this experiment insulin was loaded at 2 % w/w. We attribute the extended release of insulin to the strong electrostatic interactions between the negatively charged protein and the positively charged lumen. Interestingly, it is possible to force the insulin out of the tubes by adding cationic polyethyleneimine. This is probably due to the polymer displacing the insulin from inside of the tube.

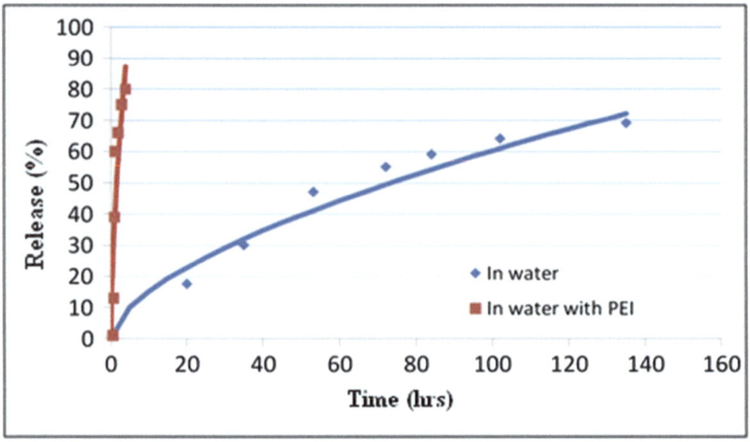

Fig. 5.24 Insulin release at two different conditions; fresh water and water containing 0.1 % cationic polyelectrolyte, polyethylenimine (*PEI*) in neutral pH

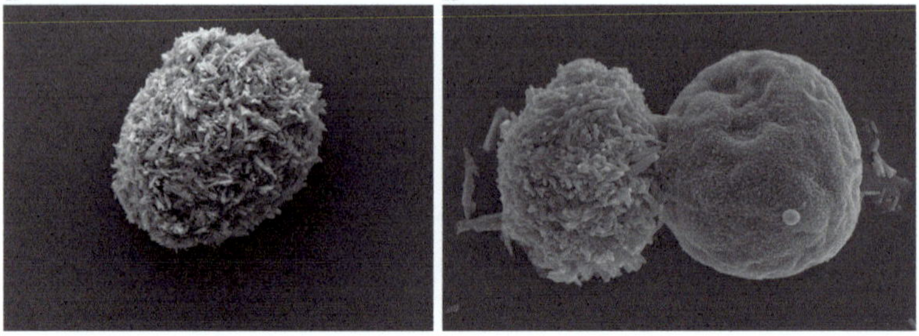

Fig. 5.25 (**a**) SEM image of the single east cell of about 4 μm diameter coated with halloysite nanotubes through layer-by-layer adsorption of polycation (polyallylamine hydrochloride) and anionic clay nanotubes. v (**b**) Proliferation of yeast cells producing daughter cell which is not coated. The cell overall size is ca. 4 × 5 μm

Nanoencapsulation of Biological Cells with Clay Nanotubes Halloysite clay nanotubes may form shells. on biological cells and bacteria. Such shells could serve a kind of armor for the cells thereby enhancing their stability in harmful media and carrying some useful compounds such as nutrition and drugs [61, 62]. Figure 5.25a–b shows a nanocoated cell and as it undergoes mitosis. The coating of nanotubes on the mother cell does not transfer to the daughter.

Fabricating biomimetic structures based on living cells in order to improve the cells' functionality and ability to survive harsh conditions is a great challenge. Genetic methods, although effective, do not allow for equipping the cells with independently produced nano-devices. Microbial cells are living entities and they have to be processed with non-harmful engineering methods in order to preserve their ability to reproduce. We proposed a method for cell surface modification to make living cells "healthier." Our method will secure their nutrition to better protect them against phagocytosis/digestion. It will also provide them with the ability to surpass a harmful media (e.g., acidic stomach on the way to the higher pH digestion tract), protect them against ultraviolet (UV) radiation with a layer of polyphenols (i.e. bacterial sun-screen), and supply cells with additional instruments for their functionality (i.e. magnetic function and mechanical robustness). We "dress" cells with a functional "cloth," preserving these newly acquired properties for one to two generations.

5.6 Conclusions

Halloysite clay nanotube based polymeric composites provide a significant enhancement in mechanical properties (5–10 % polymer doping with halloysite often double polymer strength), and the tube's empty lumen allows for loading with bioactive substances for sustained long lasting release (nanopore-limited diffusion). This enables the design of smart composite materials with synergistically improved strength and adhesivity, accompanied with controlled release of chemically active agents (antimicrobial, antifouling, flame retardant, self-healing, anticorrosion, and others). Halloysite clay is natural environmentally friendly nanomaterial available in thousands of tons which makes it perspective for scaling up polymer-clay composite products.

Halloysite nanotubes have several advantages as additives for polymeric materials: (1) General compatibility with polymers: clay nanotubes have strong surface interaction with polar biopolymers (e.g., polysaccharides, polyelectrolytes, proteins, DNA, polyacrylates), with medium polarity polymers (e.g. polyvinylchloride) and even with low polar polyethylene and polypropylene. (2) Sustained release: halloysite inner lumen can store and release molecules in a controllable manner making these nanocontainers attractive for applications in polymeric self-healing composites, in drug delivery, regenerative medicine and antimicrobial materials. Loading efficiency (20–30 % by weight), chemical and mechanical stability of halloysite is higher than conventional nanoparticles, like porous silica, alumina, titania, amphiphile vesicles and liposomes. (3) Ease of processing: halloysite has significantly less surface hydroxyl groups capable of forming intra-particle hydrogen bonds, tubule shape which makes it easy dispersable (unlike platy clay minerals stacked to each other with surface hydroxyls). (4) Biocompatibility: halloysite has no cell toxicity which allows its usage for drug delivery systems capable to release bioactive agents in sustained manner from tens hours to months. It may be useful for medical implants, dentistry, skin care, etc. (through halloysite cannot be used for intravenous injection because it is not biodegradable). (5) Protein immobilization: There are two types of immobilized proteins: about 1/3 loosely bound are released within 4–5 h but ca. 2/3 remains in the tube for a long time (probably adsorbed on the interior tube walls) preserving its biocatalytic functionality The residual proteins inside of the halloysite nanotubes allow for a new set of long lasting functionality that would otherwise not be possible.

Acknowledgements This work was performed according to the Russian Government Program of Competitive Growth of Kazan Federal University. This work was partially funded by Russian Scientific Fund grant no. 14-14-00924.

References

1. Joussein E, Petit S, Churchman J, Theng B, Righi D, Delvaux B (2005) Halloysite clay minerals – a reviewer. Clay Miner 40:383–426
2. Lvov Y, Abdullayev E (2013) Functional polymer – clay nanotube composites with sustained release of chemical agents. Prog Polym Sci 38:1690–1719
3. Du M, Guo B, Jia D (2010) Newly emerging applications of halloysite nanotubes: a review. Polym Int 59:574–595
4. Lvov Y, Shchukin D, Möhwald H, Price R (2008) Clay nanotubes for controlled release of protective agents – perspectives. ACS Nano 2:814–820
5. Price R, Gaber B, Lvov Y (2001) In-vitro release characteristics of tetracycline, khellin and nicotinamide adenine dinucleotide from halloysite; a cylindrical mineral for delivery of biologically active agents. J Microencapsul 18:713–723
6. Lvov Y, Aerov A, Fakhrullin R (2014) Clay nanotubes encapsulation for functional biocomposites. Adv Colloid Interface Sci 207:189–198. doi:10.1016/j.cis.2013.10.006
7. Abdullayev E, Lvov Y (2011) Clay nanotubes for controlled release of protective agents – a review. J Nanosci Nanotech 11:10007–10026
8. Yelleswarapu C, Gu G, Abdullayev E, Lvov Y, Rao D (2010) Nonlinear optics of nontoxic nanomaterials. Opt Commun 283:438–441
9. Abdullayev E, Price R, Shchukin D, Lvov Y (2009) Halloysite tubes as nanocontainers for anticorrosion coating with benzotriazole. Appl Mater Interfaces 2:1437–1442
10. Kirkman JH (1981) Morphology and structure of halloysite in New Zealand tephras. Clays Clay Miner 29:1–9
11. Lu D, Chen H, Wu J, Chan C (2011) Direct measurements of the Young's modulus of a single halloysite nanotube using a transmission electron microscope with a bending stage. J Nanosci Nanotechnol 11:7789–7793
12. Singh B (1996) Why does halloysite roll? A new model. Clays Clay Miner 44:191–196
13. Singh B, Mackinnon I (1996) Experimental transformation of kaolinite to halloysite. Clays Clay Miner 44:825–834
14. Abdullayev E, Shchukin D, Lvov Y (2008) Halloysite clay nanotubes as a reservoir for corrosion inhibitors and template for layer-by-layer encapsulation. Polym Mater Sci Eng 99:331–342
15. Churchman GJ, Carr RM (1975) The definition and nomenclature of halloysites. Clays Clay Miner 23:382–388
16. Bergaya F, Theng BKG, Lagaly G (2006) Handbook of clay science. Elsevier, Amsterdam
17. Vergaro V, Abdullayev E, Cingolani R, Lvov Y, Leporatti S (2010) Halloysite clay nanotubes: characterization and biocompatibility study. Biomacromolecules 11:820–828
18. Tazaki K (2005) Microbial formation of a halloysite-like mineral. Clays Clay Miner 53:224–233
19. Carr RM, Chaikum N, Patterson N (1978) Intercalation of salts in halloysite. Clays Clay Miner 26:144–152
20. Liu M, Guo B, Du M, Cai X, Jia D (2007) Properties of halloysite nanotube–epoxy resin hybrids and the interfacial reactions in the systems. Nanotechnology 18:455703

21. Suh Y, Kil D, Chung K, Abdullayev E, Lvov Y, Mongayt D (2011) Natural nanocontainer for the controlled delivery of glycerol as a moisturizing agent. J Nanosci Nanotechol 11:661–665
22. Veerabadran N, Price R, Lvov Y (2007) Clay nanotubes for encapsulation and sustained release of drugs. NANO 2:215–222
23. Lai X, Agarwal M, Lvov Y, Pachpande C, Varahramyan K, Witzmann F (2013) Proteomic profiling of halloysite clay nanotube exposure in intestinal cell co-culture. J Appl Toxicol 33:1316–1329
24. Lvov Y, Price R (2008) Halloysite nanotubules a novel substrate for the controlled delivery of bioactive molecules. In: Ruiz-Hitzky E, Ariga K, Lvov Y (eds) Bio-inorganic hybrid nanomaterials. Wiley, London/Berlin, pp 440–478
25. Lvov Y, Price R, Gaber B, Ichinose I (2002) Thin film nanofabrication via layer-by-layer adsorption of tubule halloysite, spherical silica, proteins and polycations. Coll Surf Eng 198–200:375–382
26. Liu M, Guo B, Du M, Jia D (2007) Drying induced aggregation of halloysite nanotubes in polyvinyl alcohol/halloysite nanotubes solution and its effect on properties of composite films. Appl Phys A: Mater Sci Process 88:391–395
27. Wei W, Abdullayev E, Hollister A, Mills D, Lvov Y (2012) Clay nanotube/poly(methyl methacrylate) bone cement composite with sustained antibiotic release. Macromol Mater Eng 297:645–653
28. Zhao M, Liu P (2008) Adsorption behavior of methylene blue on halloysite nanotubes. Microporous Mesoporous Mater 112:419–424
29. Luo P, Zhao Y, Zhang B, Liu J, Yang Y, Liu J (2010) Study on the adsorption of Neutral Red from aqueous solution onto halloysite nanotubes. Water Res 44:1489–1497
30. Yah W-O, Takahara A, Lvov Y (2012) Selective modification of halloysite lumen with octadecyl phosphonic acid: new inorganic tubular micelle. J Am Chem Soc 134:1853–1859
31. Shchukin D, Price R, Lvov Y (2005) Biomimetic synthesis of Vaterite in the interior of clay nanotubules. Small 1:510–513
32. Zhao Y, Abdullayev E, Vasiliev A, Lvov Y (2013) Halloysite nanotubule clay for efficient water purification. J Coll Interface 406:121–129
33. Veerabadran N, Price R, Lvov Y (2008) Tubule clay nanoreactor for template synthesis of silver nanoparticles. Polym Mater Sci Eng 99:566–577
34. Abdullayev E, Sakakibara K, Okamoto K, Wei W, Ariga K, Lvov Y (2011) Natural tubule clay template synthesis of silver nanorods for antibacterial composite coating. ACS Appl Mater Interfaces 3:4040–4048
35. Abdullayev E, Joshi A, Wei W, Lvov Y (2012) Selective lumen etching for clay nanotubes: enhanced loading capacity. ACS Nano 6:7216–7226
36. Yah W-O, Xu H, Soejima H, Ma W, Takahara T, Lvov Y (2012) Biomimetic dopamine derivative for selective polymer modification of halloysite nanotube lumen. J Am Chem Soc 134:12134–12137
37. Wei W, Minullina R, Fakhrullin R, Abdullayev E, Mills D, Lvov Y (2014) Enhanced efficiency of antiseptics with sustained release from clay nanotubes. RCS Adv 4:488–495
38. Cavallaro G, Lazzara G, Milioto S (2011) Dispersions of nanoclays of different shapes into aqueous and solid biopolymeric matrices. Ext Phys-Chem Stud Langmuir 27:1158–1163
39. Kommireddy D, Sriram S, Lvov Y, Mills D (2006) Layer-by-layer assembled nanoparticle thin films – a new surface modification approach for stem cell attachment. Biomaterials 27:4296–4303
40. Du M, Guo B, Liu M, Jia D (2007) Thermal decomposition and oxidation ageing behaviour of polypropylene/halloysite nanotube nanocomposites. Polym Polym Compos 15:321–328
41. Lecouvet B, Gutierrez J, Sclavons M, Bailly C (2011) Structure property relationships in polyamide 12/halloysite nanotube nanocomposites. Polym Degrad Stab 96:226–235
42. Voon H, Bhat R, Easa A, Liong M, Karim A (2012) Effect of addition of halloysite nanoclay and SiO_2 nanoparticles on barrier and mechanical properties of bovine gelatin films. Food Bioprocess Technol 5:1766–1774

43. Ruiz-Hitzky E, Darder M, Aranda P, Ariga K (2010) Advances in biomimetic and nanostructured biohybrid materials. Adv Mater 22:323–336
44. Hua F, Cui T, Lvov Y (2004) Ultrathin cantilevers based on polymer-ceramic nanocomposite assembled through layer-by-layer adsorption. Nano Lett 4:823–825
45. Cavallaro G, Donato D, Lazzara G, Milioto S (2011) Films of halloysite nanotubes sandwiched between two layers of biopolymer: from the morphology to the dielectric, thermal, transparency, and wettability properties. J Phys Chem C 115:20491–20498
46. Zheng Y, Wang A (2010) Enhanced adsorption of ammonium using hydrogel composites based on chitosan and halloysite. J Macromol Sci A: Pure Appl Chem 47:33–38
47. Shchukin D, Moehwald H (2007) Self-repairing coating containing active nanoreservoirs. Small 3:926–943
48. Veerabadran N, Lvov Y, Price R (2009) Organized shells on clay nanotubes for controlled release of macromolecules. Macromol Rapid Commun 24:99–103
49. Levis S, Deasy P (2003) Use of coated microtubular halloysite for the sustained release of diltiazem hydrochloride and propranolol hydrochloride. Int J Pharm 253:145–157
50. Kelly H, Deasy P, Ziaka E, Claffey N (2004) Formulation and preliminary in vivo dog studies of a novel drug delivery system for the treatment of periodontitis. Int J Pharm 274:167–183
51. Levis S, Deasy P (2002) Characterization of halloysite for use as a microtubular drug delivery system. Int J Pharm 243:125–134
52. Abdullayev E, Lvov Y (2013) Polymeric composites with ceramic nanotube endoskeleton loaded with functional chemical agents. J Mater Chem B 1:2894–2903
53. Shamsi M, Geckeler K (2008) The first biopolymer-wrapped non carbon nanotubes. Nanotechnology 19:075604
54. Ward C, Song S, Davis E (2010) Controlled release of tetracycline–HCl from halloysite–polymer composite films. J Nanosci Nanotechnol 10:6641–6649
55. Forsgren J, Jämstorp E, Bredenberg S, Engqvist H, Strømme M (2010) A ceramic drug delivery vehicle for oral administration of highly potent opioids. J Pharm Sci 99:219–226
56. Zhou W, Guo B, Liu M, Liao R, Bakr A, Rabie M, Jia D (2009) Poly(vinyl alcohol)/halloysite nanotubes bio nanocomposite films: properties and in vitro osteoblasts and fibroblasts response. J Biomed Mater Res A 93:1574–1581
57. Abdullayev E, Lvov Y (2010) Clay nanotubes for corrosion inhibitor encapsulation: release control with end stoppers. J Mater Chem 20:6681–6687
58. Zhai R, Zhang B, Liu L, Xie Y, Zhang H, Liu J (2010) Immobilization of enzyme biocatalyst on natural halloysite nanotubes. Catal Commun 12:259–263
59. Datta S, Christena LR, Rajaram YRS (2012) Enzyme immobilization: an overview on techniques and support materials. 3 Biotech 3:1–9
60. Chao C, Zhang B, Zhai R, Xiang X, Liu J, Chen R (2013) Natural nanotube-based biomimetic porous microspheres for significantly enhanced biomolecule immobilization. ACS Sustain Chem Eng 1:1145
61. Konnova S, Sharipova I, Ilinskaya O, Lvov Y, Fakhrullin R (2013) Cell-mediated three-dimensional assembly of halloysite nanotubes. Chem Commun 49:4208–4210
62. Fakhrullin R, Lvov Y (2012) "Face-lifting and make-up" for microorganisms (layer-by-layer polyelectrolyte nanocoating). ACS Nano 6:4557–4564
63. Abdullayev E, Price R, Shchukin D, Lvov Y (2009) Halloysite tubes as nanocontainers for anticorrosion coating with benzotriazole. ACS Appl Mater Interf 1:1437–1443
64. Joshi A, Abdullayev E, Vasiliev A, Volkova O, Lvov Y (2013) Interfacial modification of clay nanotubes for the sustained release of corrosion. Langmuir 29:7439–7445
65. Schomburg I, Chang A, Placzek S, Söhngen C, Rother M, Lang M, Munaretto C, Ulas S, Stelzer M, Grote A, Scheer M, Schomburg D (2013) Interfacial Modification of Clay Nanotubes for the Sustained Release of Corrosion Inhibitors. BRENDA in 2013: integrated reactions, kinetic data, enzyme function data, improved disease classification: new options and contents in BRENDA. Nucleic Acids Res 41(Database issue):D764–D772

Chapter 6
Colloidal Photonic Crystal Architectures for Advanced Light Management Applications

Martyn E. Pemble and Maria Bardosova

Abstract A brief introduction to photonic crystals and colloidal photonic crystals in particular is given followed by an outline of the two main methods of forming colloidal photonic crystals employed by the authors' research team- controlled evaporation (CE) which leads to face centred cubic structures such as are found in natural opal gemstones and the Langmuir-Blodgett (LB) method which leads to more disordered materials. The ability to control deposition on a layer-by-layer (LbL) level using the LB approach is highlighted by a discussion of the fabrication of a so-called AB heterostructured colloidal photonic crystal, made from spheres of two different sizes.

Following this the use of atomic layer deposition (ALD) to infill 3-D colloidal photonic crystals in order to be able to tune the refractive index contrast in the material is described, using as an example our work on the fabrication and characterisation of GaAs infilled colloidal photonic crystals and inverted GaAs photonic crystals that were made via infilling followed by removal of the silica host spheres. The importance of choosing the correct value for the refractive index of the infill material is highlighted and an optical Brillouin zone constructed using a free software package (MIT) which accounts for some of the observed Bragg reflections.

Finally, some examples are presented which demonstrate the potential of the use of these and other similar materials for a range of novel optical applications.

M.E. Pemble (✉)
Tyndall National Institute, University College Cork, Cork, Ireland

Department of Chemistry, University College Cork, Cork, Ireland
e-mail: martyn.pemble@tyndall.ie

M. Bardosova
Tyndall National Institute, University College Cork, Cork, Ireland

M. Bardosova, T. Wagner (eds.), *Nanomaterials and Nanoarchitectures*,
NATO Science for Peace and Security Series C: Environmental Security,
DOI 10.1007/978-94-017-9921-8_6

6.1 Introduction

6.1.1 What Is a Photonic Crystal?

At its most basic a photonic crystal is a material that can influence the motion (propagation) of photons in a manner similar to the way in which ions in crystalline materials affect the motion of electrons. Arguably the first formal report of behaviour such as this was produced by G.G. Stokes in 1884 [1] and centred on the observation of a 'peculiar internal coloured reflexion' in crystals of 'chlorate of potash'. This reflexion, meaning the observed light from the sample, disappeared twice while the sample was rotated through 360° at constant angle of incidence. In other words the colour changed as a result of some mechanism, primarily associated with the structure (and symmetry) of the crystal. In 1888, Lord Rayleigh commented on the Stokes report [2] using the results of his recently (1887) published study, which was devoted to one-dimensional (1-D) periodic multilayer dielectric structures and in which the author demonstrated that such structures exhibit a spectral range of large reflectivity known as a stop-band. In this paper [3], the original derivation of the 1-D stop band had been presented.

In this regard a number of simple schematics may be found which purport to illustrate how a photonic crystal influences the light that falls upon it. An example of such a schematic used extensively by the authors to highlight what might be seen when a photonic crystal that is active in the visible region of the spectrum is illuminated by white light is shown below, next to some images of thin film colloidal photonic crystals made in our laboratories, Fig. 6.1a, b:

Depending on the angle at which the light approaches the material, some wavelengths are forbidden to propagate through the material and as a consequence are reflected towards the viewer (the colour that is seen). Other wavelengths may propagate, but often at different angles through the material. The material therefore looks a different colour if viewed in reflection as compared to transmission.

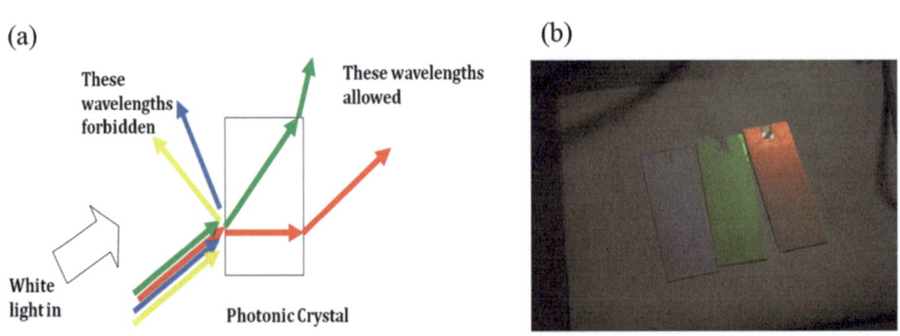

Fig. 6.1 (**a**) A schematic representation of how a photonic crystal might treat the various components of visible light and (**b**) some colloidal photonic crystals made in the authors' laboratories showing different reflected colour resulting from the different periodicities associated with each sample

Fig. 6.2 A schematic representation of materials in which the dielectric constant (and hence refractive index) varies periodically in one, two and three dimensions (Taken from Ref. [4])

Probably the best known examples of photonic crystals are natural opal gemstones and iridescent butterfly wings. Importantly, the colours seen in both reflection and transmission do not arise via the selective absorption of certain wavelengths as is typical for a normal coloured material, but rather from the structure of the material and its influence on the motion of photons- it is said to be an example of *structural colour*.

6.1.2 *Opening the Photonic Band Gap*

In general photonic crystals are much more than just 2-dimensional diffraction gratings which can be made using methods such as lithography. In fact the properties of these materials depend critically on the ordering present in all 3 dimensions although as the Fig. 6.2 above illustrates, photonic crystals can be envisaged that are periodic in one, two or three dimensions:

First proposed and demonstrated for microwave radiation by Yablonovitch in 1987 [5] and described theoretically by John in the same year [6], photonic crystals are best characterized as analogues to semiconductor materials that possess a photonic band gap, which describes a region of k-space where the electromagnetic waves may not propagate. Figure 6.3 is a schematic adapted from [7], which shows the material known as Yablonovite, in honour of its inventor. The structure, which possesses a complete photonic bandgap in the gigahertz range of the spectrum was produced by drilling an array of intersecting cylindrical holes, arranged in a diamond-like lattice, into a dielectric material.

In this analogy k is the so-called wave vector used to describe the electron in its wave form, and is usually denoted as $k = 2\pi/\lambda$, where λ is the wavelength in question. The reciprocal nature of the relationship between k and the wavelength (or more commonly energy, as so-called E vs. k diagram) can be expressed in three-dimensions via the use of the so-called Brillouin zone, which then describes the band structure of the solid in question. The reader is referred to the many textbooks and web sites where this topic is discussed in more detail. However, on an intuitive level, it may be easily appreciated that the allowed bands that exist for the electron (waves) in a particular solid will depend upon the spatial arrangement of the atoms whose

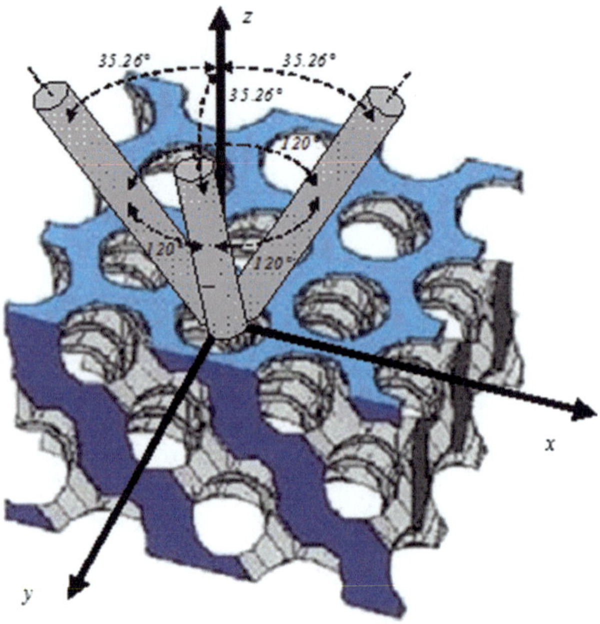

Fig. 6.3 Schematic representation of Yablonovite – the first photonic crystal produced by selective hole drilling on the millimetre scale (Adapted from Ref. [8])

atomic orbitals are contributing to the band structure in question. For this reason in any real crystalline semiconductor it is obvious that the wave vector (and hence energy) of the allowed states will vary in three-dimensions, such that the graphical depiction of this 'dispersion' is not simply a flat line across the Brillouin zone.

For electrons, the existence of discrete bands can be predicted even from a free electron perspective once boundary conditions are imposed, as a result of the discontinuities that arise when all of the Brillouin zone is 'folded' into the first zone of the diagram, defined as k lying between the limits of $\pm\pi/a$, where a is the periodicity, which we sometimes equate to lattice parameter. Folding into the first zone is permitted because of the periodic nature of the material, in that although k may increase indefinitely, the discrete zones are themselves a periodic function described by $k = \pm n\pi/a$ where n can be any integer in principle.

The physical phenomenon used to account for something otherwise simply resulting from the mathematics, is that of diffraction of the electron waves, which interrupts the propagation of the wave in question.

Accepting that this is the case, we can now stop discussing electrons in crystalline solids (!) and now extend the argument to photons in optically periodic media. Exactly the same considerations apply, but usually on much larger length scales. The structure is periodic (hence photonic crystal), but the material that the ordered structure consists of need not be crystalline. Thus an ordered array of amorphous silica or polymer particles of appropriate dimensions meets the required specification.

This argument serves to illustrate how diffraction may occur and helps the reader to imagine how the photonic band gap arises, since it is perfectly possible

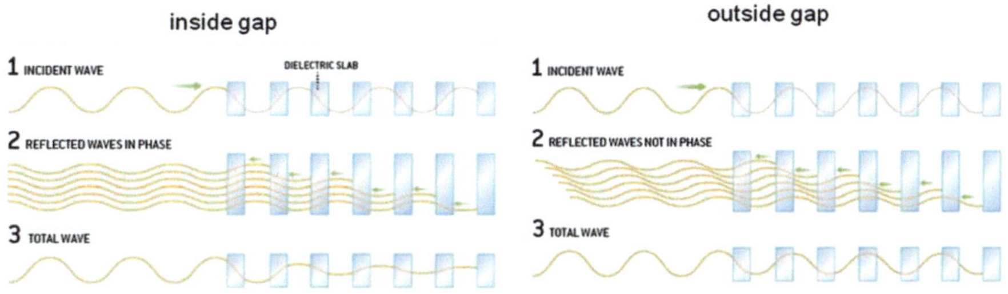

Fig. 6.4 A schematic representation of how a periodically-layered dielectric material can influence the propagation of light left, inside the forbidden gap and right, outside of the gap (Taken from reference [7])

to construct an optical Brillouin zone diagram, which shows the allowed modes in terms of an E vs. k diagram. Figure 6.4 above illustrates how reflection and interference can rapidly reduce the amplitude of a wave as it passes through a photonic crystal.

However, here the analogy breaks down because not only does the suppression of certain modes in certain directions of propagation depend on the interference phenomenon outlined above, it is also the consequence of the light wave moving from one medium, e.g. air, to another medium, e.g. an air/silica composite. The effects of reflection from the various 'layers' defined by the periodicity within the material must be accounted for together with refraction that must also occur as a result of the changing nature of the velocity of light as it passes from one medium to another, usually expressed in the form of the refractive index. Interference and refraction combine with diffraction so as to yield the photonic band gap. For diffraction it is conventional to deploy the co-called Bragg equation, $N\lambda = 2D \sin\theta$, where N is the order of diffraction and must be an integer of 1 (first order) or higher, D is the periodicity- in this case the layer spacing and θ is the angle of incidence as measured with respect to the surface normal. However this equation must then be combined with Snell's law, which defines the role of refraction across the interface between two media, 1 and 2, having refractive indices of n_1 and n_2 via the relationship $n_1/n_2 = \sin\theta_2/\sin\theta_1$.

The resulting relationship is a key one, and is known as the Bragg-Snell equation for obvious reasons,

$$N\lambda = 2D\left(n_{eff}^{2} - \sin^{2}\theta\right)^{1/2} \tag{6.1}$$

Note that Eq. 6.1 as stated here applies explicitly to the case where a light wave propagates from air ($n = 1$) into the photonic crystal, $n = n_{eff}$.

The term n_{eff} is used because it represents the effective refractive index of the photonic crystal, averaged over the medium. In reality the real refractive index on a local level varies periodically on the same length scale as that defined by D, but this variation is ignored when determining n_{eff}, which may be thought of as a composite refractive index. In the following section we discuss this further taking on board the specific example of the colloidal photonic crystals, which are the primary materials employed in our laboratories for the purpose of light management.

6.1.3 Colloidal Photonic Crystals (Including Artificial Opals)

By way of a specific example, consider a so-called 'artificial opal' made from silica spheres packed together into a face centre cubic (fcc) arrangement. Locally, n varies between that of amorphous silica (ca. 1.4) and that of the air in the spaces between the silica spheres ($n = 1$). The composite refractive index n_{eff} is then determined by the weighted average of the two refractive indices, which in the case of an fcc structure is $0.74n_{silica} + 0.26n_{air} = 1.3$, where the 0.74 and 0.26 factors are the geometrically determined weighting factors for this particular structure.

What about the D value? Again where colloidal spheres are employed and a face centred cubic arrangement constructed, we can relate the layer spacing D directly to the diameter of the spheres d, using a simple geometric relationship providing we know the orientation of the planes within the sample.

It turns out that we do indeed understand the nature of the packing of the spheres and hence the layer structure that forms. The fcc example was not introduced above without consideration of the most commonly employed arrangements of particles. For example, natural opals consist of layers of silica spheres that naturally form an fcc arrangement, and in the lab our artificial opals mimic this arrangement. Let us now ask the question, why does the fcc arrangement occur naturally? The answer to this question lies in the well-known phenomenon of close packing. Close packing is illustrated simply in Fig. 6.5, below:

This figure shows that once a close-packed layer of particles has formed, labelled layer A, then another layer (layer B) will form on top of this as a result of simple geometry. The positioning of the next layer then becomes critical in terms of the nature of the 3-D structure so formed. If the first layer is simply repeated such that the structure builds up as ABABAB etc., then a hexagonal close packed (hcp)

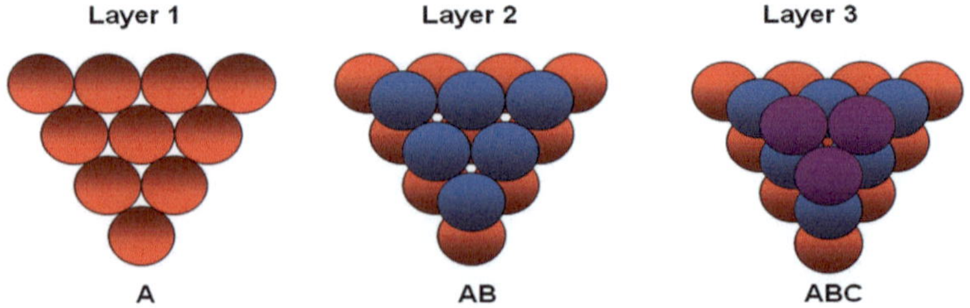

Fig. 6.5 A schematic representation of the process of close-packing of spheres. This process applies to many systems consisting of ions packing together in crystals, as well as the much larger particles used to create colloidal photonic crystals

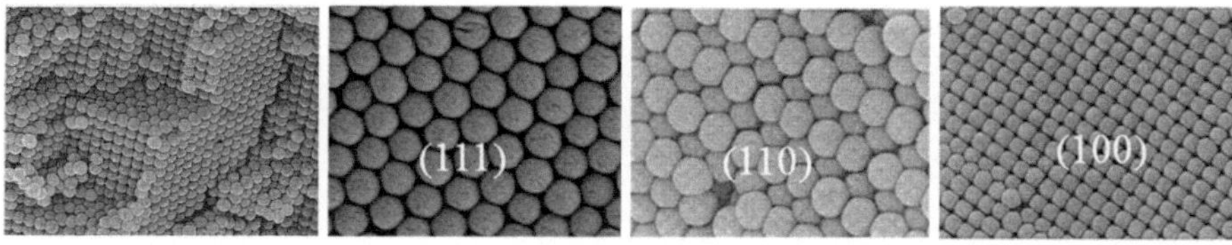

Fig. 6.6 Some SEM images of the surfaces of a range of colloidal photonic crystals made in our laboratories

arrangement results. Alternatively a third layer can form, which is distinct from the first two, layer C, such that the repeat then becomes ABCABC etc. This then is the arrangement that yields an fcc structure and it is this arrangement, which is present in natural opal gemstones. In an fcc structure each of the close packed layers shown in Fig. 6.5 would be regarded as a plane having Miller indices (111). This is because the plane cuts each of the three crystallographic axes of the face centred cube at the point equivalent to one lattice parameter. When this plane terminates the structure we obtain the most commonly viewed surface of any colloidal photonic crystal, the (111) surface in which the spheres form a close-packed surface layer having hexagonal symmetry.

Figure 6.6 above shows some images of colloidal photonic crystals made in our laboratories, which clearly illustrate the close packed nature of the top (111) surface and also reveal (side view) the existence of other low index planes such as (100) and (110) that are expected in an fcc structure.

The fabrication of so-called colloidal photonic crystals necessarily therefore involves the assembly of spherical colloidal particles of uniform size, usually consisting of amorphous silica or polymeric materials, into an ordered lattice such as the fcc structure which forms as part of assembly by close packing. The resultant structure will then possess a periodicity comparable to the wavelength of light in some easily accessible part of the electromagnetic spectrum. This so-called 'bottom-up' approach may be favourably compared to the alternative, 'top-down' approach for the fabrication of photonic crystals, which involves the use of lithographic methods developed for the electronics industry to etch a similarly periodic structure into a given substrate. Although such methods have been developed to a high degree of repeatability and precision (arguably far too high for photonic crystals given their typical dimensions and tolerances), they have not been extensively applied to the fabrication of photonic crystals because the structures possible via such an approach are somewhat limited and as such to date, there are few real examples where they have been used to create added-value optical devices.

Accordingly at the present time the bottom-up 'colloidal crystallisation' approach to making photonic crystals is still very much favoured since it offers the ability to make a wide range of interesting structures, which can then be used to fully evaluate the potential of these interesting materials via the application of cheap, simple technologies.

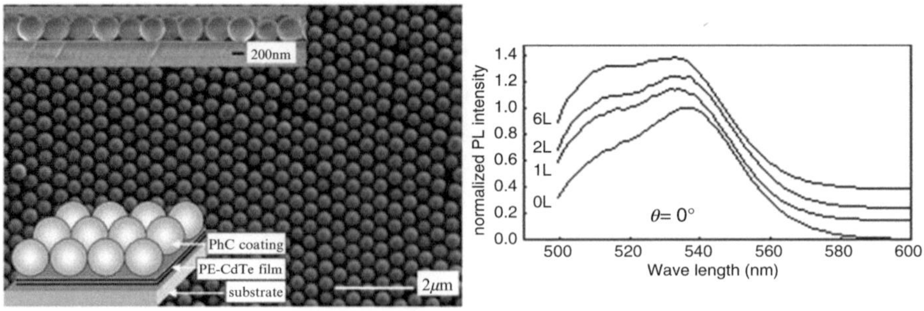

Fig. 6.7 An example of an experiment, which illustrates the power of colloidal photonic crystals as well as the added advantages associated with being able to directly control the number of layers of colloidal particles employed in a device structure, taken from Ref. [9], see text. The structures (*left*) were fabricated by one of the authors (MB) and the spectra (*right*) are courtesy of S.G. Romanov (University of Erlangen-Nuremberg; Germany)

As an example of the fundamental properties of such materials Fig. 6.7 represents the results of an experiment performed in our laboratories in 2008, in which the photoluminescence intensity from a layer of semiconductor nanoparticles excited by a laser was measured as a function of the number of layers of silica spheres deposited onto the structure [9].

This figure illustrates the power of the photonic crystal in terms of providing a more allowed 'gateway' for photons of a particular energy to flow through, than that provided by the simple device: air interface alone where a considerable fraction of the light is back reflected into the device by the interface. In terms of the particular experiment concerned, luminescent CdTe nanoparticles were embedded in a transparent matrix and excited with a laser beam. The emitted light from the nanoparticles was measured as a function of the number of layers of colloidal spheres deposited on top of the structure. As shown in Fig. 6.7, the intensity of the measured emitted light increases as the number of layers of spheres over the device increase- in other words we get more light out by 'covering' the device with a thicker layer! If the reader can understand how this occurs then they will begin to understand the power of the photonic crystal.

6.2 Fabrication of Advanced Colloidal Photonic Crystal Structures

The subject of photonic crystal fabrication via self-assembly methods has been discussed by, amongst others, Bardosova and Tredgold [10]. In general what is needed is the ability to rapidly and reproducibly produce large area defect-free colloidal photonic crystal hosts that can be structurally tunable and subsequently infiltrated in order to enhance their photonic properties. In earlier works films

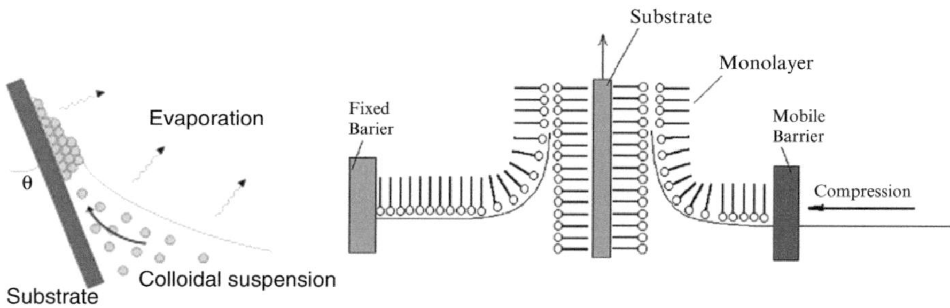

Fig. 6.8 Schematic representations of the two methods most commonly employed in our laboratories to fabricate colloidal photonic crystals

were formed by sedimentation of monodisperse particle assemblies, copying the way that opals are formed in nature, but more recently other methods have been explored with a view to obtaining a higher degree of order and to speed up the process of thin film formation, since the sedimentation approach can be very time-consuming. In the examples presented here we have employed two commonly-used methods to fabricate the desired host materials: the first of these is a method originated by Denkov et al. [11] and developed by Colvin and her collaborators [12–14] using controlled evaporation (CE) and which is known to produce an fcc arrangement of particles; the second involves application of the Langmuir-Blodgett (LB) method. The latter method has been reported previously by Bardosova et al. [15, 16] and by Reculusa et al. [17, 18] who perceived its potential as a method for engineering heterostrutures. More recently it has been determined that the LB method of fabrication, although producing high quality photonic crystals, does not produce photonic crystals with an fcc structure but with a structure that has been referred to as a $(2 + 1)$-dimensional [19]. The two different methods of assembly are depicted schematically in Fig. 6.8.

On the left the process known as controlled evaporation (CE) is illustrated. The substrate is inclined at an angle θ. As the solvent evaporates the colloidal spheres climb onto the substrate via the capillary forces inherent in the meniscus region. On the right the process of Langmuir-Blodgett (LB) deposition is illustrated. Here colloidal spheres are spread onto the subphase present in the Langmuir-Blodgett trough- usually water. The initially loose packed arrangement of spheres is then compressed into an ordered dense packed layer via a movable barrier. During this process the surface pressure is monitored in order to determine the point where the spheres form a close-packed layer, which corresponds to the steep increase in the surface pressure vs. area plot recorded. The substrate, which was immersed into subphase prior to the spreading is then drawn up via the air/water interface and out of the trough, bearing one layer of spheres deposited onto its surface. The process is then repeated as many times as required to give the desired number of layers.

As noted above, the CE method leads to the formation of true opal-like structures in which the spheres adopt an fcc arrangement. Controlling the number of layers is rather difficult using this approach although a certain degree of reproducibility can

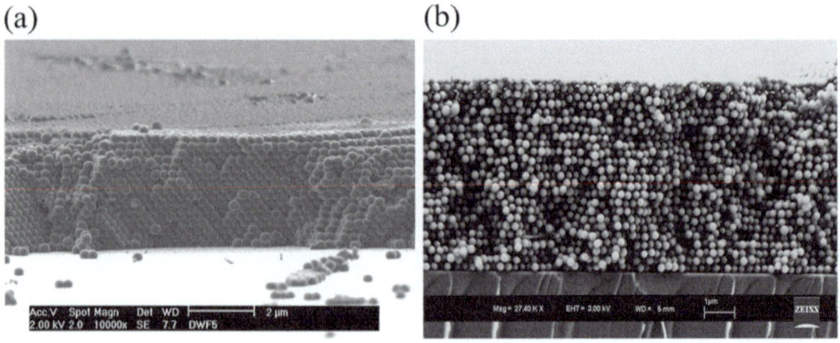

Fig. 6.9 (**a**) SEM image of a controlled evaporation sample and (**b**) a sample prepared by the Langmuir-Blodgett method in our laboratories

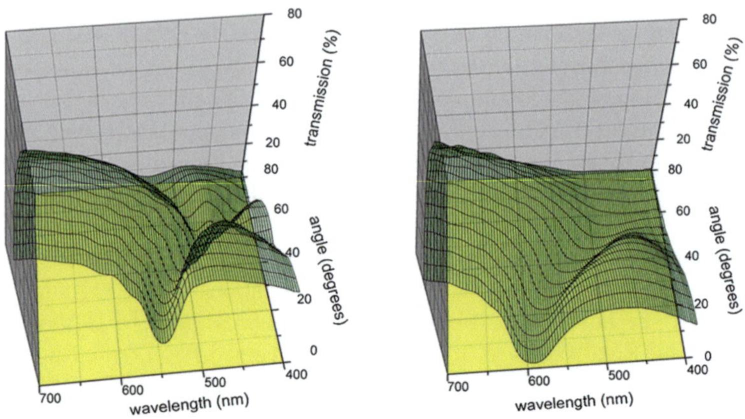

Fig. 6.10 Experimentally determined plots of transmission versus wavelength, over a range of angles of incidence, for colloidal photonic crystals made by *left*, the CE process and *right*, the LB process (Reprinted from Ref. [19] with permission from Elsevier)

be achieved by controlling the volume fraction of spheres in the suspension and the evaporation temperature. Jiang et al. produced some excellent structures this way [14]. In contrast, the LB method produces a structure, which as noted above is considerably more disordered in the third dimension. This additional disorder can be visualised as resulting from the formation of each individual close packed layer on the surface of the trough prior to it being transferred to the substrate. Sometimes the new layer falls into registry with the layer underneath and sometimes it does not. Figure 6.9 above shows the more disordered nature of this type of layer structure as compared to that formed by CE.

Romanov and co-workers compared structures made from the same size silica spheres using the two methods highlighted above [19]. This work reveals that the LB-grown samples are not true fcc opals at all, and also display different optical characteristics as demonstrated in Fig. 6.10.

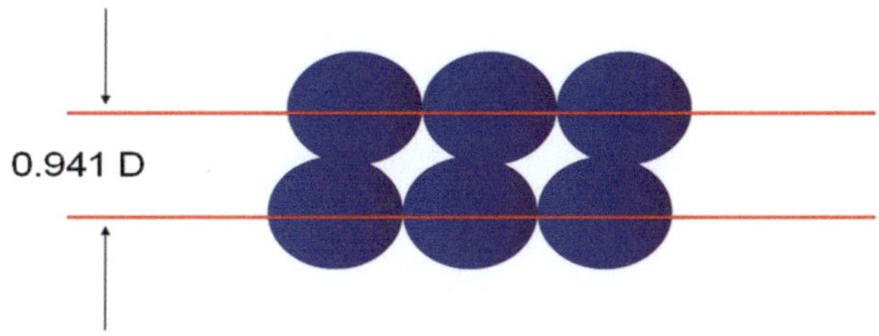

Fig. 6.11 Schematic representation of the arrangement formed between two layers of spheres via the LB deposition process. For an fcc arrangement the layer spacing would be 0.86d where d is the sphere diameter. In this real example Ref. [20] the layer spacing was measured to be 0.941d, such that the spheres in adjacent layers sit almost on top of one another

On the left of the figure we see the dispersion relation for the Bragg peaks obtained with a true fcc (opal) structure. Note the complexity between the angles of 40°–60°, which arises because of overlapping resonances that become allowed as a result of the fcc symmetry. In contrast note the simplicity of the dispersion on the right of the diagram, which was measured from an LB-grown sample. The lack of fcc symmetry results in a much simpler dispersion diagram and is clear evidence that the LB samples are not true opals.

Previously we have estimated the layer spacing that results in both types of sample, necessarily an average calculated from the optical data, using a modified form of the Bragg-Snell equation (Eq. 6.1) described earlier. If the wavelength of the stop-band is plotted as a function of angle of incidence, then one may plot a graph of the wavelength of the stop band versus the square of the angle of incidence, which is predicted to yield a straight line. Assuming $N = 1$, we may rewrite Eq. 6.1 as follows:

$$\lambda^2 = 4D^2\left({n_{eff}}^2 - \sin^2\theta\right) \tag{6.2}$$

Expanding the bracket we obtain:

$$\lambda^2 = 4D^2{n_{eff}}^2 - 4D^2\sin^2\theta \tag{6.3}$$

Thus at constant n_{eff} a plot of λ^2 versus $\sin^2\theta$ yields a straight line of slope $-4D^2$, which allows for the direct determination of the layer spacing, even when n_{eff} is not known. This is a very convenient way of directly checking the packing of the spheres in the structure since for an fcc sample D is calculated to be 0.86d, where d is the sphere diameter. For an LB sample we have shown previously that D values of up to 0.941d can be observed [20]. This means that on average for the sample the spheres are sitting almost on top of one another as shown schematically above in Fig. 6.11:

Once D has been estimated it is simple to go back to the equations above and calculate n_{eff} for the sample, which is of direct importance to experiments such as those described earlier which involve tuning of the refractive index contrast in the material.

The particular advantage of the LB technique therefore does not lie in enhanced ordering of the structure (although it was found that values for absolute reflectance for the samples prepared by the LB technique are higher than that for CE samples [20]); rather it lies in the area of thickness control, in terms of the number of layers deposited. In this regard the LB method may be thought of as almost a 'digital' process, where the number of layers of spheres deposited depends simply on the number of times the substrate is immersed into the subphase. In practice the LB trough operation is usually entirely automated and controlled by a computer interface, allowing the operator to 'dial-up' the number of layers. This is a tremendous advantage in terms of potential scale up and commercialisation of the approach, since device engineers almost exclusively work to process flows that have been developed under ISO-type guidelines which aim to ensure quality and reproducibility and it is relatively easy to see how an automated LB process could be incorporated into such a process flow.

This approach also allows for the highly controlled deposition of so-called heterostructured colloidal photonic crystals, which may be fabricated from layers of spheres of differing sizes. We first demonstrated this possibility in 2006 when we used the LB approach to grow heterostructured colloidal photonic crystals from 'size-matched' spheres, having diameters of 200 and 400 nm [21]. Figure 6.12 shows SEM images of both types of structure that we prepared which might be referred to as AB, and BA.

Figure 6.12 nicely illustrates the power of the LB approach in tailoring specific structures. The optical properties of these two systems are intriguing. Figure 6.13 illustrates the transmission spectra obtained from the two samples, which, perhaps not surprisingly, are very similar.

Figure 6.14 depicts reflectance spectra obtained at an angle of 12° from normal incidence for (a), the AB structure and (b), the BA structure. In each case spectra have been recorded from both the top of the samples (red lines) and the bottom, i.e., through the glass substrate (black lines).

Figure 6.14(a) also depicts the reflectance spectrum obtained from films of ten layers of type A particles (blue line) and B particles (green line) for direct comparison with the AB film. From Fig. 6.14 it is apparent that the peak at ca. 500 nm corresponding to the first order Bragg reflection from the type A particle layer under the type B particle layer is very intense for the case when the sample is probed through the underlying glass substrate, yet very weak when the sample is probed from the top, through the type B particle layer. Of particular note is that the peak intensity when the AB film is probed from the bottom side is greater than that for the film containing only type A particles (blue line). The possible inference that this intensity enhancement arises from a substrate effect is negated by comparison with the BA system depicted on Fig. 6.14b, where illuminating from below yields a spectrum that is comparable to that obtained by illuminating from above. For this reason the passage of the light through the glass substrate cannot in itself explain the intensity anomaly observed for the AB structure. We propose that

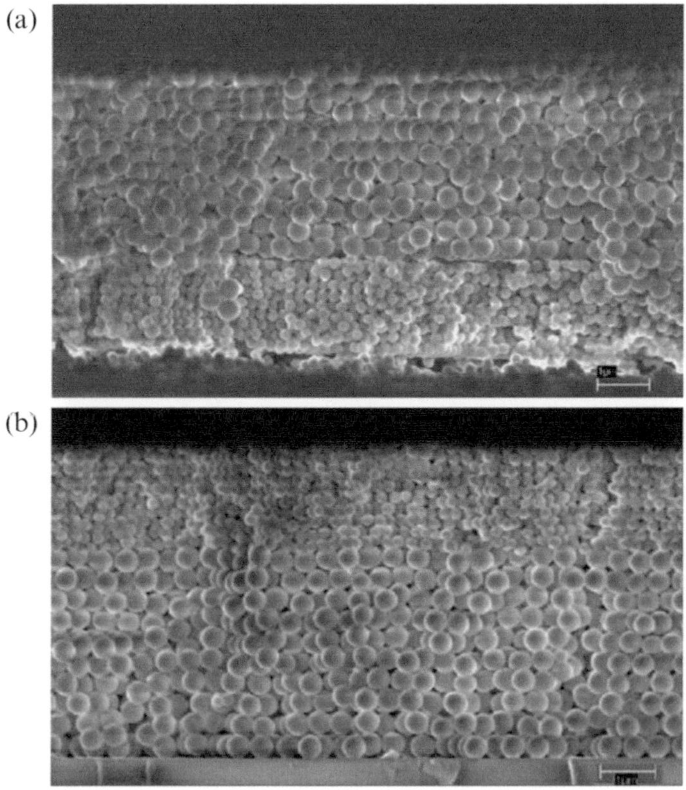

Fig. 6.12 SEM images of what we have referred to as size-matched heterostructured colloidal photonic crystals. Image (**a**) resulted from a sample in which spheres of diameter 200 nm were first deposited onto a glass substrate, followed by the deposition of layers of spheres of diameter 400 nm. Image (**b**) shows the alternate structure which now has the larger sphere layer in contact with the substrate (Reprinted from Ref. [21] with permission from Elsevier)

these results arise from the fact that the illumination of the AB structure from the top with a wavelength of ca. 500 nm (corresponding to a situation where normally there would be a first order reflection from the layers of type A particles having a diameter of 200 nm) results in the amplitude of the incident light being severely attenuated by first passing through the region of type B particles having a diameter of 400 nm, which themselves generate a strong first order Bragg reflection at about 960 nm. The weak nature of the second order Bragg reflection of the B layer, at the wavelength where the layer of type A particles will have a first order Bragg peak, i.e., ca. 500 nm, leads to the type B layer severely attenuating the light at this wavelength reaching the A layer and thus little or no first order Bragg peak from the type A particle layer will be seen. The assumption that this is an attenuation effect by the B particle layer is supported by our observation that the reflection spectrum, at this wavelength, is attenuated as a function of the B layer thickness. In contrast

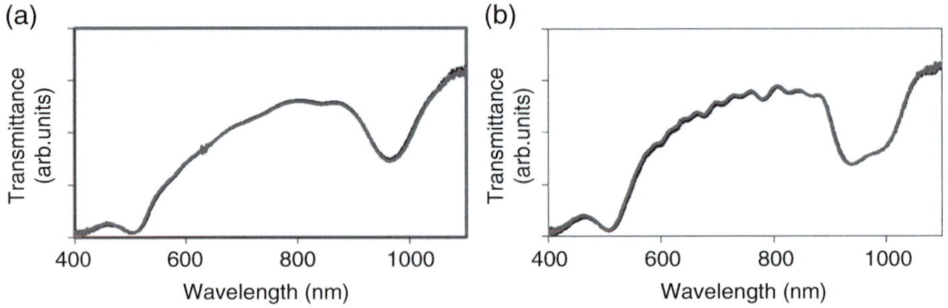

Fig. 6.13 Transmission spectra of (**a**) the AB structure and (**b**) the BA structure taken from both the A and B surfaces. As would be predicted the spectra are not dependent on the direction of illumination (Reprinted from Ref. [21] with permission from Elsevier)

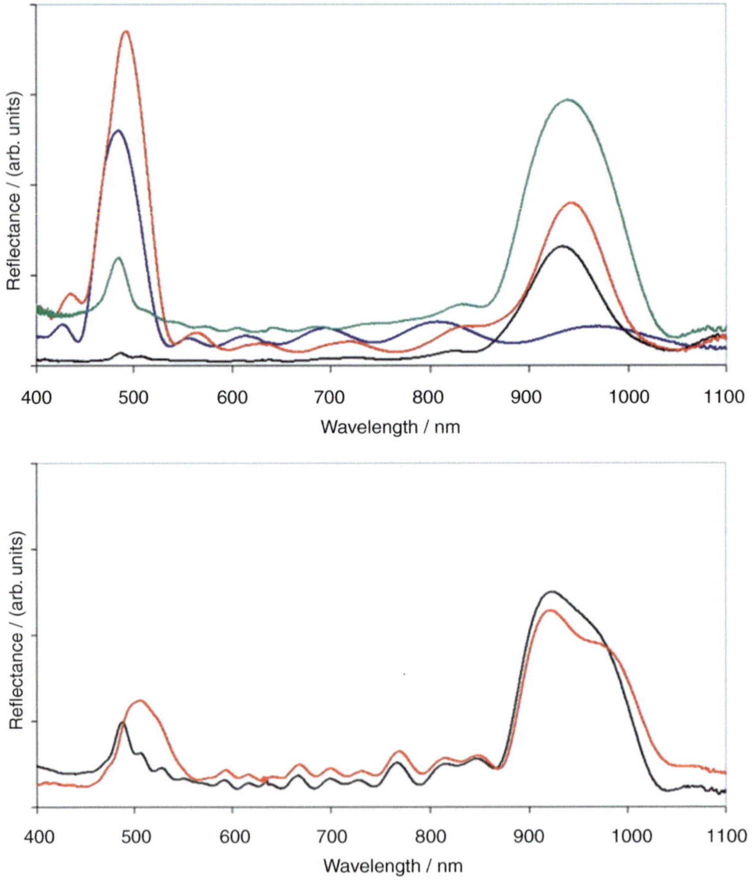

Fig. 6.14 (**a**) Reflection spectra of the AB structure taken from the A surface (*red*) and B surface (*black*). The reflection spectra of a comparable ten layer A only (*blue*) and ten layer B only (*green*) photonic crystals are given for comparison. The intensity of the ca. 500 nm peak is clearly enhanced when illuminated from the A surface. (**b**) Reflection spectra of the BA structure taken from the A surface (*red*) and B surface (*black*). Enhancement is not observed and the spectra are complicated by an interface effect (Reprinted from Ref. [21] with permission from Elsevier)

when the AB sample is illuminated from the bottom, such that the type A particle layer sees the light first with no prior attenuation, the first order Bragg peak from the type A particle layer is very intense and enhanced by the overlap with the weaker second order peak from the type B particle layer, i.e., an *additive process occurs*. Similar behavior might be expected to occur for the BA structure, in that some enhancement of the first order Bragg peak for the type A particle layer might be expected, this time under illumination from the top. Figure 6.14b reveals that while this peak is certainly more intense than its counterpart recorded via illumination from the bottom, it is much weaker than the enhanced peak depicted in Fig. 6.14a. We attribute this observation to the disruption of the quality of the photonic crystal formed from the BA structure due to the interface effects referred to earlier.

What then is the significance of these observations? The main finding presented here is that in reflection mode, size-matched heterostructured photonic crystals may exhibit *cooperative effects* whereby the first order Bragg reflections from one part of the structure mix with the second order Bragg reflections from other parts of the structure. We suggest that the resulting differences in reflectance for the AB samples for the different modes of illumination may find exploitation in a range of optical filtration devices.

6.3 Tuning the Properties of Colloidal Photonic Crystals

In general the nature of the band gap in a colloidal photonic crystal is influenced by the contrast in the dielectric constant between the host structure and its voids, which in the first instance can simply be equated to the difference in the real part of the dielectric constant, the refractive index contrast Δn, as well as the lattice constant [5, 6]. In order to engineer colloidal photonic crystal structures with a significant degree of functionality it is necessary to be able to exert a high degree of control over both of these parameters. The lattice constant can be selected by utilising an appropriate sphere diameter, which necessarily involves the development of techniques for the synthesis of colloidal spheres of a given diameter. For opaline materials where the spheres are made of silica, the well-known Stöber synthesis [22] or a variant of this may be used to prepare suspensions of colloidal silica spheres having a variance in diameter of less than 5 % [23, 24], Examples of a range of synthetic routes to the production of polymeric spheres, usually of polymethyl methacrylate or polystyrene are also readily available in the literature.

With all of these syntheses the primary target is to produce spheres essentially all having the same diameter, since as is apparent from Figs. 6.9, 6.11 and 6.12 this will facilitate the assembly process.

The properties of the material also depend on choice of crystal structure. Colloidal crystallisation is very limited in this regard, being only able to make fcc structures via close packing, or closely related systems (see later), with growth occurring mainly in the (111) direction unless another low index plane is deliberately induced via the use of a templated substrate [25, 26]. Such templates

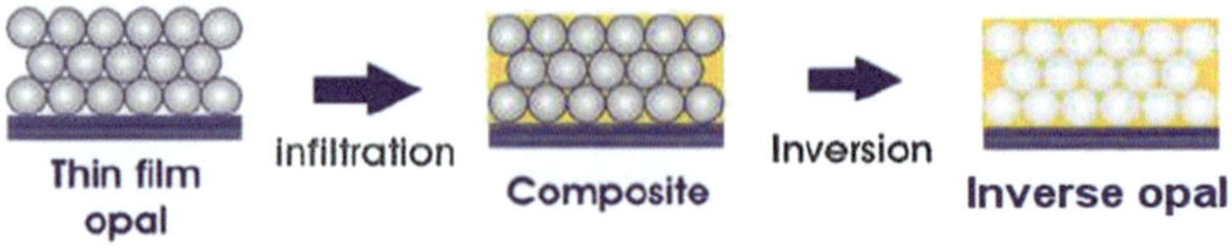

Fig. 6.15 A schematic representation of the infilling and subsequent inversion of a thin film colloidal photonic crystal, in this case an opaline structure (Adapted from Ref. [31] with the permission from IOP Science)

may be readily prepared using the lithographic methods referred to earlier. For optical materials the dimensions and tolerances associated with the fabrication of such templates are not at all challenging.

Interestingly colloidal photonic crystals are also perfect materials in terms of being able to tune the contrast in refractive index or more precisely dielectric constant since it is possible to infill the voids in the host structure with a material of differing refractive index. This process has been demonstrated with a wide range of methods such as wet chemical impregnation [27] and chemical vapour deposition (CVD) [28–30], One very notable example of CVD infilling by Blanco et al. [24] employed silane gas to infill a colloidal photonic crystal with silicon. The silica spheres of the host material were then removed by chemical etching to produce what is known as an 'inverted' structure and in shorthand form for an opal host the resulting material is often referred to as opal^{-1}.

This process may be described schematically as shown in Fig. 6.15 above.

In the particular example presented by Blanco et al. [24] the inversion process necessarily increased the refractive index contrast from that of silica to air to that of air to silicon. When this occurred, a so-called complete band gap opened up in the material. Here the term complete band gap represents a region across the optical Brillouin zone where light waves of certain energy may not propagate in *any* direction. This is an extraordinary condition where apart from scattering losses which may be predicted and quantified, the material behaves as a perfect mirror for the light in question, which cannot formally 'exist' in the bulk.

Figure 6.16 shows some images taken from reference [24] which illustrate (a) the inverted silicon photonic crystal and (b) the resulting optical Brillouin zone which also clearly displays the complete gap at around 1.54 μm, in the communications region of the spectrum.

A further example of a complete gap was achieved using a CVD-like infiltration process by Miguez et al., who also observed a complete photonic band gap after infilling an opal template with germanium [32].

Another approach that we and others have often employed in order to provide a highly uniform and conformal infilling of a 3-D colloidal photonic crystal host is that of atomic layer deposition (ALD). The ALD technique has been shown by several groups to be an extremely versatile technique for the infiltration of 3-dimensional photonic crystals producing uniform and highly conformal infill with a wide range of materials including ZnO [33], WN [34], ZnS [35], TiO_2, ZnS:Mn [36] and the III/V materials GaP [37] and GaAs [38].

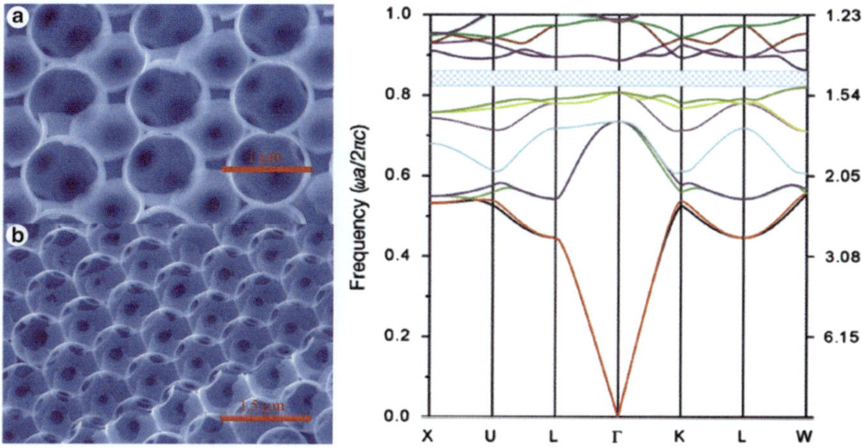

Fig. 6.16 Images taken from the work of Blanco et al. [24], who first demonstrated the appearance of a complete photonic band gap in the communications wavelength region, which resulted from the infilling of a colloidal photonic crystal with silicon, followed by inversion to increase the refractive index contrast still further. *Left*, SEM images of the inverted structure and *right*, the optical Brillouin zone that resulted (Reprinted with permission from *Nature*)

ALD is a variant of the CVD approach in which a layer is formed via the surface reaction of at least two discrete chemical precursors, which are introduced into the deposition chamber sequentially, as compared to being introduced simultaneously as in CVD. Growth is then limited by surface adsorption and reaction such that only a single monolayer or less is capable of forming during each growth ‘cycle’. The reader is referred to several reviews of the ALD method for further reading [39–41]. While ALD is a slow method entirely unsuitable for the growth of thick layers (e.g. several microns) it is perhaps the most conformal and uniform growth method known, and is particularly suitable for the infilling of complex 3-D architectures such as may be found in colloidal photonic crystals, as is demonstrated in the following section.

6.4 Atomic Layer Deposition for the Modification of Colloidal Photonic Crystals

Figure 6.17 below illustrates the ALD approach that we and others have taken in order to modify colloidal photonic crystals and also illustrates perfectly the ability of the ALD technique to infill a complex 3-D structure fully via self-limiting surface reactions. This figure was recorded from a colloidal photonic crystal that had been modified by filling the voids between the silica spheres with Al_2O_3, grown using trimethyl aluminium and water as precursors.

The reaction in question is really the ‘standard’ of all known ALD processes [42] and is known to occur from room temperature through to around 300 °C, at a

Fig. 6.17 SEM image of a colloidal photonic crystal following ALD-infill with Al_2O_3 (Reprinted from Ref. [42] with permission from Elsevier)

growth rate of ca. 0.1 nm per cycle- thus granting the operator phenomenal control over layer thickness. In the preparation of the sample for the recording of the image shown in this figure some of the photonic crystal slab has broken away from the substrate (foreground of the image). In this region a pattern can clearly be seen on the substrate, which has arisen as a result of the growth occurring all the way through the slab, to the substrate, except where the first layer of spheres has acted as a mask. This observation highlights the power of the ALD approach very well and also illustrates that nanosphere lithography can be used to prepare patterned substrates.

From the point of view of enhanced refractive index contrast there is little to be gained by infilling a silica colloidal photonic crystal with alumina because the two materials are quite similar optically. In contrast, our first real venture into this area in 2006 involved infilling a silica opal with GaAs [38]. We selected GaAs as the material with which to infill the colloidal photonic crystals because of its relatively high refractive index across the visible region of the spectrum. Just as Blanco et al. and Miguez et al. demonstrated that a full photonic band gap could be created by infilling with silicon and germanium respectively [24, 32], we postulated that similar effects could be seen with GaAs infilling. In addition GaAs-based photonic crystals are of particular interest in terms of both the nature of the photonic gap produced in a high refractive index contrast medium and the possibility of coupling these effects with both linear and nonlinear optical processes such as second harmonic generation, which would be allowed in a material such as GaAs as a result of the direct band gap and the non-centrosymmetric nature of the bulk unit cell, respectively [43]. To date while several two dimensional photonic band gap structures have been fabricated using lithographic methods, reports of fabrication of three dimensional GaAs photonic crystals are limited [44] and to the best of our knowledge there are no other reports of growth by CVD or related methods. Figure 6.18 below shows electron micrographs of the infilled opal at various stages of the procedures employed.

(a) (b) (c)

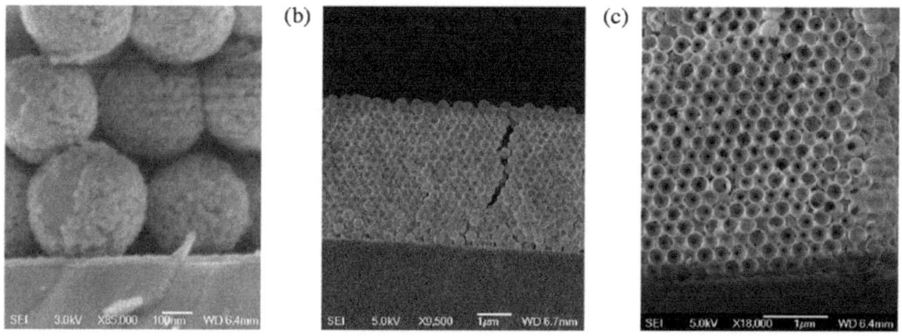

Fig. 6.18 SEM images of cleaved synthetic opals after (**a**) partial infill after 200 ALD cycles, (**b**) nearly complete after 800 ALD cycles and (**c**) the subsequent inverted structure (Reprinted from Ref. [38] with permission from the American Institute of Physics (AIP))

It is noteworthy that the GaAs ALD growth was performed in a standard horizontal metalorganic vapour phase epitaxy (MOVPE) reactor system (Thomas Swan Scientific Equipment) using a process which represented only a slight modification of the normal conditions employed for GaAs MOVPE yet still allowed for the self-limiting growth mechanism inherent to ALD to operate. The reactor was maintained at 400 °C and 200 Torr using trimethylgallium (TMGa) and arsine AsH_3 as precursors. Alternating gas pulses of TMGa (3 s) and AsH_3 (8 s) at flow rates of 5.3×10^{-5} and 4.5×10^{-3} mol min^{-1}, respectively, were separated by a hydrogen purge (2 s) to prevent the reactant gases from mixing in the gas phase. Abrupt switching of precursors was achieved using a pressure-balanced run-and-vent gas manifold. The flow rate of hydrogen through the reactor was 16 dm^3 min^{-1} throughout precursor pulsing and purging. Inversion of infilled opal structures was achieved by immersion in 1 % HF solution for 20 min.

Figure 6.18 depicts the GaAs coated silica photonic crystal at two different stages of growth as well as the inverted structure that was formed after removal of the host silica spheres so as to increase further the refractive index contrast. Figure 6.18b and c in particular show that near complete infill was achieved using the approach described, despite the fact that the ALD precursor gases had to penetrate over 4 μm through the photonic crystal in order to achieve this. Figure 6.18c demonstrates that a self-supporting inverted structure consisting almost entirely of GaAs and air could also be formed. The optical changes induced by the infilling process were particularly interesting. Figure 6.19 shows reflectance spectra recorded at an angle of incidence of 12°, from the bare opal sample (blue line) and for increasing levels of infill being 15 % (green), 26 % (grey), 67 % (black) and 91 % (red).

Several features that appear in Fig. 6.19 are noteworthy. Firstly, as predicted by the Bragg-Snell equation increasing infill with a higher refractive index material such as GaAs leads to an increasingly red shift in the position of the stop band. Secondly, for the black and red traces two peaks are apparent while the blue and green traces show only one peak and the gray trace shows no peaks other than the small Fabry-Perot oscillations which arise due to interference within any transparent thin film, which themselves may be used simply to estimate the overall

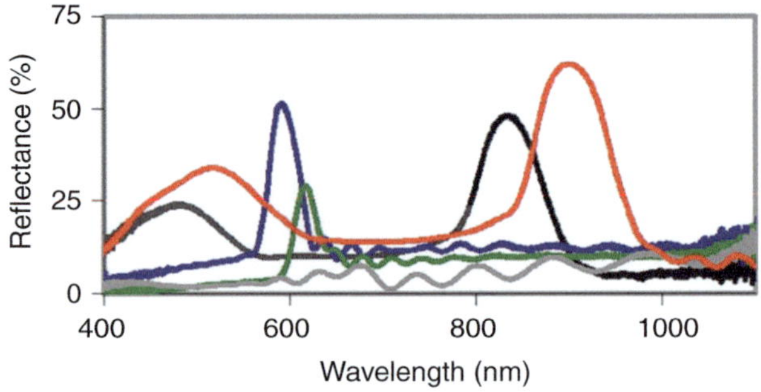

Fig. 6.19 Diffuse reflectance spectra of opal thin films for (**a**) a bare opal (0 % infill, *blue line*), (**b**) 15 % infill (*green line*), (**c**) 26 % infill (*grey line*), (**d**) 67 % infill (*black line*), and (**e**) 91 % infill of GaAs (*red line*). Note the progressive red shift of the first order Bragg peaks as the infill level increases. At higher levels of infill the partially unassigned secondary features shift into the accessible spectral window (Reprinted from Ref. [38] with permission from the American Institute of Physics (AIP))

film thickness and may be regarded as a measure of thickness uniformity [45]. The second peaks at shorter wavelengths shown in the black and red traces are the result of second order effects, i.e. where N in the Bragg-Snell equation = 2, rather than 1. The corresponding peak for the blue trace (bare opal) and the green trace (15 % infill) would appear at wavelengths which lie outside of the range of the detector employed. Why then are there no Bragg peaks for the grey trace?

The answer to this question perfectly illustrates the working of the infill material in that at this level of infill (26 %), the value of n_{eff}, being the weighted average of the material filling the voids between the silica spheres (now 74 % air and 26 % GaAs), almost exactly equals the refractive index of the silica spheres themselves. Thus there is no refractive index contrast and hence no stop band is observed. We have turned off the photonic band gap by using just the right amount of infilling!

Possibly one of the most significant points to arise from this work involves its consideration of potential absorption processes, hitherto specifically ignored. GaAs has a direct band gap of 1.42 eV, which means that it absorbs light of energy greater than this value, which equates to a wavelength of ca. 875 nm. Thus all wavelengths below this value may be absorbed by the GaAs.

We addressed this issue by considering not just the real part of the refractive index, which would normally be sufficient at least to a first approximation for a 'transparent' material, but now take into account the full value of the dielectric constant for the material, ε.

We introduced the concept that the refractive index now must be described by a complex number such that $\mathbf{n} = n + ik$, where k represents the imaginary part of the refractive index and may be equated to the absorption coefficient of the medium.

Using this formalism it is possible to equate the complex refractive index directly to dielectric constant (over the frequency range in question) using $\varepsilon = |\mathbf{n}|^2$. Under this condition the Bragg-Snell relationship may be rewritten as

$$N\lambda = 2D\left(\varepsilon_{eff} - \sin^2\theta\right)^{1/2} \tag{6.4}$$

Furthermore by substituting the fill factors (f) for each material, as well as the n and k values for each material, we obtain:

$$\varepsilon_{eff} = f\left(n_{SiO_2} + ik_{SiO_2}\right)^2 + f'(n_{infill} + ik_{infill})^2 + f''(n_{air} + ik_{air})^2 \tag{6.5}$$

To a first approximation Eq. 6.5 can be shown to simplify to:

$$\left(n^2 - k^2\right)_{SiO_2} + \left(n^2 - k^2\right)_{infill} + \left(n^2 - k^2\right)_{air} \tag{6.6}$$

Note that in order to generate Eq. 6.6 from Eq. 6.5 we introduced the relationship $f + f' + f'' = 1$, in order to eliminate the fill factors from the expression for ε.

For GaAs, the value of n varies from 4.4 to 3.5 while k varies from 2.0 to 0.0 over the wavelength range 400–950 nm [46, 47].

The estimates of infill stated earlier were made using Eqs. 6.5 and 6.6 above. While these expressions are approximate in several regards, this formalism serves well in terms of predicting the behaviour of the GaAs infilled opal and the resulting inverted structure [38].

We were able to take this approach one step further, and calculate the complete photonic band structure for this material. Figure 6.20 shows the Brillouin zone for a GaAs inverted structure calculated employing the MIT photonic band package [48] using a face-centered-cubic lattice with 260 nm spheres, $n = 3.62$, and $k = 0.0$. The singular refractive index value used in the calculations was chosen to match the Bragg reflection peak observed for the inverted structure. The calculated Brillouin zone can be directly compared to measured reflection data displayed in Fig. 6.20 from which it is evident that there is good agreement with the position of the first

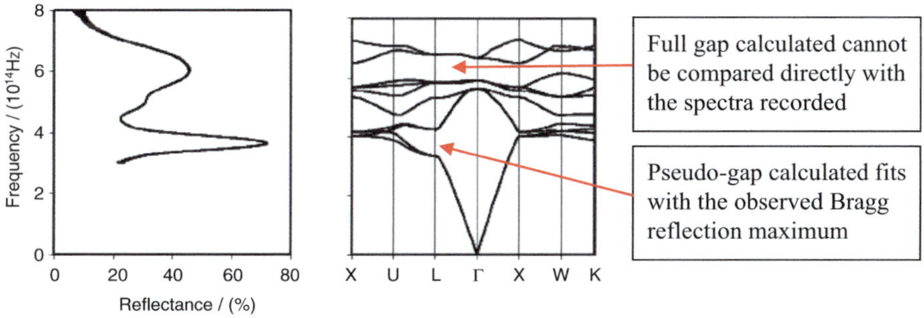

Fig. 6.20 Brillouin zone calculated using the MIT photonic band package software [47] for an inverted GaAs: air photonic crystal fabricated in our laboratories, together with the measured reflectance spectrum (Adapted from Ref. [38] with permission from the American Institute of Physics (AIP))

order Bragg reflection maximum and the pseudo-gap between the second and third bands (counting upwards) between the *U* and *L* points in the Brillouin zone.

Similarly it can be argued that there could be a genuine full gap response associated with the feature around 6×10^{14} Hz between the eighth and ninth bands but one must be cautious when making such assignments because there may be a contribution from second order effects while the refractive index selected for these calculations does not allow for any dispersive effects which will undoubtedly influence the real band diagram.

This example serves to demonstrate how colloidal photonic crystal architectures may be subsequently exploited as convenient templates for the fabrication of a wide range (limited only by the available ALD or other infilling processes) of interesting structures.

6.5 Colloidal Photonic Crystals for Photovoltaic Applications

6.5.1 Towards Colloidal Photonic Crystal-based Transparent Conducting Materials

Transparent conducting materials, most commonly oxides (TCOs) are an integral part of most photovoltaic devices. Being the first functional layer in the device that is exposed to the incident light the purpose of the TCO is to allow/direct as much light into the absorber layers as possible while also providing electrical contact. The issues faced with the use of many commonly-used TCO materials such as indium tin oxide, fluorine-doped tin oxide and aluminium-doped zinc oxide are electrical conductivity versus transparency- there is always a trade off between these two parameters, and reflection losses. It is this latter problem that has received considerable attention in recent years, mainly in the form of texturing. TCO layers can be grown having a highly textured surface for example. Some CVD processes naturally yield textured, columnar deposits [49], which greatly reduce reflection. Wet chemical and plasma etching have also been employed to alter the morphology of the TCO again with a view to reducing reflection losses [50–52].

Our venture into this field has centred on the use of colloidal photonic crystal thin films not only because they can act as superb anti-reflection layers, but they can actively enhance light propagation into the absorber layers. How then to add the conductive properties required of a true TCO? We have already seen how this may be achieved- via the use of ALD.

Figure 6.21 shows an opal sample in which we have deposited ZnO and Al-doped ZnO using an ALD process developed in our laboratories whereby the dopant Al ions are introduced at well-defined points during the ALD cycles, so as to create a nanolaminate.

Such materials are currently under further development in our laboratories and the example shown illustrates the potential power of combining colloidal

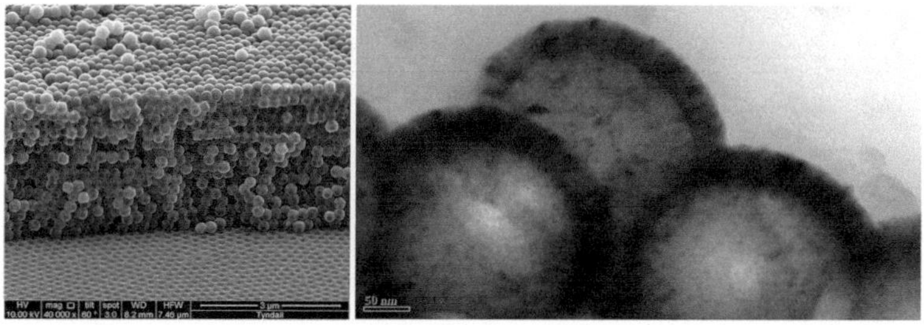

Fig. 6.21 *Left*, an SEM image very similar to that shown in Fig. 6.17, but this time for a colloidal photonic crystal that has been infilled with Al-doped ZnO using an ALD process developed in our laboratories [53]. Note the substrate patterning effect which like Fig. 6.17, demonstrates the ability of the ALD approach to infill a 3-D structure completely. *Right*, a TEM image of the SiO_2 particles after ALD coating. The columnar ZnO coating may clearly be seen. The scale bar is 50 nm

crystallisation with ALD infilling to produce a conducting film, transparent over a large spectral range (away from the Bragg reflection), which is also tailored to facilitate light propagation.

6.5.2 Colloidal Photonic Crystal-Based Solar Cell Architectures

In general, it is well-known that colloidal photonic crystals can be used as hosts for the absorber material itself since they represent essentially transparent (again away from the Bragg peak) materials which possess a high internal surface area and high scattering ability [54–56]. This concept has been investigated by Bermel and co-workers who produced a seminal article on the subject [57].

Figure 6.22 shows how using photonic crystals of various designs might influence the performance of a solar cell.

In this paper, Bermel et al. calculate that the use of the 3-D inverse opal architecture shown in the third panel of Fig. 6.22 which consists of eight layers or periods of an inverse PhCs, can result in an enhancement of the overall efficiency of a crystalline silicon (c-Si) solar cell by 26.5 %, which compares very favourably with other potential light harvesting structures such as gratings and distributed Bragg reflectors.

In our own work we have studied the use of colloidal photonic crystals as potential light trapping structures and also for a specific type of solar concentrator system that may be employed in building-integrated photovoltaics.

The first example presented here concerns our development of what we refer to as omni-directional light trapping structures [58]. Figure 6.23 shows schematically a structure which we have fabricated in our laboratories.

The system consists of the two distinct forms of colloidal photonic crystal that we have described earlier- an fcc opal type slab made by controlled evaporation and a more disordered LB grown slab. Between these two photonic crystal slabs a

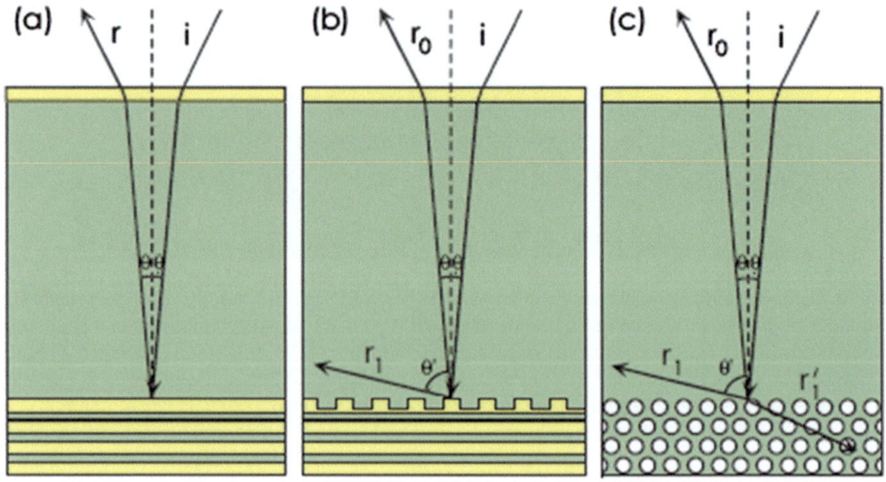

Fig. 6.22 shows designs aimed at enhancement of efficiency for a solar cell based on increased path length and light trapping by gratings and photonic crystals. (**a**) a distributed Bragg reflector (DBR), (**b**) a DBR plus an etched grating, (**c**) an inverse opal structure (Reprinted from Ref. [57] with permission from OSA (The Optical Society))

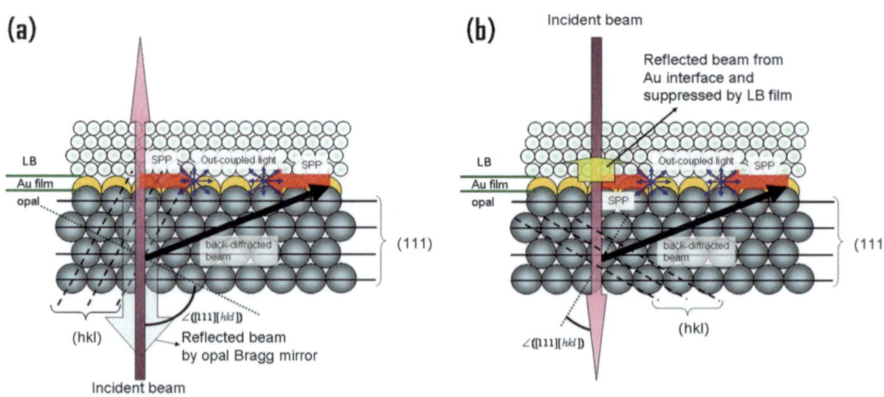

Fig. 6.23 A schematic representation of a so-called omni-directional light trapping structure that would be suitable for application to a PV system where tracking of the sun over the course of a day is either not feasible or too expensive in terms of energy costs [59]

gold layer has been introduced. The various light pathways through the structure are illustrated. Note that in this work we necessarily had to consider the role of surface plasmon polaritons (SPPs) in the gold layer because the natural corrugation that exists will allow these to be readily excited. Crucial to the operation of this layer is the LB slab. The disorder inherent in this layer results in the overall light trap becoming essentially polarization insensitive. In addition the losses only weakly depend on the angle of light incidence, i.e. this trap becomes almost omni-directional. Finally we calculate that the intensity of losses exceeds 90 % in the broad spectral range $\lambda < 1{,}000$ nm. Thus the trap architecture covers most of the solar spectrum.

A further example concerns the light trapping ability of a much simpler structure which has been examined in more detail than the structure described above. This is simply an LB-grown sample, in which the number of layers of colloidal spheres is increased and the subsequent response of a somewhat specialised PV unit evaluated. The application under consideration is that known as building-integrated photovoltaics. Using the concept of a window as a potential concentrator, capable of collecting light and forcing it to propagate in a particular direction via the process of total internal reflection, we have demonstrated the added advantage of using a colloidal photonic crystal overlayer [60].

In this case, the concentrators are integrated with the windows of a building and represent a way of balancing contributions towards natural lighting as well as creating electrically active building elements. The plastic slab impregnated with luminescent dye illustrates concept of total internal reflection. Note how the edges of the slab appear to glow as light is collected from the top face and directed towards the edges.

As may be seen from Fig. 6.24, the overall concept is rather novel. The solar cells are placed in the window frame and do not face the sun directly. The photonic crystal overlayer facilitates light scattering/trapping into the window material and its subsequent direction towards the frame via total internal reflection. The benefits arising directly from the presence of the photonic crystal layer were then examined by varying the number of layers of colloidal spheres deposited, figure 6.25:

Figure 6.25 reveals the increase in light detected at 90° to the incident light beam employed as a function of the number of layers, with the maximum being observed in this particular experiment at eight layers, which is close to the number of layers generally required in order to achieve full development of the photonic band gap, typically between 10 and 15 layers [45].

Figure 6.26 then shows 1–V characteristics of an Si solar cell with different concentrators attached and how the structure was tested.

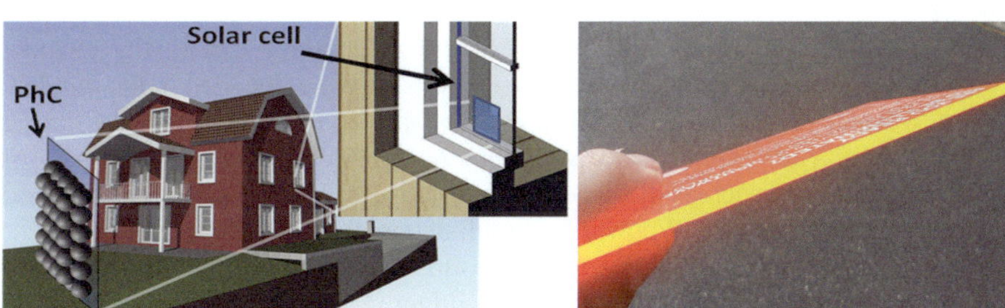

Fig. 6.24 *Left*, a schematic of possible building integration of solar concentrators taken from Ref. [60]; *right*, a plastic slab impregnated with luminescent dye

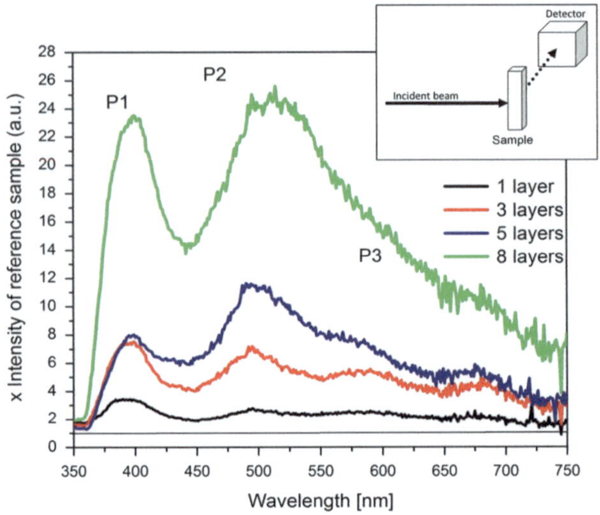

Fig. 6.25 Scattering normalized to the transmission from a blank PMMA sheet (*straight line* at "1" represents the transmission from the reference sample). Peaks *P1* and *P2* are a result of forward scattering of incident light into angles greater than 42° and subsequent back-scattering. Insert shows the measurement geometry (Reprinted from Ref. [60] with permission from Elsevier)

Except for the sample with only one layer of silica particles, the short circuit current I_{sc} increases with the number of deposited layers leading to an increase in light power reaching the solar cell at the edge of the concentrator. There is an optimum number of layers (estimated to be between 20 and 30) exceeding which the gain from beneficial effects of diffraction and scattering will be overcome by the losses introduced.

So if the concept described in Figs. 6.24, 6.25 and 6.26 were to become a reality, what might we expect to see? For example, would the window into the building be completely transparent? The answer to this question is no. If one viewed the window employing the particular angle of incidence and wavelength combination needed to match the condition described by Eq. 6.1 then one would see an opaque coloured slab- just like the colours seen at the start of this chapter on Fig. 6.16. However, once the angle of viewing changes, the condition would be broken and the sample would again appear to possess some degree of transparency. This concept is illustrated in Fig. 6.27.

Thus while full transparency may be sacrificed to some extent, the window would become the concentrator and the solar cells, being embedded in the window frames, would essentially be an almost maintenance free source of energy during daylight hours. The same concept could also be applied to other types of passive photovoltaic devices such as those designed to harvest ambient indoor lighting and provide background levels of power for building energy control purposes.

This type of passive energy harvesting using colloidal photonic crystals could well become very significant in terms of building design and construction in the future.

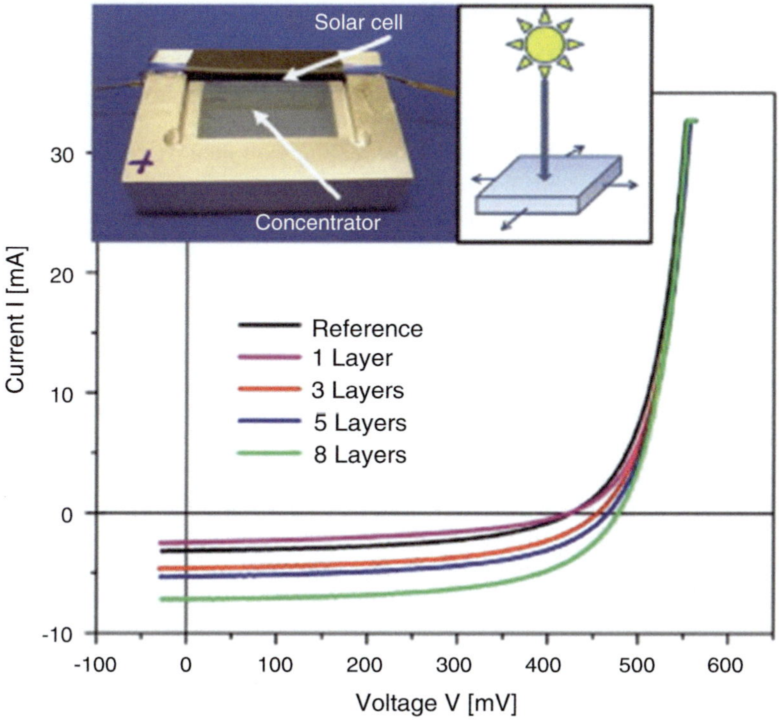

Fig. 6.26 I-V curves for a silicon solar cell attached to one face of the planar concentrators in an aluminium frame. Insert shows the experimental set up used in simulations (Reprinted from Ref. [60] with permission from Elsevier)

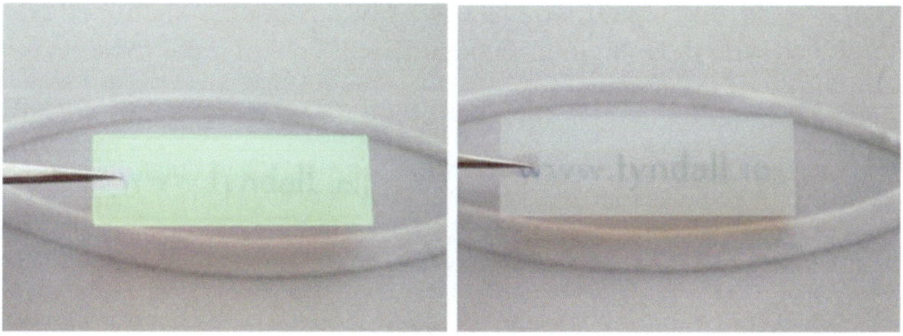

Fig. 6.27 *Left*, a colloidal photonic crystal deposited onto a microscope slide and held at the correct angle such that the camera catches the Bragg reflection. *Right*, the same sample after the slide has been tilted away from the Bragg reflection angle. Note that the transparency of the sample is partially restored simply by tilting

6.6 Conclusions and Outlook

In this chapter the concept of the colloidal photonic crystal has been described with emphasis on how such materials may be fabricated and modified, or tuned, so as to give the desired response. Two methods for making colloidal photonic crystals have been described, but emphasis has been placed on the use of the Langmuir-Blodgett method because of the inherent ability to control precisely the number of layers formed.

The technique of atomic layer deposition has been shown to be a particularly powerful tool for modifying these materials and also for facilitating the fabrication of inverted structures, which provide additional scope in terms of the available parameter space.

Finally, some potential applications of these materials have been highlighted, focusing on energy harvesting photovoltaic devices.

Photonic crystals have been under investigation for some considerable time now. Why then have they not become widely employed in a range of devices? The answer to this is twofold: firstly, while the examples of colloidal photonic crystals presented here and elsewhere in the scientific literature often do an excellent job in terms of what they were designed to do, observers generally balk at images of these materials which show the presence of defects such as cracks. We and others have developed a number of methods designed to minimise cracking in such materials [61] but we have to point out that sometimes cracks just do not matter! In terms of the desired optical response a degree of cracking or other forms of defect often makes little or no difference to the performance of the material. However, this is really a topic for another chapter.

Secondly, methods are needed whereby colloidal photonic crystals may be reliably and reproducibly manufactured in high volumes, perhaps using roll-to-roll type processing. In this regard we are currently developing such methods which can be combined with the Langmuir-Blodgett approach, in order to demonstrate the manufacturability of these materials.

Acknowledgements This work was supported by Science Foundation Ireland Grant no. 11/PI/1117 and EU FP7 project no 295182 PhANTASY. We would like to thank the members of our research team, in particular Dr. I.M. Povey, Dr. J. McGrath, Dr M. Schmidt and Ms C. Ryan for fruitful discussions and help with experiments.

References

1. Stokes GG (1884) On a remarkable phenomenon of crystalline reflection. Proc R Soc Lond 38:174
2. Strutt JW, Lord Rayleigh (1888) On the remarkable phenomenon of crystalline reflexion described by Professor Stokes. Phil Mag Ser 26(160):256
3. Strutt JW, Lord Rayleigh (1887) On the maintenance of vibrations by forces of double frequency, and on the propagation of waves through a medium endowed with a periodic structure. Phil Mag Ser 24(147):145

4. After Johnson SG, http://ab-initio.mit.edu/photons/tutorial/
5. Yablonovitch E (1987) Inhibited spontaneous emission in solid state physics and electronics. Phys Rev Lett 58:2059
6. John S (1987) Strong localization of photons in certain disordered dielectric superlattices. Phys Rev Lett 58:2486
7. Yablonovitch E (2001) Photonic crystals: semiconductors of light. Sci Am 285(6):47
8. Joannopulos JD, Johnson SG, Winn JN, Meade RD (2008) Photonic crystals: molding the flow of light. Princeton University Press, Princeton, 286 pp
9. Romanov SG, Bardosova M, Povey IM, Sotomayor Torres CM, Pemble ME, Gaponik N, Eychmueller A (2008) Modification of emission of CdTe nanocrystals by the local field of Langmuir-Blodgett colloidal photonic crystals. J Appl Phys 104:103118
10. Bardosova M, Tredgold RH (2002) Ordered layers of monodispersive colloids. J Mater Chem 12:2835
11. Denkov ND, Velev OD, Kralchevsky PA, Ivanov IB, Yoshimura H, Nagayama K (1992) Mechanism of formation of 2-dimensional crystals from latex particles on substrates. Langmuir 8:3183
12. Bertone JF, Jiang P, Hwang KS, Mittleman DM, Colvin VL (1999) Thickness dependence of the optical properties of ordered silica-air and air-polymer photonic crystals. Phys Rev Lett 83:300
13. Colvin VL (2001) From opals to optics: colloidal photonic crystals. MRS Bull 26:637
14. Jiang P, Bertone JF, Hwang KS, Colvin VL (1999) Single-crystal colloidal multilayers of controlled thickness. Chem Mater 11(8):2132
15. Bardosova M, Hodge P, Pach L, Pemble ME, Smatko V, Tredgold RH, Whitehead D (2003) Synthetic opals made by the Langmuir-Blodgett method. Thin Solid Films 437:276
16. Bardosova M, Hodge P, Smatko V, Tredgold RH, Whitehead D (2004) A new method of forming synthetic opals. Acta Phys Slov 54(4):409
17. Reculusa S, Ravaine S (2003) Synthesis of colloidal crystals of controllable thickness through the Langmuir-Blodgett technique. Chem Mater 15:598
18. Reculusa S, Masse P, Ravaine S (2004) Three-dimensional colloidal crystals with a well-defined architecture. J Colloid Interface Sci 279:471
19. Romanov SG, Bardosova M, Pemble M, Sotomayor Torres CM (2006) (2+1)-dimensional photonic crystals from Langmuir-Blodgett colloidal multilayers. Appl Phys Lett 89(4):043105
20. Pemble ME, Bardosova M, Povey IM, Tredgold RH, Whitehead D (2007) Novel photonic crystal thin films using the Langmuir-Blodgett approach. Phys B 394:233
21. Bardosova M, Pemble ME, Povey IM, Tredgold RH, Whitehead D (2006) Enhanced Bragg reflections from size-matched heterostructure photonic crystal thin films prepared by the Langmuir-Blodgett method. Appl Phys Lett 89(9):093116
22. Stöber W, Fink A, Bohn E (1968) Controlled growth of monodisperse silica spheres in the micron size range. J Colloid Interface Sci 26:62
23. Bardosova M, Dillon FC, Pemble ME, Povey IM, Tredgold RH (2009) Langmuir-Blodgett assembly of colloidal photonic crystals using silica particles prepared without the use of surfactant molecules. J Colloid Interface Sci 333:816
24. Blanco A, Chomski E, Grabtchak S, Ibisate M, John S, Leonard SW, Lopez C, Meseguer F, Miguez H, Mondia JP, Ozin GA, Toader O, van Driel HM (2000) Large-scale synthesis of a silicon photonic crystal with a complete three-dimensional bandgap near 1.5 micrometres. Nature 405:437
25. van Blaaderen A, Ruel R, Wiltzius P (1997) Template-directed colloidal crystallization. Nature 385:321
26. Hoogenboom JP, Retif C, de Bres E, van de Boer M, van Langen-Suurling AK, Romijn J, van Blaaderen A (2004) Template-induced growth of close-packed and non-close-packed colloidal crystals during solvent evaporation. Nanoletters 4(2):205
27. Lee YC, Kuo TJ, Hsu CJ, Su YW, Chen CC (2002) Fabrication of 3D macroporous structures of II-VI and III-V semiconductors using electrochemical deposition. Langmuir 18:9942

28. Yates HM, Pemble ME, Blanco A, Miguez H, Lopez C, Meseguer F (2000) Growth of tin oxide in opal. Chem Vap Depos 6:283
29. Yates HM, Pemble ME, Miguez H, Blanco A, Lopez C, Meseguer F, Vazquez L (1998) Atmospheric pressure MOCVD growth of crystalline InP in opals. J Cryst Growth 193:9
30. Tan CH, Fan GH, Zhou TM, Li ST, Sun HQ (2005) Preparation of lnP-SiO2 3D photonic crystals. Phys B 363:1
31. Galisteo JF, Garcia-Santamaria F, Golmayo D, Juarez BH, Lopez C, Palacios E (2005) Self-assembly approach to optical metamaterials. J Opt A: Pure Appl Opt 7:S244
32. Miguez H, Chomski E, Garcia-Santamaria F, Ibisate M, John S, Lopez C, Meseguer F, Mondia JP, Ozin GA, Toader O, van Driel HM (2001) Photonic bandgap engineering in germanium inverse opals by chemical vapor deposition. Adv Mater 13(21):1634
33. Scharrer M, Wu X, Yamilov A, Cao H, Chang RPH (2005) Fabrication of inverted opal ZnO photonic crystals by atomic layer deposition. Appl Phys Lett 86:151113
34. Rugge A, Becker JS, Gordon RG, Tolbert SH (2003) Tungsten nitride inverse opals by atomic layer deposition. Nano Lett 3(9):1293
35. King JS, Neff CW, Summers CJ, Park W, Blomquist S, Forsythe E, Morton D (2003) High-filling-fraction inverted ZnS opals fabricated by atomic layer deposition. Appl Phys Lett 83:2566
36. King JS, Heineman D, Graugnard E, Summers CJ (2005) Atomic layer deposition in porous structures: 3D photonic crystals. Appl Surf Sci 244:511
37. Graugnard E, Chawla V, Lorang D, Summers CJ (2006) High filling fraction gallium phosphide inverse opals by atomic layer deposition. Appl Phys Lett 89:211102
38. Povey IM, Whitehead D, Thomas K, Pemble ME, Bardosova M, Renard J (2006) Photonic crystal thin films of GaAs prepared by atomic layer deposition. Appl Phys Lett 89:104103
39. Knez M, Niesch K, Niinistoe L (2007) Synthesis and surface engineering of complex nanostructures by atomic layer deposition. Adv Mater 19(21):3425
40. Choy KL (2003) Chemical vapour deposition of coatings. Prog Mater Sci 48(2):57, PII S0079-6425(01) 00009–3
41. Ritala M, Leskela M (1999) Atomic layer epitaxy – a valuable tool for nanotechnology? Nanotechnology 10:19
42. Povey IM, Bardosova M, Chalvet F, Pemble ME, Yates HM (2007) Atomic layer deposition for the fabrication of 3D photonic crystals structures: growth of Al_2O_3 and VO_2 photonic crystal systems. Surf Coat Technol 201(22–23):9345
43. Yariv A (1989) Quantum electronics, 3rd edn. Wiley, New York, p 305
44. Lee YC, Kuo TJ, Hsu CJ, Su YW, Chen CC (2002) Fabrication of 3D macroporous structures of II-VI and III-V semiconductors using electrochemical deposition. Langmuir 18:9942
45. Bardosova M, Pemble ME, Povey IM, Tredgold RH (2010) The Langmuir-Blodgett approach to making colloidal photonic crystals from silica spheres. Adv Mater 22(29):3104
46. Burkhard H, Dinges HW, Kuphal E (1982) Optical properties of In_{1-x} Ga_x P_{1-y} As_y, InP, GaAs, and GaP determined by ellipsometry. J Appl Phys 53:655
47. Zollner S (2001) Optical constants and critical-point parameters of GaAs from 0.73 to 6.60 eV. J Appl Phys 90:515
48. Johnson SG, Joannopoulos JD (2001) Block-iterative frequency-domain methods for Maxwell's equations in a planewave basis. Opt Express 8:173
49. Fay S, Steinhauser J, Oliveira N, Vallat-Sauvain E, Ballif C (2007) Opto-electronic properties of rough LP-CVD ZnO: B for use as TCO in thin-film silicon solar cells. Thin Solid Films 515(24):8558
50. Thomson M, Hodgkinson JL, Sheel DW (2013) Control of zinc oxide surface structure using combined atmospheric pressure-based CVD growth and plasma etching. Surf Coat Technol 230:190
51. Hodgkinson JL, Thomson M, Cook I, Sheel DW (2011) Non-thermal atmospheric pressure plasma etching of F:SnO2 for thin film photovoltaics. J Nanosci Nanotechnol 11(9):8403

52. Hongsingthong A, Aino A, Sichanugrist P, Konogai M, Kuramochi H, Akiike R, Iigusa H, Utsumi K, Shibutami T (2012) Development of novel Al-doped zinc oxide films fabricated on etched glass and their application to solar cells. Jap J Appl Phys 51(10/2):SI 10NBO9
53. Povey IM (2014) Unpublished personal communication
54. Gratzel M (2005) Solar energy conversion by dye-sensitized photovoltaic cells. Inorg Chem 44:6841
55. Zhang QF, Chou TR, Russo B, Jenekhe SA, Cao GZ (2008) Aggregation of ZnO nanocrystallites for high conversion efficiency in dye-sensitized solar cells. Angewandte Chemie – Int Ed 47(13):2402
56. Fujita M, Takahashi S, Tanaka Y, Asano T, Noda S (2005) Simultaneous inhibition and redistribution of spontaneous light emission in photonic crystals. Science 308(5726):1296
57. Bermel P, Luo C, Zeng L, Kimerling LC, Joannopoulos JD (2007) Improving thin-film crystalline silicon solar cell efficiencies with photonic crystals. Opt Express 15:16986
58. Ding B, Bardosova M, Pemble ME, Peschel U, Romanov SG (2010) Omnidirectional broadband scattering in metal-dielectric colloidal photonic hetero-crystals. Proceedings of Optical Society of America (OSA), paper PTuC2
59. Ding B (2010) Development of heterogeneous and hybrid opal-based 3-dimensional photonic crystals for extended light moulding and trapping. PhD thesis, UCC Cork
60. Kocher-Oberlehner G, Bardosova M, Pemble ME, Richards BS (2012) Planar photonic solar concentrators for building-integrated photovoltaics. Sol Energy Mater Sol Cells 104(53)
61. McGrath J (2014) Development of large scale colloidal crystallisation methods for the production of photonic crystals. PhD thesis, UCC Cork

Chapter 7
Crystalline and Amorphous Chalcogenides, High-Tech Materials with Structural Disorder and Many Important Applications

M. Frumar, T. Wagner, K. Shimakawa, and B. Frumarova

Abstract The paper reviews and discusses crystalline, nanocrystalline, glassy and amorphous chalcogenides and their thin films and fibres, their preparation, structure, properties, changes and applications in optics, optoelectronics and electronics, data storage and sensors with accent on recent data and progress. The area of interest is so broad that only some materials and processes could be discussed in detail, and a lot of data were chosen for the sake of illustration only.

Various ways of the preparation of crystalline, glassy and amorphous chalcogenides, their crystals, thin films and nanoparticles are mentioned; the structure, properties and applications of individual groups of materials are discussed. The applications are numerous: in the infrared technique, data storage, in light transformation and emission, in sensors, X-ray sensors, Xerox facilities, luminophors, lasers, ferroelectrics, thermoelectrics, catalysers, plasmonics materials; in topological insulators, fibres, in light up-conversion, nanodots and nanomaterials; in one-, two-, and three-dimensional systems, planar optical circuits, waveguides, non-linear optical and optomechanical devices, in surgical instruments, etc. Some of them are discussed in detail.

The review paper is based on several lectures presented on the occasion of "Nanomaterials and Nanoarchitectures International Summer School, Advanced

M. Frumar (✉) • T. Wagner
Department of General and Inorganic Chemistry, University of Pardubice, 532 10 Pardubice, Czech Republic
e-mail: miloslav.frumar@upce.cz; tomas.wagner@upce.cz

K. Shimakawa
Department of General and Inorganic Chemistry, University of Pardubice, 532 10 Pardubice, Czech Republic

Center of Innovative Photovoltaic Systems, Gifu University, Gifu 501-1193, Japan
e-mail: koichi.shimakava@upce.cz; koichi@gifu-u.ac.jp

B. Frumarova
Institute of Macromolecular Chemistry of Academy of Sciences of the Czech Republic, v.v.i., 162 06 Prague, Czech Republic
e-mail: bozena.frumarova@upce.cz

M. Bardosova, T. Wagner (eds.), *Nanomaterials and Nanoarchitectures*, NATO Science for Peace and Security Series C: Environmental Security, DOI 10.1007/978-94-017-9921-8_7

Study Institute of NATO", held in June/July 2013, in Cork, Ireland. The paper is primarily devoted to MSc. and Ph.D. students and postdoctoral students of solid state physics, solid state chemistry, material science and material engineering, but also to researchers as well as to general audience interested in science and technical progress, in order to help them understand and apply some of these materials and data.

Abbreviations

ACh	Amorphous chalcogenide(s)
ACCh	Amorphous and crystalline chalcogenide(s)
AChF	Amorphous chalcogenide films
AGCh	Amorphous and glassy chalcogenide(s)
AIST	Argentum-Indium-Antimony-Telluride(s)
CCh	Crystalline chalcogenide(s)
CD	Compact disc
Ch	Chalcogenide(s)
CTAB	Cetyl trimethyl ammonium bromide
CVD	Chemical vapour deposition
1D	One-dimensional
2D	Two-dimensional
3D	Three-dimensional
DVD	Digital video disc
ECM	Electrochemical metallization memories
ECB RAM	Electrochemical conductive bridge random access memory
E_g	Energy gap
ESA	Excited state absorption
ETU	Energy transfer up-conversion
EXAFS	Extended X-ray Absorption Fine Structure
GCh	Glassy chalcogenide(s)
GSA	Ground state absorption
HRTEM	High-resolution transmission electron microscopy
IR	Infrared region
LED	Light emitting diode(s)
M	Metal or metalloid
MID IR	Mid infrared region
MOCVD	Metal organic chemical vapour deposition
MOF	Microstructured optical fibres
n	Index of refraction
NIR	Near-infrared region
OUM	Ovshinsky Universal Memories
PCM	Phase change memory
PCMM	Phase change memory materials
PC RAM	Phase change random access memory

PL	Photoluminescence
RE	Rare earth element(s)
SERS	Surface enhanced Raman spectroscopy
SHG	Second harmonic generation
SSM	Solid state (non-volatile) memory
T_c	Temperature of crystallization
T_g	Temperature of glass transition
T_m	Melting temperature
X	Halogen (Cl Br, I)
XANES	X-ray Absorption Near Edge Structure
Y	Chalcogen

7.1 Introduction

Chalcogenide(s) (Ch) form a large group of compounds, crystals, glasses, thin films and nanomaterials and colloids. Many of them can be found in nature as minerals and ores, such as PbS, PbSe, PbTe, HgS, ZnS, CdS, CdSe, Ag_2S, Ag_2Te, Cu_2S, Cu_2Se, As_2S_3, As_4S_4, Sb_2S_3, Bi_2S_3, Bi_2Te_2S, $Pb_2Sb_2S_4$, $Pb_5Sb_4S_{11}$, Cu_3AsS_4, GeTe, MoS_2, etc. [1]. The larger part of chalcogenides was synthesized and prepared artificially, often in very high purity, e.g. for semiconductors, optical elements and optical fibres. Many of crystalline or amorphous chalcogenides (CCh and ACh, respectively) have been used as important materials in electronics and in optics (e.g. crystals of CaS, PbS, PbSe, PbTe, HgTe, CdTe, Se, As_2Se_3, As_2Se_3, Ge-Sb-Se, Ge-Sb-S and Ge-La-S systems, Sb_2S_3, Sb_2Te_3, Sb_2Se_3, Bi_2S_3, Bi_2Se_3, Bi_2Te_3, Ge-Sb-Se, Ge-Bi-S, Ge-Bi-Se, $Ge_2Sb_2Te_5$, $Ge_8Sb_2Te_{11}$, $Ge_8Bi_2Te_{11}$ and many others). They have also been applied as active materials for reconfigurable electronics, phase change random access memories (PC RAM), multilevel memories, cognitive devices, as Ovshinsky universal memories (OUM), in ionic memories (electrochemical conductive bridge random access memories, ECB RAM), etc. They have been applied in other various areas, e.g. CaS was the first man-made luminophor, selenium was used as the first photoconductor, photovoltaic element and semiconducting rectifier, ZnS and CdS have been used in earlier television (TV) tubes as luminophors for the visible part of spectrum; sulphides, selenides and tellurides (binary and ternary) have been used for their good transmittance in the infrared (IR) region as optical elements, and for their luminescence as lasers and light amplifiers. Many Ch have been applied as light and composition sensors, e.g. PbS, PbSe, HgS, HgSe, HgTe, PbTe, CdTe, Sb_2S_3, for spectroscopy (e.g. of alcohol in wine by evanescent spectroscopy), night vision, remote temperature detection, security and defense applications.

Some Ch or chalcohalides (also contain halogen atoms) are important ferroelectric materials with ultra-high permittivity and with low temperature of transition from the ferroelectric to the paraelectric state (e.g. SbSI, BiSeI, SbSeI) [2]; some have low glass-transition temperature (T_g) and have been used as immersion liquids in optical elements for the IR region.

A significant group of selenides and tellurides (e.g. Sb_2Te_3, Bi_2Te_3, Sb_2Se_3, Bi_2Se_3 and their solid solutions) have been applied as important thermoelectric materials for room and not very high temperatures, some have been exploited as Peltier cells for cooling, e.g. in portable automotive fridges, for cooling of sensors, or for cooling of bio-medical samples in microscopy, etc. [3, 4].

Some chalcogenides are important catalysts in chemical industry, e.g. MoS_2, $MoSe_2$, WSe_2 and many others (see e.g. [5]). Some CCh, such as MoS_2, $MoSe_2$, or WSe_2, have a layered structure similar to graphite and have recently been intensively studied as 2D graphen-like materials, as topological insulators and as potential materials for light emitting diodes (LED). Other CCh, such as Sb_2Te_3, Bi_2Te_3, and Bi_2Se_3, also form a layered type of crystals and have been recently studied as topological insulators and optoelectronic materials [6, 7]. They can also be used for the preparation of field-effect transistors, e.g. from MoS_2 [8].

Amorphous and glassy chalcogenides (AGCh) also have numerous present and potential applications as materials for optics and optoelectronics, mostly for the IR region [9–11]: Here we only list a few of the applications: optical elements, e.g. lenses, micro lenses, filters, coatings, antireflection coatings, waveguides, planar optical devices and circuits, IR lasers, acousto-optical elements, optical signal processing (light up-conversion, signal couplers, frequency mixing); materials for holography, diffractive elements and gratings, active materials for non-volatile phase change memories (DVD, Blu-Ray discs, etc.). They are used in non-linear optics, as the second and third harmonic generators, IR and X-ray sensors. They are used in Xerox facilities and laser printers, photoresists of very high resolution, fibres and micro structured optical holey fibres, as plasmonic materials.

The most important element out of the group of chalcogens is Se that is chemically simple; however, its structure is very complex. Selenium photocells (photoconductive or photovoltaic) have been used for many years as sensors; lately it has been Se which has been used as a Xerox-sensitive material, in X-ray sensitive panels and as a material for very sensitive TV pickup tubes, which can work very well even in the moonlight [11]. Formerly, Se was also applied as a rectifier, catalyst and as a colouring element for oxide glasses [12].

Rare-earth doped chalcogenides are applied in luminescence in the near-infrared region (NIR) and the mid-infrared region (MID IR) of spectra as light amplifiers, generators, eye-safe lasers, IR lasers for tissue coagulation, non-bleeding medical operations; for tissue excision, removal of the arterial plaque, cutting bone and drilling teeth, and environmental and other sensors.

Glassy chalcogenides, GCh, are applied as optical materials, in the preparation of fibres in power delivery, mechanical sensing, temperature, chemical, and remote monitoring (in medicine, in vivo spectroscopy, etc.); as thin solid electrolytes in solid batteries, ionic selective sensors, sensitive materials for Xerox facilities and laser printers, and potentially, in materials for *all optical computing*.

In comparison with oxides, amorphous and crystalline chalcogenides (ACCh) contain heavier atoms, posses high optical transmittivity in the IR region, a higher index of optical refraction, higher optical nonlinearity and lower chemical bonds' energies. A possibility to form stable homopolar chemical bonds is higher in

AGCh than in oxidic materials. AGCh have greater free volume [13] and larger composition areas for the formation of homogeneous solids as compared with corresponding crystals. Some AGCh have low barriers for structural changes and their crystallization can be very fast, at least at higher temperatures.

Ch can be prepared in polycrystalline or in single-crystal forms, as glasses or thin amorphous or crystalline films and also as nanoparticles. Many of them have been prepared and studied as 1D, 2D and 3D nanoparticles, as colloids or quantum dots, where D stands for dimension (see e.g. [6, 14]).

Because of their interesting and sometimes peculiar composition, structure and properties, and because of the importance of Ch in numerous areas of science and technology, Ch have recently been described and discussed not only in many original papers, but also in many books, monographs and many review papers (see e.g. [3–5, 13, 15–37]).

The area of science and technology of Ch is very extensive. This review paper focuses on and presents recent results of research, new or potential applications of chalcogenides' properties. We also discuss data which are important for basic understanding of chalcogenide structure, properties and applications especially in optics and in electronics. The review papers also mention elemental selenium; some of its properties are closer to chalcogenides (Ch) than to the elements of the 16th group of the periodic table. Its applications are similar to those of many selenides, some applications are exclusive [9–12, 19, 25].

The other applications of Ch are also discussed in Sect. 7.6.

7.2 Chemical Composition, Stoichiometry, Synthesis, Glass Formation, Thin Films, Nanoparticles and Fibres

Chalcogen(s) (Y) react with almost all elements of the periodic table, excluding noble gases. They form many molecular, polymeric or solid compounds, alloys, glasses, amorphous solids and nanomaterials. Our attention is paid on solid Ch (CCh and AGCh) and elements, compounds and materials that have present or potential application in electronics, optics and in optoelectronics, as already mentioned.

7.2.1 Crystalline Chalcogenides

The number of CCh is enormous. Many binary, ternary or multinary compounds are formed (see e.g. above and also Figs. 7.1, 7.2, 7.3, 7.4, 7.5, 7.6, and 7.7 [12, 36–43]; Fig. 7.8 [44] is included for the sake of comparison). The chemical composition of many CCh is not confined only to one formula like in daltonides; the area of possible compositions can be broad (Figs. 7.3, 7.4, and 7.5); even in many individual and simple systems, there can be found several stoichiometric and/or non-stoichiometric (Fig. 7.7), or homologous compounds (Figs. 7.1 and 7.7).

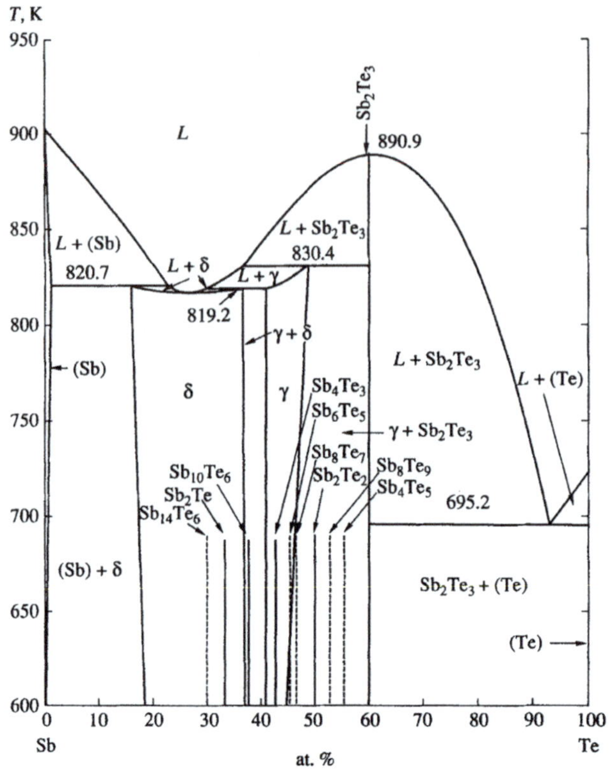

Fig. 7.1 T-x diagram of Sb-Te system [37]. In the equilibrium, many so-called homologous compounds are formed of the general formulae $(Sb_2)_n(Sb_2Te_3)_m$. Such compounds are formed in annealed samples, not under the conditions with a fast cooling rate. In such a case, metastable compounds are formed. The individual metastable compounds are denoted by the *vertical lines*

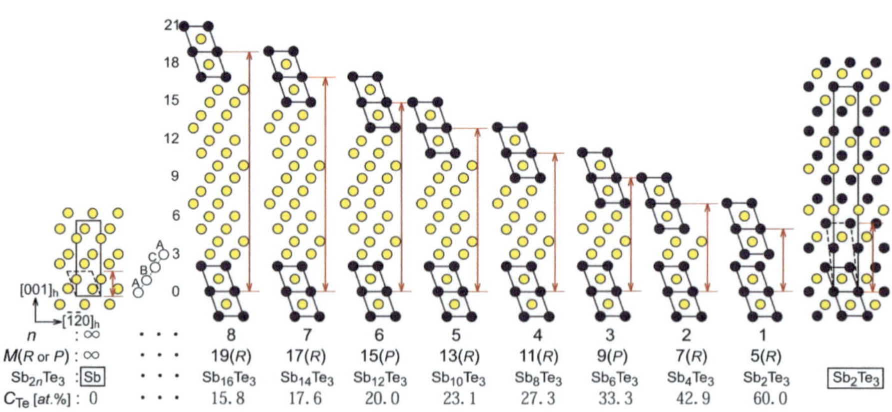

Fig. 7.2 The [100] projection of a crystal structure model of a Sb-Te binary system [38]. The *dark blue circles* represent Te atoms. In the structure of Sb and Sb_2Te_3, each atom shows a projection of a real atomic position and the *rectangles* represent each unit cell. The stacking period M, chemical formula and Te concentration are shown under each model

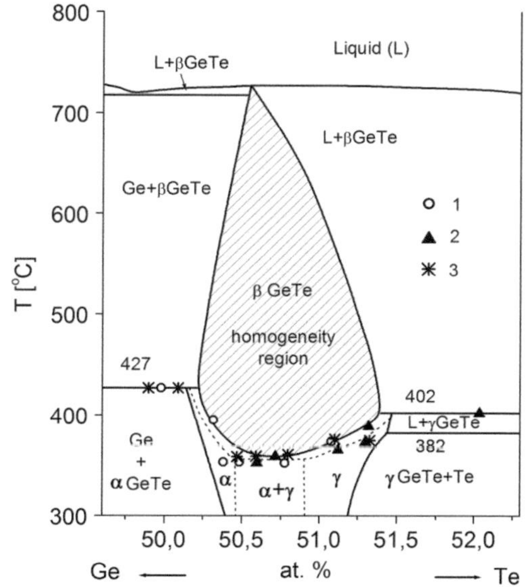

Fig. 7.3 The homogeneity region of germanium telluride (phase β) [39]. Data: *1* from dilatometry, *2* from electrophysical measurements, *3* differential thermal analysis (DTA)

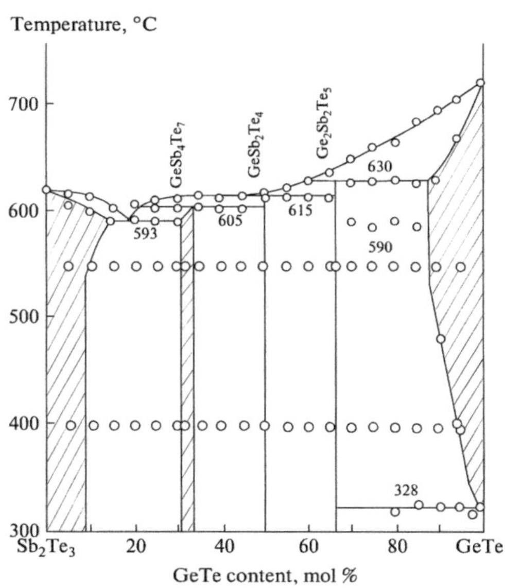

Fig. 7.4 An equilibrium T-x diagram of a Sb_2Te_3-GeTe system [40]. The cross-hatched regions are formed by solid solutions. According to [37], these regions are formed by many homologous compounds analogically to Figs. 7.1, 7.2

The phase diagrams of many Ch systems show not only narrow sharp regions of formation of stoichiometric compounds, but also broad regions (see e.g. the crosshatched regions of nonstoichiometric compounds (solid solutions) in Figs. 7.4 and 7.5) [36, 39–41]. Recently it has been revealed that in these nonstoichiometric regions, there can be found several phases (an example is given in Figs. 7.1 and 7.7) [37, 43]. Such phases (compounds) are formed e.g. in the Sb-rich region of the Sb-Te system and in the GeTe – Sb_2Te_3 systems (Figs. 7.1 and 7.7); their

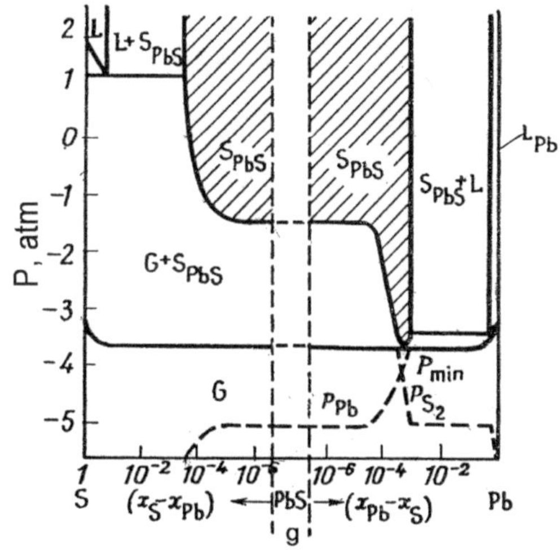

Fig. 7.5 The p-x phase diagram of a Pb-S system [41]. The cross-hatched regions are formed by a solid PbS with Pb excess (*right-hand side*) and with excess of sulphur (*left-hand side*). The minimum pressure is in the Pb excess region ($\Delta x \approx 10^{-3}$). The PbS is a typical example of Ch with small non-stoichiometry. It can be prepared with Pb excess (the n-type conductivity) or with S excess (the p-type conductivity)

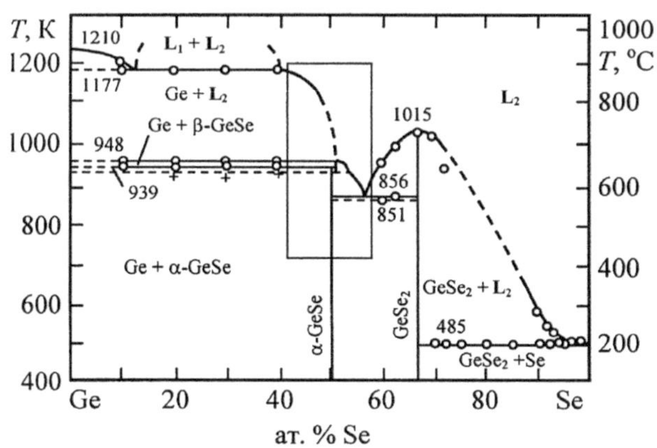

Fig. 7.6 T-x phase diagram of a Ge-Se system [42]. L is for melts (liquids)

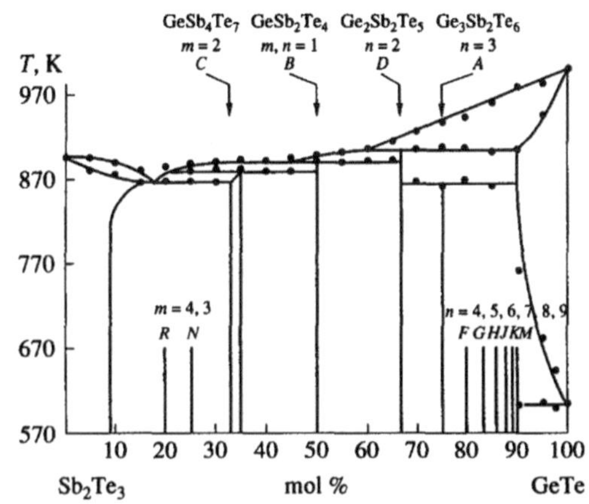

Fig. 7.7 T-x phase diagram of a GeTe-Sb_2Te_3 system [43]. *A–D* are compounds found earlier in equilibrium diagrams. Homologous compounds found recently are denoted by *H* to *R*. Compare with Fig. 7.5

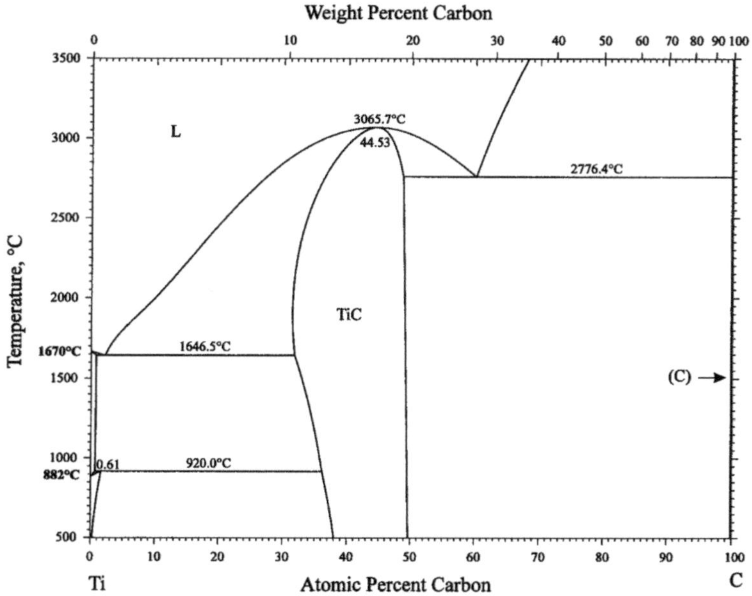

Fig. 7.8 T-x phase diagram of a Ti-C system [44]. The Ti-C system represents so-called "intermediate compounds", which are something in between alloys and compounds. They have a large homogeneity region and the maximum melting points are often outside of the stoichiometric composition. In this case it is near 4.5 at.% of C. Carbon occupies interstitial positions in this crystal. Similar compounds are formed in phase change memory materials (PCMM)

composition can be expressed by a general formula $(Sb_2)_n(Sb_2Te_3)_m$, for the Sb-Te system, and $(GeTe)_n(Sb_2Te_3)_m$ for the GeTe-Sb_2Te_3 system, where m and n are integers [37, 38, 43]. In analogy with organic chemistry, such phases are called *homologous compounds*. Homologous compounds of such systems can be prepared by the annealing of samples of a proper composition at higher temperatures [37, 43]. As an example, in a Sb-rich part of the Sb-Te system, there can be found many crystalline homologous compounds, e.g. Sb_2Te_3, Sb_4Te_3, Sb_6Te_3, Sb_8Te_3, $Sb_{10}Te_3$, etc., up to $Sb_{72}Te_3$, and possibly further (Fig. 7.2). The whole region of composition from Sb_2Te_3 to Sb, including β and γ phases [36], forms nearly continuous a series of crystalline compounds (Figs. 7.1 and 7.2). The structure of the Sb-rich system of Sb-Te reminds of an intercalated compound with one or more bilayers of Sb_2 between Sb_2Te_3 layers (Fig. 7.2 [37]). The compounds from this system and the system Ge-Sb-Te (Fig. 7.7) are important as phase change memory materials (PCMM) and are discussed in Sects. 7.5 and 7.6.

The properties of such homologous phases change nearly smoothly with composition change without any abrupt changes near the individual compounds formation. Metastable states of such compounds (alloys) can be formed in a broad range of compositions – oversaturated solid solutions or structural vacancies can be formed during quick phase changes, as discussed in Sects. 7.5 and 7.6.

Many CCh and even their single crystals can be non-stoichiometric, containing a different ratio of chalcogens and other atoms in the compound [36, 41, 43, 45] (Figs. 7.3, 7.4, and 7.5). It is valid for both binary and multinary compounds. Some of them are close to intermediate compounds or alloys by composition and structure. Intermediate compounds are compounds with a significant possible deflection of composition from a stoichiometric one, e.g. of the type given in [41, 45–47]. These compounds are intermediate because their structure and composition is in between solid solutions and classic compounds. Intermediate compounds are similar to some metallic alloys and they form phases easily outside of the stoichiometric composition (Fig. 7.8). The maximum melting point of such systems is very often outside of the stoichiometric composition, too. They are formed by many metals or heavy elements, sometimes in combination with small elements like H, C, Si, etc. that are placed in interstitial positions. Their crystalline structures are often determined by the size of atoms and density of electrons, etc. Significant changes of stoichiometry and of structure of these types of materials influence their electrical, optical and other properties and allow for their numerous important applications (hard materials, superconductors, emissive materials, etc.). In case of PCMM, interstitial defects are not formed as the size of all atoms in PCMM is large. Only substitution defects and crystal vacancies are evidently formed.

Formation *of metastable phases* can be seen not only in amorphous and glassy systems but also in many crystals. The GeTe-Sb_2Te_3 system crystals crystallize in the fcc metastable form and, at higher temperatures, at the equilibrium hexagonal lattice. The reasons for the formation of metastable structures can be kinetic (the formation of a stable structure needs a long transport of atoms), energetic, or their combination. Individual metastable and stable states (compounds) can be departed by relatively high energy barriers between the individual states. At a given temperature, some barriers between them cannot be easily overcome, they are then relatively stable. While the metastable states of glasses are well known and have been widely studied, the metastable states of crystalline solids are almost unknown and have recently been named as "*Terra incognita awaiting discovery*" [48]. The energetic scheme of phase transitions via metastable states is also provided in Fig. 7.24, Sect. 7.4.2.

7.2.2 Amorphous and Glassy Chalcogenides

A large group of chalcogenide materials can form glasses and amorphous systems. Materials usually considered being glasses are those received by fast cooling of their melts (liquids). The amorphous state can also be formed by fast cooling of vapours (e.g. in thin films), or by fast precipitation or formation of chemical elements or compounds. Amorphous solids can also be prepared by intensive mechanical milling of crystalline powders. The number of defects in milled crystalline particles grows steeply until it results in a full loss of the long-range order ($\approx$ which corresponds to amorphization).

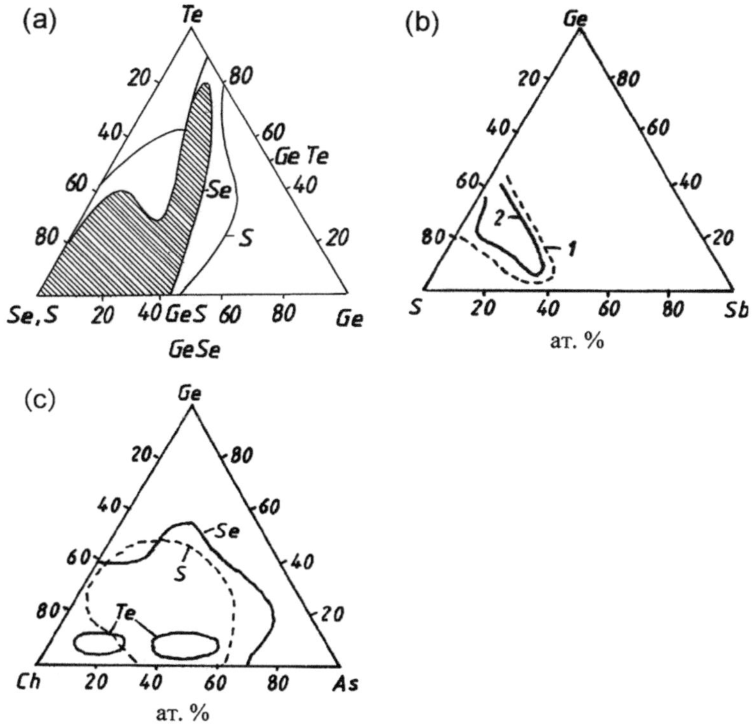

Fig. 7.9 Glass-forming regions for Ge-Se(S)-Te (**a**), Ge-Sb-S (**b**) and Ge-As-S(Se,Te) (**c**) [29]

The glass-forming ability of many chalcogenides is significant, glasses and amorphous state are formed very easily (Fig. 7.9 [29]). Se and S, and partly also Te, are important components of AGCh; they can have their coordination number typically close to two, forming two covalent bonds. Such pairs of bonds are easily deformable and can follow the disorder of the melt and form the amorphous or glassy state without requiring a greater amount of energy. For example, the stability of As_2S_3 glass is so high that even its annealing for several months near the glass transition temperature, T_g, is not sufficient for crystallization. Chalcogen atoms can also form homopolar chemical bonds between two chalcogens (e.g. -S-S-, or -Se-Se- bonds) which can be twisted or deformed very easily, too. Such bonds (as well as As-As, Sb-Sb or Ge-Ge bonds) can compensate for a local disorder in AGCh without a significant change of energy. The bonds between like atoms can also compensate for the non-stoichiometry of the glasses. The important fact is that the energies of many different bonds in typical chalcogenide glasses do not differ from each other very much (Table 7.1) [29]; the changes of bond statistics in the system do not change the energy of the system very much. In AGCh, homopolar bonds between the like atoms are considered being the "wrong bonds" (e.g. S-S, Se-Se and As-As in As-Y systems, Y stands for chalcogen).

Table 7.1 Bond energies of different chalcogenide bonds ([29], p. 108)

Bond	Energy, kJ/mol	Bond	Energy, kJ/mol	Bond	Energy, kJ/mol
S-S	280	As-As	200	Si-Si	225
Se-Se	225	Sb-Sb	378.7	Ge-Ge	185
S-Se	255	As-S	260	Ge-S	265
Se-Te	195	As-Se	230	Ge-Se	230
Sb-Te	195	As-Te	205	Ge-Te	200
Sb-Sb	175	Sb-Se		Si-Te	220
Sb-S	230			Si-S	225

Glass-forming areas of Ch are dependent on the cooling rates of the melts and they are generally broadened with increasing cooling rates. Generally, for the formation of a glass free of any crystal, cooling rates need to be faster than the formation of crystal nuclei and the following growth of crystals. When the cooling rates of the melts are sufficiently high (in thin films of phase change memory materials (PCMM); they are evaluated to be as large as $\sim 10^9$ to 10^{11} K/s), almost any melt can produce a glass.

Non-crystalline systems are also formed by fast precipitation of some compounds or solids from liquid solutions (melts). In such a case, the obtained particles can be of nano-size dimensions, or even colloidal, and can be embedded in melt, in solvent, in crystal, or in glass.

Many ACh and CCh can be prepared by precipitation from aqueous or non-aqueous solutions, from soluble alkali- or ammonium- chalcogenides (e.g. from $(NH_4)_2S$) and from the corresponding metal or metalloid salts (such as $GeCl_4$, $AsCl_3$, $PbCl_2$). When the formed Ch are poorly soluble in such solutions, the precipitation can be fast and amorphous (colloidal) solids or amorphous nanoparticles of different size and form can be prepared [14]. ACCh can also be prepared by the chemical vapour deposition (CVD) method and by the ion implantation of elements into a glass.

7.2.3 Synthesis of Chalcogenides

Synthetic methods are numerous and they vary depending on the composition of the compound or material to be prepared. The simplest method, which is often used for preparation of semiconducting crystalline and amorphous chalcogenides of high purity, is the direct synthesis from elements of 5 N or higher purity (5 N corresponds to 5 nines, 99.999 at. %) in evacuated silica ampoules. The purity of precursors for optical fibre preparation needs to be much higher ([49] in [50]). In order to avoid possible crystallization of prepared glasses, the melts shall be cooled relatively fast.

Methods which are often used for the single crystal preparation are methods of growing from their own melt (its composition is the same as the composition of the crystal), or from foreign melts of a composition different than the crystal one (Czochralski, Bridgman, flux methods, epitaxial or atomic layer growth). Single crystals can also be prepared from the gas phase (sublimation, chemical transport reactions, chemical vapour deposition, epitaxy, etc. (see e.g. [51])). In transport reactions, a volatile element or compound often containing a halogen, is usually applied as a transporter. It reacts easily with the transported compound. The transport and the growth of single crystals is based on a thermal shift of the equilibrium of chemical reactions, e.g. of the type given in (Eq. 7.1), where M is metal or metalloid, Y is chalcogen and I is iodine or another halogen. The T_1 and T_2 are the temperatures of warmer and cooler parts of the reaction vessel, respectively. Depending on the value and sign of enthalpy ΔH of the transport reaction (negative or positive), the shift of the reaction equilibrium and the transport of the compound goes from the colder to the warmer part of the growing vessel or vice versa [52, 53]. According to (Eq. 7.1), the compound MY is transported by molecules of iodine (I_2) forming two molecules of iodide MI_n and a molecule of chalcogen Y_2. The M stands for a metal or metalloid. I_2, Y_2 and MI_n are volatile.

$$2MY + nI_2 \underset{T_2}{\overset{T_1}{\rightleftarrows}} 2MI_n + Y_2 \tag{7.1}$$

The equilibrium constant of the reaction (7.1) is given by (Eq. 7.2). K is temperature-dependent.

$$K = \frac{[MI_n]^2\,[Y_2]}{[MY]^2[I_2]^n} \ . \tag{7.2}$$

The symbols in square brackets correspond to the activities ($\sim$ concentrations) of the compounds (elements) in Eq. 7.1. Also some single crystals of today's PCMM were prepared using chemical transport reactions [53].

A relatively common method for the preparation of crystalline or amorphous chalcogenides is the chemical vapour deposition (CVD) method with its several modifications, such as metal organic vapour phase epitaxy, MOVPE, etc. The compound or material is formed by a chemical reaction from volatile vapours or gases at higher temperatures. These volatile compounds are usually dissociated at higher temperatures and form (also at higher temperatures) the compound in demand, e.g.

$$2AsH_3 + 3H_2S \leftrightarrow As_2S_3 + 6H_2 \tag{7.3}$$

Often-used sources of metallic or metalloid elements are organometallic or organometalloid hydrides, such as derivates of AsH_3, or SbH_3, or other organometallic or organometalloid compounds. Possible chalcogen sources can be the following: chalcogen vapours, or chalcogen hydrides H_2S, H_2Se, or some volatile chalcogen containing compounds, including organo-element ones that can be dissociated at higher temperatures creating sulphur, selenium or tellurium vapours; or a compound

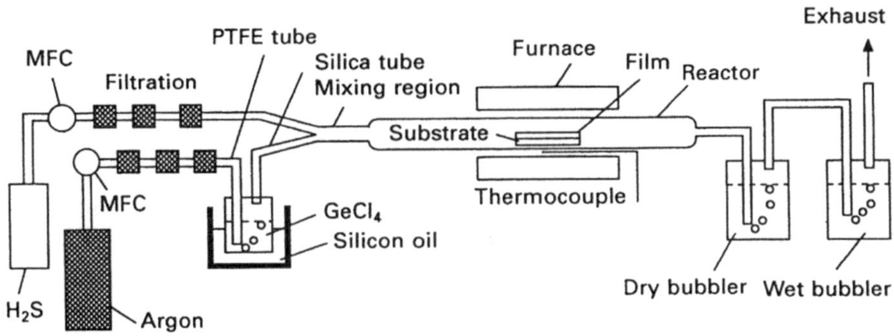

Fig. 7.10 Scheme of a CVD facility used for the Ge-S deposition, *MFC* represents a mass flow controller [54], p. 290

that can react with a metal source forming a chalcogenide [54] (Fig. 7.10). Many of organometallic or organoelement compounds can be prepared in very high purity with very low content of different impurities. The CVD method is also often used for thin film preparation (see also Sect. 2.4).

The preparation of Ge-Sb-Te thin films from $Ge(C_3H_5)_4$, $Sb(C_3H_7)_3$ and $Te(C_3H_7)_2$ precursors can be mentioned as an example. The $Ge_2Sb_2Te_5$ film was prepared at 350 °C under 25 % of hydrogen flow [55]. The metal organic chemical vapour deposition (MOCVD) methods have also been successfully applied for the preparation of many chalcogenide thin films (e.g. PbS, CdS, SnS and also $Ge_2Sb_2Te_5$).

7.2.4 Thin Films

The research and many applications of Ch need not only bulk crystals, polycrystals or bulk glasses; they also need thin or thick crystalline or amorphous films. For thin film preparation, many methods can be used. The physical methods, such as the vacuum- and flash- evaporation, use the evaporation of a substance by heating to higher temperatures in a high vacuum. The difference is that in the classical vacuum evaporations, the whole batch of material from one or several boats is evaporated, while the flash evaporation uses a finely powdered material; its small grains are slowly dropped on a preheated boat and every grain is immediately evaporated and its vapour condensed on a substrate. In such a case, the homogeneity of the sample obtained is much higher. In the classical vacuum evaporation of a material with components of different volatility, fractionation can take place and nonhomogeneous films with a gradient of composition can be obtained.

Important and commonly used methods for laboratory and also for industrial thin film preparation are radio frequency (rf), or direct current (dc) sputtering and

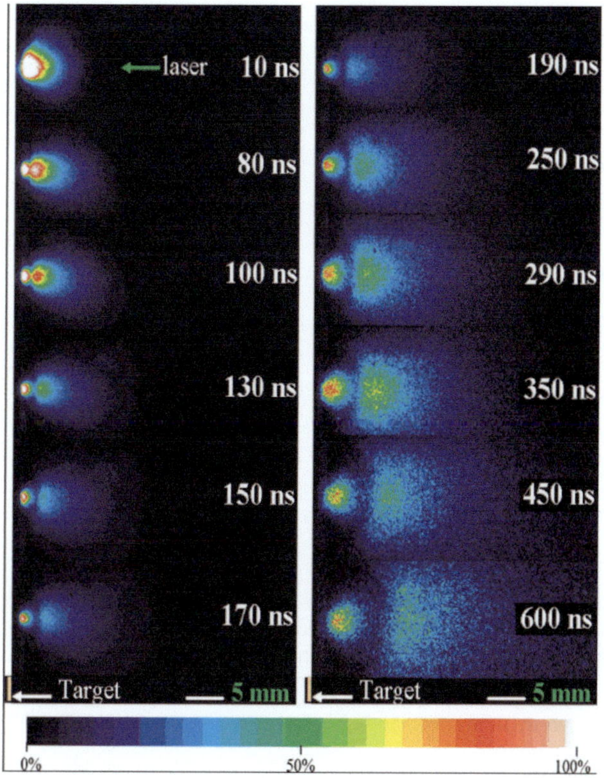

Fig. 7.11 Temporal evolution of the total optical emission from a plasma plume obtained by laser ablation of an $As_{40}Se_{60}$ sample. Successive laser pulses of equal energy (50 mJ) were used to record the shots at different delays. For visibility reasons, the colour scale represents relative intensities, i.e. with respect to the maximum value (note that the maximum value is saturated up to 130 ns). For comparison, in the absolute intensity, the maximum of the 150 ns picture is about 15 times more intense than that of the 600 ns shot [58]

deposition of a material. The bombardment of the surface of a target by electrons, Ar ions or other inert gas ions at lower pressure, is used here. A very important method is also pulsed laser deposition (PLD, [56, 57]), in which a short intensive laser pulse evaporates the material. As an example, the evaluation of time dependence of temperature in a plasma plume during the As-Se system samples evaporation is given in (Fig. 7.11 [58]).

The PLD films of As_2Se_3 contain fewer 'wrong' As-As and Se-Se bonds and less As_4Se_4, As_4Se_3, Se_8 or Se_n particles (Figs. 7.12 and 7.13 [59]) than the flash- and thermally- evaporated ones. Such films have higher density, a higher index of refraction, probably due to higher kinetic energy of the plasma-plume particles. It allows for their interaction in the hot plume or during the deposition, the content of As_4Se_4, As_4Se_3, Se_n decreases, e.g. due to reactions (7.4, 7.5, and 7.6)

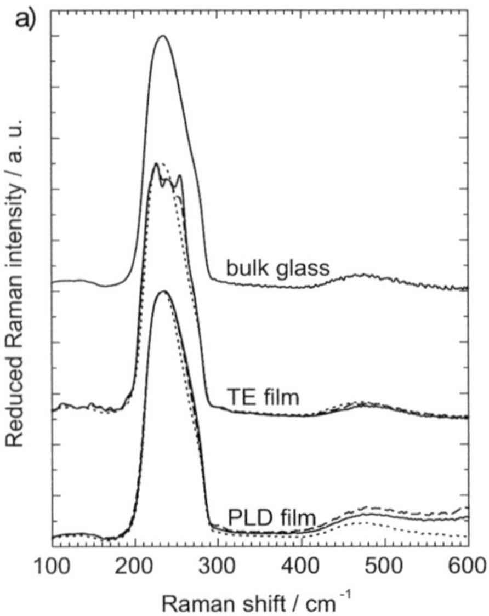

Fig. 7.12 Raman spectra of a bulk As_2Se_3 glass [59]. (**a**) region 100–600 cm^{-1};(**b**) region 100–190 cm^{-1}; as deposited, exposed, and annealed TE, and PLD As-Se thin films: the *full line* – as-deposited films, the *dashed line* – exposed films, and the *dotted line* – annealed films

Fig. 7.13 Raman spectra of a bulk $As_{60}Se_{40}$ glass (*the full line*); PLD $As_{60}Se_{40}$ thin films: as-deposited (*the dashed line*), and exposed (*the dotted line*) [56]

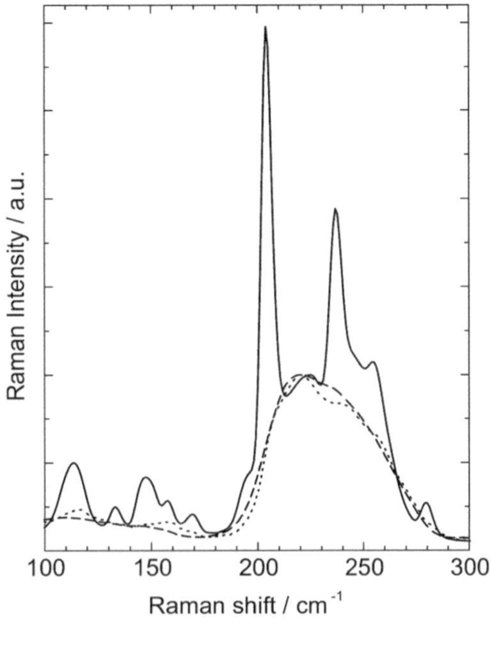

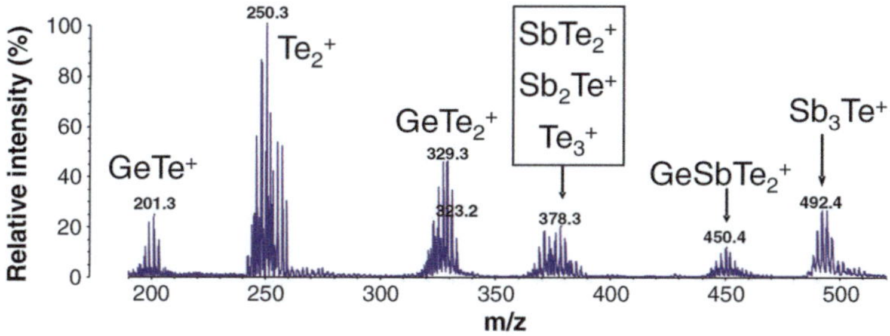

Fig. 7.14 A mass spectrum of ionized vapours of $Ge_2Sb_2Te_5$ evaporated by high intensity laser pulses [60]. The stoichiometry of negatively or positively charged fragments/clusters typical of laser-generated plasma decomposition was determined in vapours of a laser-evaporated $Ge_2Sb_2Te_5$ compound. In the mass spectra of the $Ge_2Sb_2Te_5$ thin films identified the $GeTe^+$, $Te_2{}^+$, $GeTe_2{}^+$, $GeSbTe_2{}^+$, $SbTe_2{}^+$, Sb_2Te^+ and Sb_3Te^+ clusters

Fig. 7.15 A mass spectrum of ionized vapours of GeTe evaporated by high intensity laser pulses [60]. In the vapours of the laser-evaporated GeTe, there were found the ions of $GeTe^+$, Ge_2Te^+, $GeTe^{2+}$ and Ge_2Te^{2+} ions

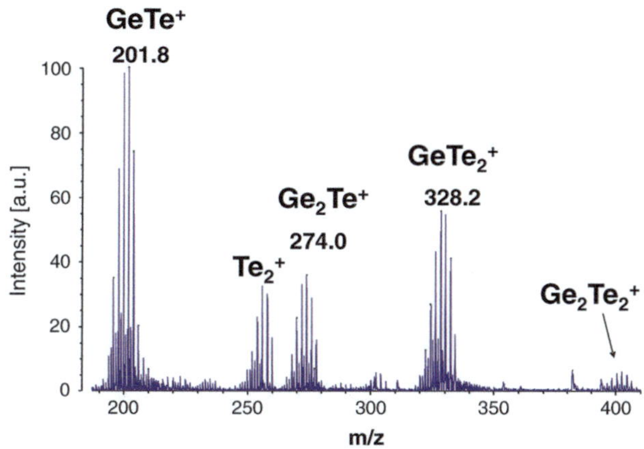

$$As_4Se_4 + (2/n)\ Se_n = 2\ As_2Se_3 \tag{7.4}$$

$$As_4Se_3 + (3/n)\ Se_n = 2\ As_2Se_3, \tag{7.5}$$

$$|As - As| + |Se - Se| = 2\,|As - Se|\,. \tag{7.6}$$

"Physical" methods of the evaporation of compounds (sputtering, thermal-, vacuum-, or flash evaporation, PLD, sputtering by electron or ionic bombardment of a target surface) are, in the end, also chemical methods. The evaporation e.g. by increased temperature, by light pulses, or by ionic or electron bombardment is connected with the dissociation of the evaporated material into molecular or ionic fragments (Figs. 7.14 and 7.15) [60] and needs chemical synthetic reactions for the preparation of the necessary compound. The compounds and their fragments that are evaporated by PLD by sputtering or by vacuum evaporation are synthesized again, either during the deposition in cooling vapours, inside the plume when the temperature decreases or after the deposition on the surface of the substrate. The Raman spectrum of PLD films is similar to the spectrum of bulk glasses in spite of the fact that the evaluated temperature in the plume was 8,000–10,000 K [56]. The backward synthetic chemical reaction of the wanted compound from its fragments can be, however, slow in thermally evaporated films and is not fully completed if the substrate is cold (non-heated). The prepared films are then non-homogeneous, containing also the fragments of the original material. This problem is not very important for PLD films; the temperature in the plume is very high and also kinetic energy of particles in the plasma plume is very high [56, 59], so the reactions (Eqs. 7.4, 7.5, and 7.6) are relatively fast and the content of dissociation products (Figs. 7.12 and 7.13) is very low. The formation of several minor Ge-H clusters was observed for a pure germanium and for germanium telluride. Almost no Ge-H or Sb-H clusters were observed in the mass spectra of PLD-deposited materials.

The homogenization of fresh evaporated amorphous films can be received by the *annealing* of films at higher temperatures, e.g. in the *vicinity of the glass transition temperature,* T_g. At around this temperature, atoms or structural groups become relatively mobile and reactive; their diffusion and the chemical reactions between them progress in the film relatively fast (Eqs. 7.7 and 7.8), a backward reaction.

$$2\left|M-Y\right| \overset{h\nu_i, I_i, T_i}{\longleftrightarrow} \left|M-M\right| + \left|Y-Y\right|. \tag{7.7}$$

M stands for a metal or metalloid, Y stands for a chalcogen (mostly S or Se), $h\nu_i$ and I_i stand for the energy and intensity of exposing light, respectively. The value of $i=1$ stands for a forward reaction (from the left-hand side to the right-hand side), and $i=2$ is for a backward reaction.

In As chalcogenides, As_4Y_4 and As_4Y_3 molecules are formed (here Y stands for S or Se). The reactions (7.7, and 7.8) can be shifted with temperature forward and backward.

$$2(As_2S_3)_n \longleftrightarrow nAs_4S_4 + nS_2; \tag{7.8}$$

The dissociation reaction for the evaporation of As_2Se_3 can be written, e.g. as

$$2As_2Se_3 = As_4Se_4 + 2Se, \tag{7.9}$$

As_4Se_4 can be further dissociated, in the PLD plume probably to elements. When the temperature of the plume is lowered, the reactions of the type of Eq. (7.4, 7.5, 7.6, 7.7, and 7.8) go back and the elements or fragments are forming the original compound again. The vapours of the PLD process have very high kinetic energy [56] that supports the backward synthetic reaction between fragments [59]; the structure of PLD films is similar to the structure of bulk materials (Figs. 7.12 and 7.13), as already mentioned.

In the case of the $As_{60}Se_{40}$ glass, the PLD films can be more “amorphous” than the corresponding bulk glasses (Fig. 7.13). The bulk glass is partly dissociated, with strong narrow Raman lines of As_4Se_4, while the PLD film only has one broad Raman band of a fully amorphous solid (Fig. 7.13).

The structure of annealed films can be close to the structure of glasses before evaporation. The as-evaporated films prepared by different methods of evaporation can have a different structure, composition and properties when compared with bulk samples and with annealed films.

Thin films prepared by physical methods from materials with higher volatility of some of their components (e.g. chalcogens or halogens) can be not only dissociated, but their composition can also be changed. Chalcogens, especially S and Se, generally have higher tension of vapours than most of the other components of a compound (material). In the films prepared by classic thermal evaporation, or by sputtering of chalcogenides, there can be found either a gradient of composition through the film thickness due to fractional evaporation, or the films contain sub-stoichiometric content of the more volatile chalcogens. The stoichiometric composition can be received, or at least the stoichiometry improved, either by the heating of the thin film in vapours of chalcogen, by the evaporation of materials containing the overstoichiometric excess of the more volatile parts (e.g. chalcogen), or by the evaporation of the material from two boats of different temperature according to the volatility of the evaporated compounds. The pulsed laser deposition (PLD) method can also be used, too. In the PLD method, the temperature of the evaporated surface is very high (evaluated value is $\approx$8,000 K, see above), all components are evaporated at once without any fractionation [56, 58–60], as already mentioned. This process is illustrated in Figs. 7.12, 7.13, 7.14, and 7.15.

The *spray pyrolysis* can also be used for film preparation. A solution of the wanted material or its thermally unstable compound is sprayed on a heated substrate. When a cold substrate is used, it is consequently heated together with the spray-deposited film in order to remove the solvent and, if necessary, to remove the volatile part of the compound. Synthetic reactions among the precursors can also take place during the heating. A similar process proceeds in the solution processing of chalcogenides ([61] and the papers cited).

Another very important method for chalcogenide film preparation is a so called *spin-coating* method. It is based on an application of a solution of AGCh in a proper solvent (e.g. in alkaline amines or other alkaline solvents). Such a solution often forms a viscous liquid, which is dropped or poured on a substrate placed in the centre of a rotating disc. The liquid is spread under centrifugal forces all over the substrate. The solvent can be removed (or the temporarily formed compound dissociated) after drying and heating. For thick film preparation, the process can be repeated several times [62]. The spin-coating method can be used for the preparation of thin as well as thick films and also for the preparation of waveguides, gratings, and photoresists of very high resolution. Using this method, homogeneous large-area thin or thick films with thickness up to several micrometres can be prepared. The temporary compounds formed between the solvent and the chalcogenide should not be very stable. They shall be dissociated by annealing when the solvent and volatile parts of the solutions are removed. Good quality large-area films of the original chalcogenide can be then obtained; their surface can be modified e.g. by the embossing method [61]. The major problem very often is to find a proper solvent for a given chalcogenide. While chalcogenides based on the As-S or As-Se system can be dissolved very easily in alkaline amines or alkaline solutions, it is very difficult to dissolve tellurides or some complex chalcogenides.

7.2.5 Nanoparticles

Nanocrystalline materials display many optical, electronic and structural properties that are present neither in isolated molecules nor in micro- or macroscopic samples. The forbidden band gap of semiconductors widens when the dimension of a material becomes of nano-size values.

Nanoparticles can be prepared using different methods. The hydrothermal technique in a water solution or in a water-organic solvent mixture is often used for that purpose. Many sulphides, e.g. SnS, SnS_2, CoS_2, CdS, CdSe, Bi_2S_3, etc., have been prepared using this method [63, 64]. There have also been attempts to prepare and then use nanorods or nanowires for phase change memory (PCM) cells [65]. Further information can be found in [66].

Also quantum nanodots are very important, formed in many materials by the inclusion of metals or semiconductors into chalcogenides. In fact, metallic nanodots are not very new. They were already used, without any knowledge of the principle, for colouring glasses even in Ancient Rome (Au), e.g. in the famous Lycurgus Cup [67]. They were, in fact, the first plasmonic materials prepared without understanding the principle. Many nanocrystalline Ch (CdS, PbS, PbSe, ZnSe, CdSe, etc.) have also been studied, see e.g. [68].

Thin 2D films of a nano-size thickness can be prepared from every Ch. They have been used and studied for a long time. Recently they have also been intensively studied and applied in PCMM.

7.3 Chemical Reactivity, Etching, Oxidation and Hydrolysis

The most of Ch are relatively sensitive to oxidation (O_2) and to hydrolysis by water vapours. The sensitivity differs with composition, temperature and the state of individual samples. Most of Ch shall be studied and applied either in an atmosphere free of oxygen and water vapours, or covered by a protective layer. Their higher reactivity with some substances can be used for the etching of photoresists, gratings, planar optical circuits and waveguides. Ch generally have higher absorption of UV light than organic photoresists; they can be applied in a deep UV region for smaller details preparation with very sharp edges for production of high density electronic elements, memories or integrated circuits [69].

Many Ch (e.g. from the As-S, As-Se, Ge-S, Ge-Se systems) are relatively well soluble in alkaline solutions of NH_3, NaOH, amines, diamines, alkyl-ammonium compounds, hydrazine, etc. [69–71]. The solution of such compounds can dissolve or etch Ch; the solutions can also be used for the preparation of their thin or thick films by spin coating and by spraying methods (see above). The chemical reactivity of ACh can be strongly changed by the doping of the surface of ACh by some elements or compounds, e.g. by Ag, which sharply increases their resistivity against etching [33–35]. The reactivity of many Ch can also be changed by the exposure of the material by a band-gap, or even by sub-band-gap light, or by the sorption of some specific compounds on the surface of chalcogenides [34]. The problem is discussed in Sect. 5.3 in connection with photoinduced effects.

The doping of the layers or the bulk sample surface can be done by classic methods and also by the *ion implantation* of the surface or of the whole thin film. Almost all elements can be implanted. The surface of the implanted matrix is distorted by the ion impact and by their penetration into the sample volume. It should be annealed for healing the defects [72]. The system is shifted towards the equilibrium due to such annealing.

7.4 The Structure of Chalcogenides and Its Changes

7.4.1 Crystalline State

The structure of CCh covers a wide area of various arrangements of atoms from a highly symmetric diamond-like structure (e.g. the cubic ZnS (Fig. 7.16), the cubic fcc $Ge_2Sb_2Te_5$ (Fig. 7.17 [73]), the cubic $Au_{25}Ge_4Sn_{11}Te_{60}$ (Fig. 7.18 [74])), via a hexagonal symmetry of many tellurides, selenides and sulphides (e.g. CdS, Sb_2Te_3, Bi_2Se_3 and Ge-Sb-Te equilibrium crystals ([75] Fig. 7.19) to the rhombohedral GeTe.

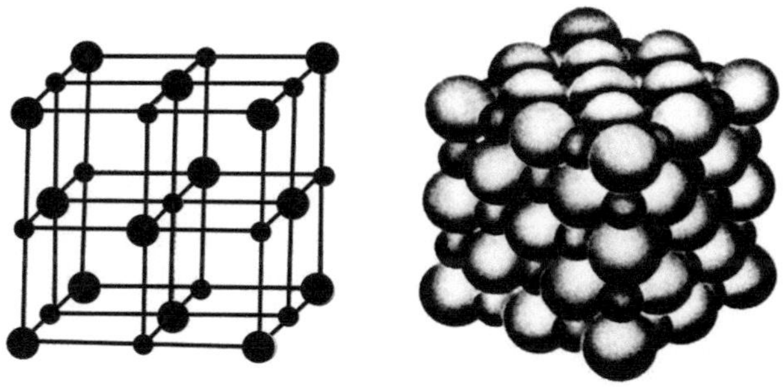

Fig. 7.16 Structure of a cubic ZnS. *Larger (balls) circles* are for S, *smaller ones* for Zn

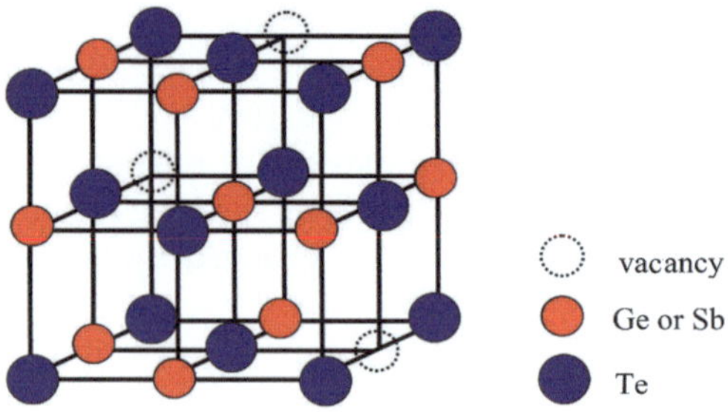

Fig. 7.17 A model of a metastable cubic modification of $Ge_2Sb_2Te_5$ [73], p. 214, adapted. The *dashed circles* represent vacancies, the *larger blue circles* represent Te, the *red* positions are randomly occupied by Ge or Sb

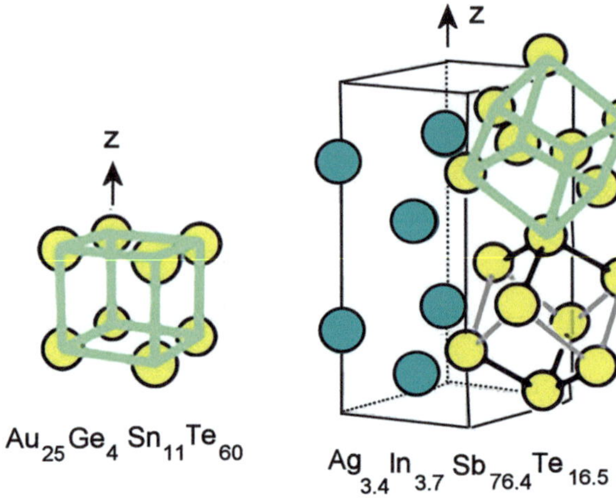

Fig. 7.18 Structure of two PCMM ($Au_{25}Ge_4Sn_{11}Te_{60}$, $Ag_{3.4}In_{3.7}$ $Sb_{76.4}Te_{16.5}$) [74]. All crystal positions are occupied randomly. The first alloy ($Au_{25}Ge_4Sn_{11}Te_{60}$) was applied for a short time formerly, the second one (AIST) is still used in rewritable DVD discs. The AIST crystallizes in a rhombohedral (slightly distored cubic) lattice. The crystal positions are occupied statistically by all atoms of their formula. By this behavior they are close to metallic alloys

The structure of many CCh contains some crystal positions that are randomly (statistically) occupied by two or several atoms. It is either a result of similar properties and size of some atoms, e.g. of Te and Sb (see Fig. 7.18, the Ag-In-Sb-Te structure), or it can also be caused by difficulties in distinguishing two different atoms by the used method, which is the case of Sb and Te from X-ray diffraction or from their X-ray spectra. Their intensive X-ray bands are overlapping. The third reason can be a kinetic one; the phase change in phase change memory materials (PCMM) is so fast that the atoms have not enough time to take the equilibrium positions and they form metastable structures.

The statistical occupation of some crystal positions has been found in many PCMM (see, e.g. AIST, $Au_{25}Ge_4Sn_{11}Te_{60}$, Fig. 7.18), partly in $Ge_2Sb_2Te_5$ (Fig. 7.17), at least in its metastable modifications. Random occupation of some (or

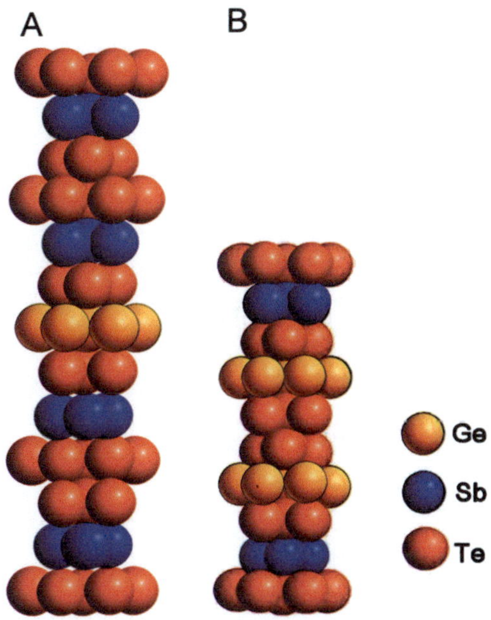

Fig. 7.19 A model of the structure of hexagonal crystals of $Ge_2Sb_2Te_5$ (**a**) and $GeSb_4Te_7$ (**b**) [75]

all) crystal positions can make the crystallization of an amorphous film easier and faster (a shorter transport of atoms) as it has been found e.g. in AIST (Fig. 7.18), which forms very fast PCM. The movement of atoms during the crystal formation need not the diffusion for a long distance. One atomic jump is often enough for most of atoms.

Many CCh change the symmetry when heated and have several crystallographic modifications, separated by energy barriers of different height. Not only stable but also metastable compounds or crystal structures can be formed. A typical example can be seen in the GeTe-Sb_2Te_3 system, where the fast or low temperature crystallization of amorphous thin films results in metastable cubic fcc crystals (Fig. 7.17), while further heating of fcc crystals to higher temperatures gives the equilibrium a hexagonal modification (Fig. 7.19). In the cubic fcc structure of $Ge_2Sb_2Te_5$ (Fig. 7.18), the positions in "cationic" sublattice of $Ge_2Sb_2Te_5$ are statistically occupied by Ge, Sb atoms and by crystal vacancies, as already mentioned. The statistical occupation has also been found in some layers of hexagonal Pb-Bi-Te system crystals which are similar to PCMM [76, 77] (Fig. 7.20). Contrary to the systems Sb-Te and Ge-Sb-Te, all atoms can be distinguished in Pb-Bi-Te system.

Crystals of some ternary sulphides and selenides, e.g. of $Pb_5Sb_4S_{11}$, $PbSb_2S_4$ [78, 79] (Figs. 7.21 and 7.22) have low symmetry. The elemental cell of the last mentioned crystals is very large and the repetition of the structural motifs (long range order) is only realized after a large number of atomic distances. The movement of conductivity electrons (holes) is similar to the one in amorphous solids which have the elemental cell of the size of the whole sample.

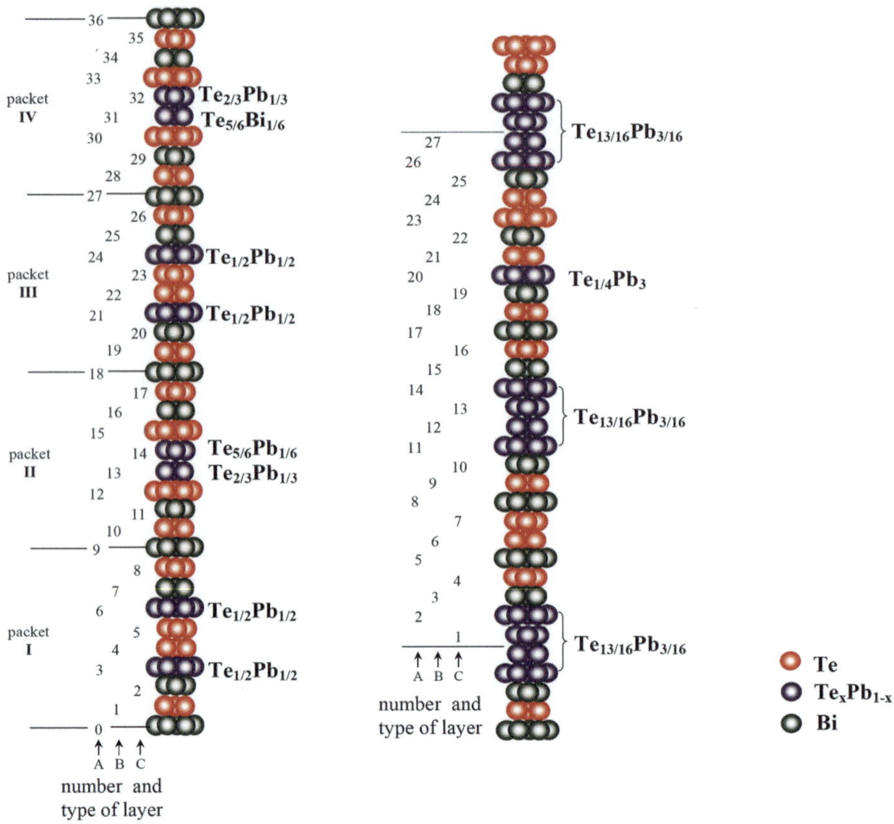

Fig. 7.20 A crystal structure of hexagonal modification of a compound $PbBi_4Te_7$ from the system Pb-Bi-Te [76, 77]. Some crystal positions are occupied statistically by Pb and Te, mostly with excess of Te. Such behaviour allows for a large nonstoichiometry

Crystalline chalcogenides (CCh) have very different properties, depending on their structure, composition and band gap energy E_g. They form dielectrics, wide-gap semiconductors, semiconductors, narrow-gap semiconductors, metals and superconductors. A detailed discussion of all these groups of crystalline chalcogenides is beyond the scope of this review paper.

7.4.2 Amorphous and Glassy State

The composition and structure of AGCh covers a broader range than those of crystalline ones; the solubility of excess components and/or other elements or compounds is higher in glasses and amorphous solids than in the corresponding crystals. The energies of chemical bonds between a chalcogen and a metal or a

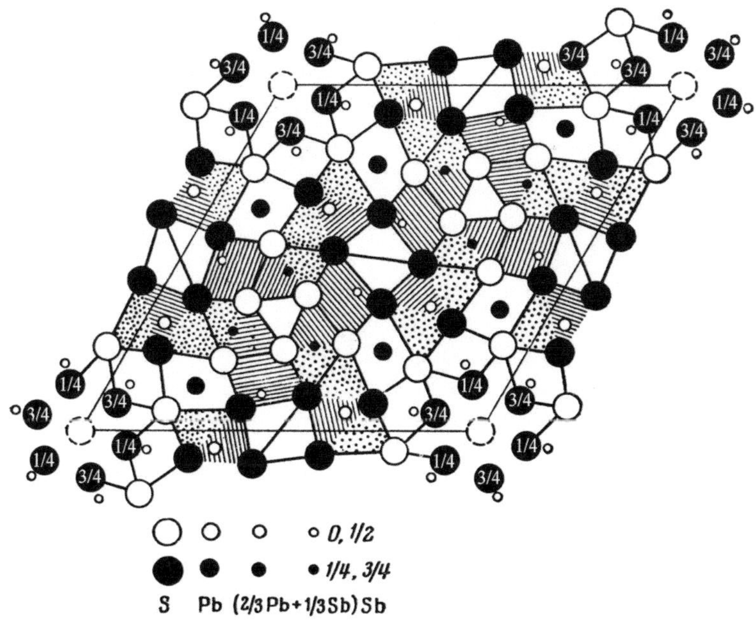

Fig. 7.21 The projection of the structure of $PbSb_2S_4$ in the direction x-y. The *dashed* areas correspond to trigonal pyramids, the *dotted* ones are tetragonal pyramids [78]

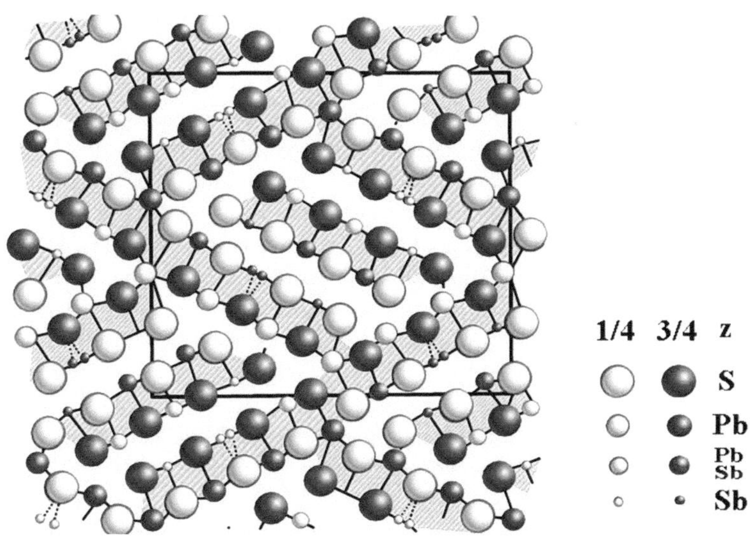

Fig. 7.22 Structure of $Pb_5Sb_4S_{11}$ in the projection of (001) [79]. Parts of the structure that are similar to PbS are *dashed*. Parts of crystal positions are occupied again statistically, in this case by Sb and Pb

Fig. 7.23 A picture of Pt/C replica of a fracture face of the glass $Ge_5Sb_{24}S_{71}$ from an electron microscope. Magnification 10^4 [80] (The figure was obtained at the University of Jena with kind permission of the late prof. Vogel and prof. Linke)

chalcogen and a metalloid are lower than those in oxides. They often are not very different from the energies of homopolar bonds between like atoms (e.g. -S-S-, -Ge-Ge-, -Sb-Sb-, etc.), as we have already mentioned (Table 7.1). This fact allows for a large nonstoichiometry of many glasses without any phase separation. In spite of that, some glasses can become phase separated not only by crystallization, but also by the process of micro-liquation, when several glasses of different composition are formed during the cooling of their melt of composition from the non-miscibility area (Fig. 7.23) [80].

The structure of GCh, AChF and amorphous nanoparticles depends on the composition, temperature of synthesis, the way of preparation and the rate of cooling. In fact, every AGCh is an original solid with its own structure. The amorphous state is a disordered and metastable one. It can have many local minima in dependence of the volume or Gibbs energy on transition coordinates (Fig. 7.24 [81]). They depend on the temperature from which the melt was cooled and from which the metastable structure of the glass was frozen. The barriers between individual metastable states (E_1 to E_4, Fig. 7.24) can be lower than the barrier between the starting and the final states. As a result, the transition from the M_A to M_E states does not go directly, but more probably via the M_1, M_2 and M_3 metastable states.

Fortunately, the influence of the above-mentioned factors (the rate of cooling, the way of preparation of AGCh) can often be kept in reasonable limits, and the samples with relatively low fluctuations of structure and properties can be prepared with some precautions. The frozen structure of AGCh can relax by annealing at a temperature close to T_g. The atoms or structural groups of several atoms are relatively mobile at this temperature. The structure is then shifted towards the equilibrium one; the value of Gibbs energy decreases (Fig. 7.24). The structure can also be changed by other outer forces, e.g. by exposure to absorbing light (see Sect. 5.2).

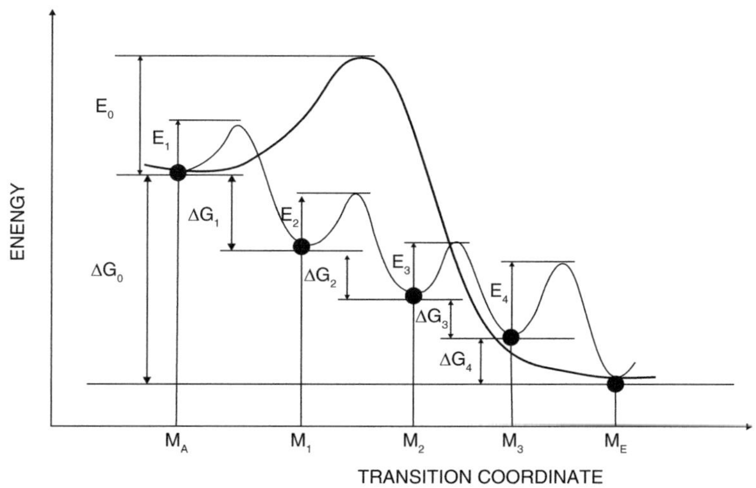

Fig. 7.24 Two paths for the transition to the equilibrium state. ΔG represents Gibbs energy change, E_i barriers for the metastable or amorphous state ($i = 1 - 4$), M_A is original state, E_0 is thermodinamic barrier for the transition from original to equilibrium state, M_E the equilibrium state, M_1–M_3 are metastable states [81]

The structure of the amorphous state does not possess the long-range order, and also the mid-range order is rather limited. S. R. Ovshinsky, who has contributed fundamentally to the application of amorphous chalcogenides in the area of electrical switches and PCM, did not like the negative determination of the absence of the long-range order (i.e. "no long-range order"). He coined a positive formulation of the great randomness of the structure by saying that "the amorphous solids possess larger freedom of composition and structure", contrary to crystals which are of a "totalitarian structure and composition".

In spite of easier dissociation of many chemical bonds in Ch, the fundamental structural units, such as AsY_3, SbY_3 pyramids and SiY_4, GeY_4 tetrahedra, Y_2-As-As-Y_2 bipyramids, where Y stands for chalcogen, are preserved not only in ChG but also in their thin films (Figs. 7.25 and 7.26) [82]. In the As-S system, there can be found e.g. AsS_3 pyramids, but also the -S-S-, -As-As- structural units, the As_4S_4 and As_4S_3 molecules (Figs. 7.25 and 7.26). In the Ge-S system, there can be found $GeS_{4/2}$ tetrahedra, Ge_2S_6 bitetrahedra and also the -S-S- and Ge-Ge bonds.

Contrary to crystals, the energies of identical chemical bonds in different structural units of AGCh are not exactly equal in every structural unit; the length of individual bonds and the angles between individual bonds are dispersed (Fig. 7.27). As a result, there are no sharp lines on their X-ray diffraction diagrams. Instead of them, there are broad bands corresponding to the distribution of local diffraction distances.

Complex glassy structures have been found in some telluride and selenide binary and ternary glasses and in PCMM, which are generally very bad glass-formers (Sects. 7.5 and 7.6). For the glass formation in PCMM, their melt should be cooled

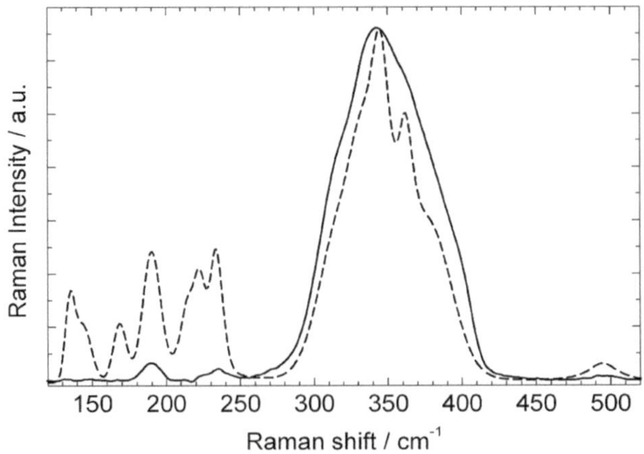

Fig. 7.25 Raman spectra of an $As_{38}S_{62}$ bulk glass (the *full line*) and a freshly evaporated thin film (the *dashed line*). The broad line near 345 cm^{-1} corresponds to the vibration of AsS_3 pyramids, the *narrow band* near 233 cm^{-1} to As-As vibrations. The 494 cm^{-1} band corresponds to S-S bonds vibrations, the other narrow bands correspond to As_4S_4 and As_4S_3 molecule vibrations [82], adapted

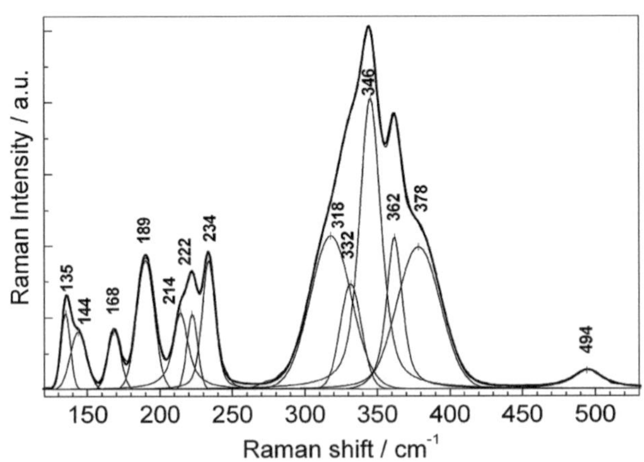

Fig. 7.26 Raman spectra of an $As_{40}S_{60}$ freshly evaporated thin film [82], adapted. The spectrum was deconvoluted into individual bands. They can be assigned to the vibrations of bonds in molecular features with narrow bands given in Fig. 7.25

by the rate up to 10^9–10^{11} K/s [83], otherwise crystals are obtained. The glassy state of these materials can be obtained only in very thin films (several nm or several tens of nm), usually melted by nanosecond optical or electrical pulses. The Raman spectra of such PCMM in PCM cells can only be measured with great difficulties, as the vibration frequencies of their atoms, mostly heavy, are low and their vibrations are often overlapped by the vibrations of other groups of atoms. The real PCM films also are very thin. The measurements of their X-ray absorption near edge structure

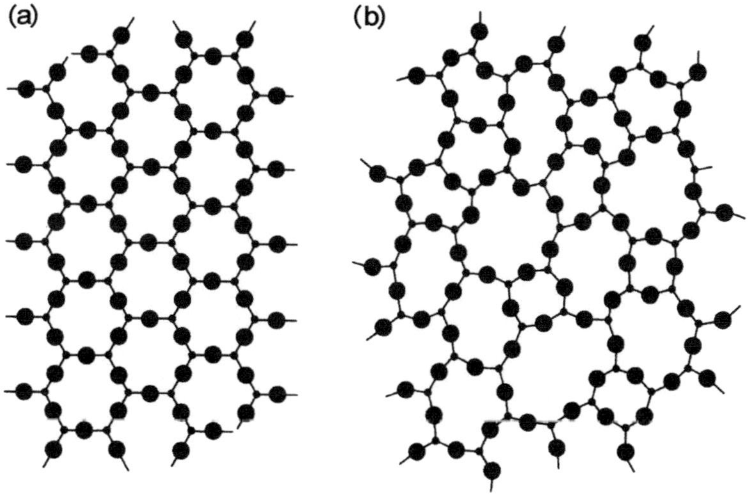

Fig. 7.27 A 2D scheme of the structure of crystalline (**a**) and amorphous solid (**b**). The changes in bond lengths and bond angles between crystalline and amorphous state are evaluated as ±5 % and ±10–20 %, respectively. The structure of the amorphous solid in Fig. 7.30b does not contain any homopolar bonds, so this structure corresponds more to oxide glasses than to chalcogenide ones

(XANES) and extended X-ray absorption fine structure (EXAFS) spectra are very useful. In the case of the Ge-Sb-Te and Sb-Te systems, it is difficult to distinguish between the Sb and Te atoms that are neighbours in the periodic table and have almost identical a structure of their electron configuration.

The amorphous and glassy states are metastable ones. By heating above the crystallization temperature, T_c, they crystallize and lower their Gibbs energy. The crystallization can be induced not only by heating, but also by intensive light or electrical pulses. Crystallized thin films can be melted again by more intensive light or electrical pulses. The glassy (amorphous) state can be obtained repeatedly by fast cooling of such a melt. For some systems, the process can be repeated many times, e.g. for PCMM up to 10^{12}, usually up to 10^6, which is enough for memories of mobile phones, electronic books, etc. Such principle of crystallization-amorphization, and vice versa, is used in phase change memories (PCM, Sects. 7.5 and 7.6).

Contrary to the scheme of the structure of AGCh given in Fig. 7.27, the Figs. 7.28 and 7.29 [84] show experimentally obtained pictures of the real structure of thin films of $Ge_2Sb_2Te_5$, in which individual atoms can be seen. The figures were measured on the films using high-resolution transmission electron microscopy (HRTEM). Both samples were amorphous according to X-ray diffraction and optical transmission. However, Fig. 7.29 shows the areas of the size of several nanometres which are partly ordered [84].

The area of the research and application of amorphous and glassy solids, including Ch, is very broad and important and its development is still in progress. The two men who played an outstanding role in the research in amorphous and glassy

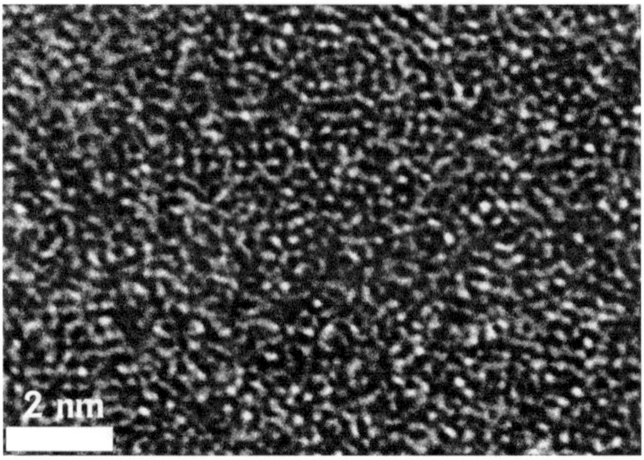

Fig. 7.28 A view of a real structure of an amorphous $Ge_2Sb_2Te_5$ [84]. The individual particles correspond to individual atoms. The picture was obtained on a very thin amorphous film using a high-resolution transmission electron microscope

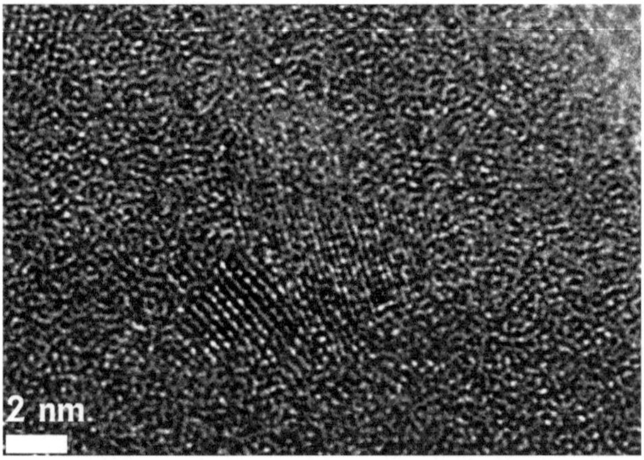

Fig. 7.29 A view of a structure of a thin $Ge_2Sb_2Te_5$film [84]. In the middle of this figure can be seen a small area which has a partly ordered structure. The size of such baby-nuclei is, however, so small that it can be detected neither by X-rays diffraction, nor by optical transparency measurement

semiconductors especially in their early days, were Sir N. F. Mott (Fig. 7.30), who was awarded the Nobel Prize for his research in non-crystalline semiconductors, and S. R. Ovshinsky (Fig. 7.31), who launched the intensive research of switching effects and of PCM, which were successfully applied later and led to the production of many billions of CD and DVD memory discs and to electrical phase change memories. Not only these two men contributed fundamentally to the science and technology of AGCh; there have been many other important and successful pioneers in the area, such as N. A. Goryunova and B. T. Kolomiets, who started the study of glassy and amorphous semiconducting chalcogenides. We shall also name

Fig. 7.30 Sir N. F. Mott, winner of the 1977 Nobel Prize in Physics (With kind permission of E. A. Davis)

Fig. 7.31 S. R. Ovshinsky, Dr. h. c. multi, Energy Conversion Devices, Ovonyx, chalcogenide-based switches, solar elements, amorphous Si, NiMH batteries (With kind permission of Stan Ovshinsky widow)

fundamental contributions by J. Tauc, H. Fritzsche, E. A. Davis, D. Adler, V. Lyubin, H. K. Henisch, A. E. Owen, W. Spear, A. Feltz, R. Grigorovici, Z. Borisova, V. F. Kokorina, S. R. Elliott, N. Yamada, A. Kolobov, K. Tanioka, K. Shimakawa, K. Tanaka, S. Kasap, M. Wuttig, T. Ohta, B. Hyot, T. Matsunaga, S. Raoux, M. Popescu, and many, many others.

7.5 Properties of Amorphous and Glassy Chalcogenides

Due to a great structural disorder in amorphous chalcogenides and different length and angles of chemical bonds, there are, contrary to crystals, many new electron levels formed inside the forbidden energy-gap, E_g, near the bottom of conductivity and near the top of valence bands (Fig. 7.32); the electron mobilities in these states are low (Fig. 7.33) [85, 86].

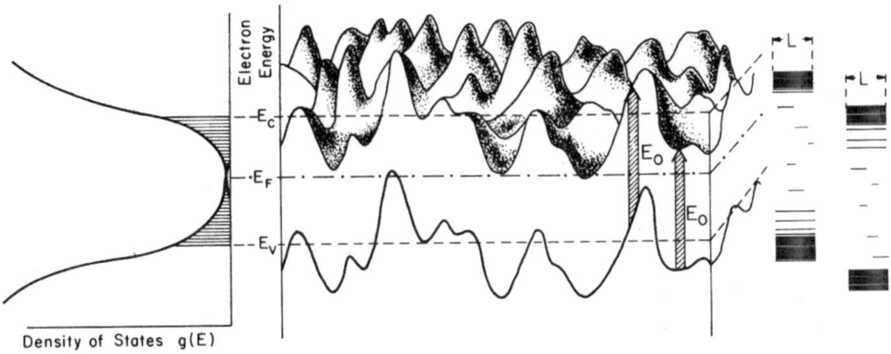

Fig. 7.32 A distortion of the *bottom* of the conduction band and of the *top* of the valence band due to a structural disorder of glasses and thin films [85]. The disorder causes local changes of the *bottom* of the conduction band and of the *top* of the valence band. As a result, the edges of these bands are not sharp as in crystals and many local levels for electrons and holes are formed near both edges. The electrons and holes can become localized on these levels; the electrical conductivity and free path of the carriers are drastically lowered. The very short free path of free carriers is important for the application of amorphous solids in Xerox, laser printers, X-ray panels, etc. Due to that, very sharp lines and other details can be obtained. The heights of individual electrons and holes are formed near both edges

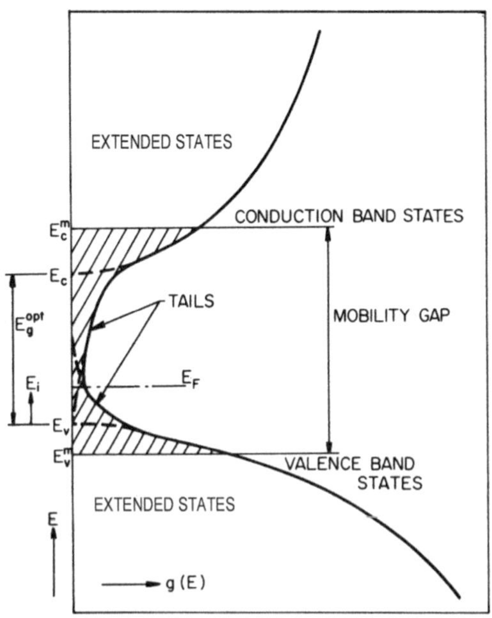

Fig. 7.33 The density of electronic states g(E) in dependence on energy E of electrons in amorphous semiconductors according to the Mott-Cohen-Fritzsche-Ovshinsky model [86], p. 177. Energy gap E_g^{opt} is obtained by extrapolation of extended states to g(E) = 0

The free carriers can be temporarily localized near the structural defects and the electrons in their movement shall overlap many local energy barriers. They can move much more quickly outside of the mobility gap in the extended electron/hole states (Fig. 7.33) which are above the tops of individual crests (Fig. 7.32). This fact is partly behind the significant increase in conductivity in the switching effect, when the energy of free carriers increases due to a high electrical field applied on the material. The effect also plays a significant role in electrically or optically induced phase changes; the excitation of carriers by intensive electrical or optical pulses is strong, the bonding electrons are excited into an antibonding orbital (conductive band); the order of chemical bonds is lowered and the changes in the structure become easier and faster.

7.5.1 Optical Properties

Optical transmission of chalcogenides is high in the IR region; some sulphides are transparent also in the visible part of the spectrum. The short-wavelength-absorption edge is sharp in CCh (Fig. 7.34, [86], p. 173); it is not so sharp in AGCh. In AGCh it can be divided into three main parts; A stands for the band gap edge, B stands for the Urbach edge, and C stands for the weak absorption edge ([86], p. 172, Fig. 7.35).

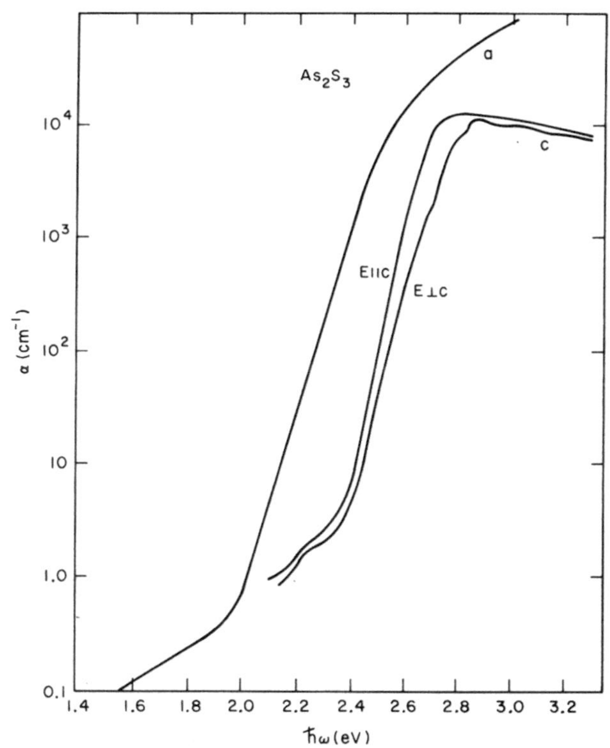

Fig. 7.34 The energy dependence of the optical absorption coefficient of an As_2S_3 glass (*a*) and As_2S_3 crystal (*c*) in the region of a short-wavelength absorption edge [86], p. 173

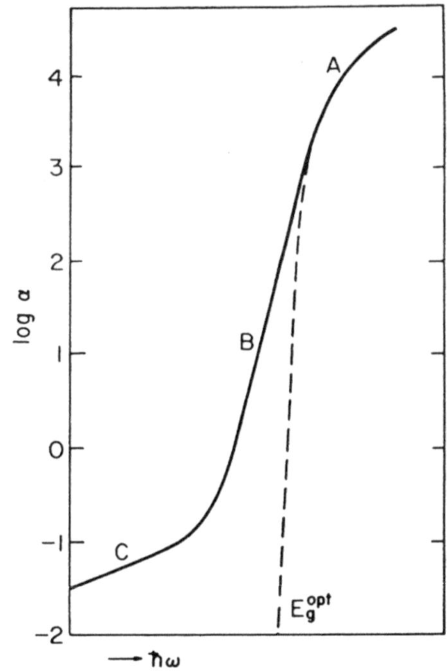

Fig. 7.35 A schematic view of three parts of a short-wavelength absorption edge of glassy or amorphous solids [86], p. 172. Part *A* is connected with a band to the band electronic transition; part *B* is the Urbach edge; part and *C* is caused by absorption on impurities, nonhomogenities in the glass composition and on the fluctuation of the index of refraction

The spectral dependence of the absorption coefficient α at a high absorption region ($\alpha \geq 10^4$ cm^{-1}) follows the relation

$$\hbar\omega.\alpha(\omega) \sim \left(\hbar\omega - E_g^{opt}\right)^r, \tag{7.10}$$

where r is a constant (for As_2Se_3 and for many AGCh $r = 2$) [86].

The α is given by Eq. 7.11 in the Urbach part (B) of the absorption edge

$$\alpha(\omega) \sim \exp\left[\frac{\sigma\left(\hbar\omega - E_g(T)\right)}{kT^*}\right], \; \hbar\omega < E_g(T). \tag{7.11}$$

σ is a constant, $E_g(T)$ is the temperature-dependent energy gap and T^* is the effective temperature that is constant at low temperatures and proportional to T at higher temperatures [86]. The spectral dependence of α is linear in the $log\ \alpha - \hbar\omega$ scale.

In a pure AGCh, the weak absorption tail (C, Fig. 7.35) is induced by fluctuations of the local structure, local composition and of local changes of the refraction index. The value of the absorption coefficient α in the weak absorption part of the spectrum is increased by the presence of impurities in AGCh. In this part (C), the optical absorption coefficient is higher in AGCh than that in pure silica fibres used in optical data transmission. Because of that, the GCh fibres are not good candidates for a long-distance transport of high intensity of light and also for a long-distance transport of signal. Chalcogenide fibres become very important for short-distance applications, e.g. in medicine, IR sensing, environmental areas, etc. (see also Sects. 7.5 and 7.6). Chalcogenide fibres have already been prepared from several chalcogenides, such as As-S, As-Se, As-S-Se, Ge-As-Se, Ga-Ge-Sb-S, Ge-Sb-S, Ge-Sb-Se, etc. [49, 50, 87]. The losses of light intensity in chalcogenide glasses are higher than those in silica glasses and fibres due to the lower purity and higher compositional disorder of GCh. Higher absorption is critical mainly for the delivery of high-intensity light. This problem can be solved by the application of hollow fibres in which light is delivered in air inside the fibre hole. Very low losses (1.7 dB/km) have been obtained. Several hollow Ch fibres can be melted together and so-called micro-structured optical fibres (MOF) with properties of photonic crystals can be prepared, mainly for IR applications (Fig. 7.36 [87]). The topic is discussed in Sect. 7.6.

The transparency of Ch at the long-wavelength IR part of the spectrum is limited by multiphonon absorption (Fig. 7.37, [87]). It depends on composition; for tellurides it can spread in the IR region up to 25 μm.

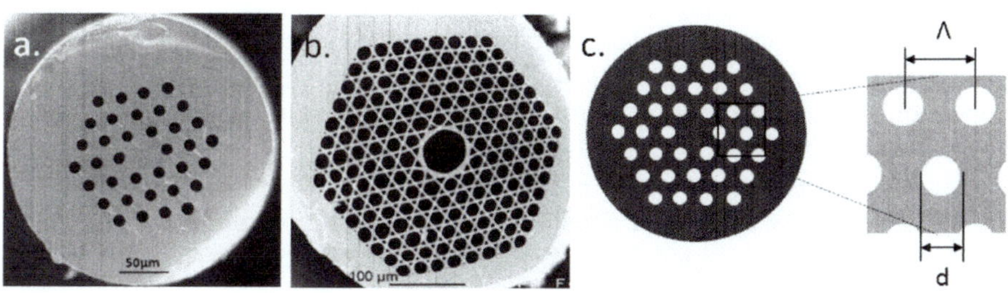

Fig. 7.36 Microstructured optical fibres. They are prepared by melting several hollow fibres together and by cladding their surface. The light proceeds through the air and the attenuation is only caused by the roughness of the inner walls of the fibres. The MOF have properties of photonic crystals [87]

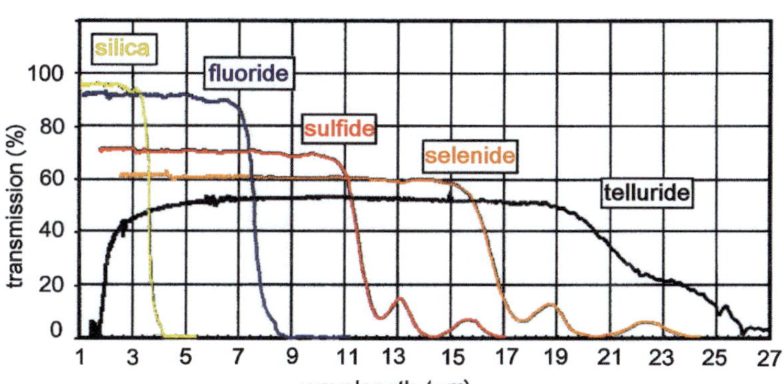

Fig. 7.37 Transparency regions of silica, fluoride, sulphide, selenide and telluride glasses in the IR spectral region [87]

The transparency and other optical constants in the far-IR part of the spectrum and beyond can be measured also in the THz region (see e.g. [88]).

Optical reflectivity of chalcogenides spans from 0.1 to <1. It depends on the index of refraction n and the index of absorption k

$$R = \frac{(n-1)^2 + k^2}{(n+1)^2 + k^2}. \tag{7.12}$$

Equation 7.13 can be simplified for low values of k

$$R = \frac{(n-1)^2}{(n+1)^2}. \tag{7.13}$$

The optical reflectivity and the index of refraction values increase in the succession from sulphides via selenides to tellurides. Tellurides of heavy metals have a metallic lustre in the visible part of the spectrum. Crystals of heavier tellurides (e.g. Bi, Te, Hg, Cd, Ag,) are narrow gap semiconductors or possess semimetallic or metallic properties.

The reflectivity of some crystalline tellurides, such as GeTe, Sb_2Te_3, $Ge_2Sb_2Te_5$ and others, is much higher than that of the corresponding glasses. This property is important for their application as materials for non-volatile optical PCM used in CD, DVD and Blu-Ray discs, see also Sect. 7.6.

The index of optical refraction in chalcogenides is higher than in oxides due to higher polarizibility of heavier atoms of chalcogens. In amorphous sulphides, its value is typically 2–3, in tellurides it can be near 6 for the amorphous form, and near 10 for the crystalline ones.

7.5.1.1 Non-linear Optical Effects

Very important and promising for many new applications are high coefficients of optical non-linearity of AGCh [89–101], which often are of 2 or 3 orders higher than those in silica. The non-linear effects can be applied in ultra-fast optical switching devices (the expected switching time is in the order of femtoseconds, 10^{-15} s.), frequency converters, electro-optic modulators and devices, all-optical circuits and all-optical signal processors (see e.g. [89, 90, 101] and the papers quoted). The non-linear effects have been found in many crystals and glasses; the advantage of glasses is their lower cost, easier processing, and the possibility to include them easily into optical planar circuits and connect them with optical fibres.

The polarization P induced by an electric field E in a material can be written as

$$P = \chi^{(1)}E + \chi^{(2)}E^2 + \chi^{(3)}E^3, \tag{7.14}$$

where the first term describes the conventional linear response; the other two terms correspond to the non-linear response; $\chi^{(n)}$ are susceptibility constants, which depend on the material illuminated. $\chi^{(1)}$ is connected with the linear refractive index n_o and with the absorption coefficient k. As glass is isotropic, it cannot provide even-order nonlinearities, or they are very little [96]. The third term with $\chi^{(3)}$ becomes important for isotropic glasses. The index of refraction is a constant at low intensity of the electric field. At higher intensities of the electric field E ($E > 10^7$ V/cm), the index of refraction starts to be dependent on E; $n = n_0 + n_2 E^2$, where n_0 and n_2 do not depend on the intensity of light, and $n_0 >> n_2$. E^2 is the mean square of the electric field. The part of the non-linear index of refraction, n_2, is connected with non-linear electron polarizibility.

The value of $\chi^{(3)}$ can be measured by different methods, such as Z scan, non-linear imaging technique [97], by the non-linear imaging technique modified as the non-linear imaging technique with a phase object (see e.g. [98], and its variation in [101]). The experimental determination of non-linear parameters is not an easy task, and several semi-empirical relationships have been proposed for $\chi^{(3)}$ and n_2 evaluation [90, 91].

Non-linear optical parameters can also be evaluated, e.g. from the linear refractive index *(n_0)*,

$$\chi^{(1)} = \left(n_0^2 - 1\right)/4\pi, \tag{7.15}$$

or from optical susceptibility $\chi^{(1)}$ (see e.g. [90, 101] and the papers cited therein).

The physically based Miller's rule is one of the most convenient evaluation methods especially for visible and near-infrared frequencies [90]. The accuracy of the determination of non-linear parameters using Miller's rule is generally better than the order of magnitude for many covalent or ionic compounds; for sulphides it agrees with experimental values within a factor of 2 [90].

For the region far from resonance, $\chi^{(3)}$ is approximately equal [101] to

$$\chi^{(3)} \approx A\left(\chi^{(1)}\right)^4, \tag{7.16}$$

where A is a constant. The value of $\chi^{(3)}$ can also be evaluated from the Wemple and Di Domenico parameters E_d and E_0 [102]

$$\chi^{(3)} = A\left[E_d E_0/4\pi\left(E_0^2 - \hbar\omega\right)^2\right]^4, \tag{7.17}$$

where E_d is so-called dispersion energy, ω is energy of light and $E_o \approx 2E_g$ [102]. For $\omega \to 0$, one obtains

$$\chi^{(3)} == \frac{A}{(4\pi)^4}\left(\frac{E_d}{E_0}\right)^4 = \frac{A}{(4\pi)^4}\left(n_0^2 - 1\right)^4. \tag{7.18}$$

The mean value of the constant, A, as evaluated from 97 experimentally found values [91] is approximately $A \sim 1.7 \times 10^{-10}$ (for $\chi^{(3)}$ in esu).

In several chalcogenides, the non-linear susceptibility $\chi^{(3)}$ was found to be several orders higher than the one of a SiO_2 glass (see above) in particular; the chalcogenide glasses were found to have the greatest non-resonant third-order non-linearity and thus become promising candidates for optical switching and other applications [90]. As an example, the value of $\chi^{(3)}$ for an As_2S_3 glass was found to be $\chi^{(3)} = (1.48 \div 2.2) \times 10^{-12}$ esu (measured experimentally by different methods), while for a GeS_2 glass, $\chi^{(3)} = 1 \times 10^{-12}$ esu, and for a SiO_2 glass, $\chi^{(3)} = 2.8 \times 10^{-14}$ (esu) for $\lambda = 1{,}900$ nm [103]. A higher $\chi^{(3)}$ value does mean a lower necessary power, a shorter interaction lengths and higher probability of a fully optical signal processing.

The $\chi^{(3)}$ values obtained by above given method are very close to the values of Nasu [104]. In thin films, the values of $\chi^{(3)}$ are influenced by the history of the films. In As_2S_3 films (Fig. 7.38), $\chi^{(3)}$ of fresh evaporated films is increased by the exposure to the band gap light and even more enhanced by the following annealing of the films. It is supposed that this fact is connected with an increase of homogeneity and with decreased content of fragments of As_2S_3 which are formed during the evaporation.

The high values of $\chi^{(3)}$ were obtained in several other chalcogenides, e.g. in [103, 104].

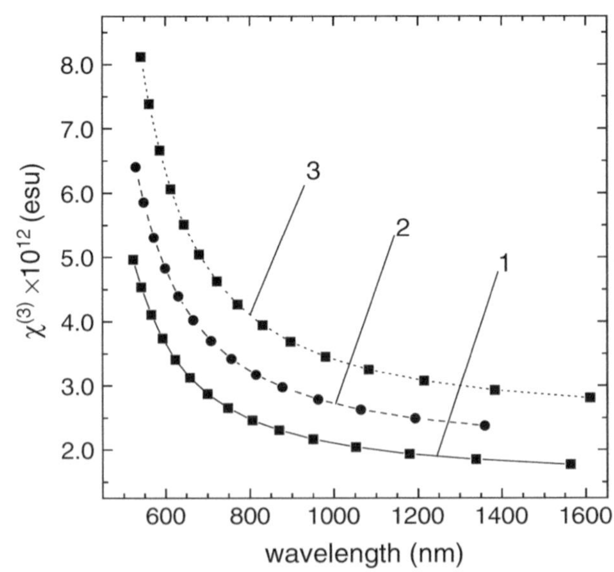

Fig. 7.38 Spectral dependences of non-linear coefficient $\chi^{(3)}$ values of TE $As_{40}S_{60}$ films [90]. *1* a fresh-evaporated film, *2* an exposed film, *3* an annealed film. Curves *1–3* were calculated by Eq. 7.18

The value of $\chi^{(3)}$ can be increased by poling. The electrical field of the value E ~ (10^6 V/cm) induces $\chi^{(2)} = 3E_{dc}.\ \chi^{(3)}$ [105], where E_{dc} is the applied DC field. The E_{dc} can also be induced inside the glass by the migration of mobile ions, e.g. Ag^+, Na^+ in the electrical field. At low temperatures, the electrical field created by the migration of mobile ions decays relatively slowly due to their lower mobility.

The high non-linearity of the index of refraction of many glasses can also be caused by the presence of nanocrystals embedded in the glass matrix (Au, CdS, and Ag [95, 96]).

7.5.1.2 Luminescence of Chalcogenides

Luminescence can be defined as the excess emission over and above the thermal emission of a body, if this emission has a duration considerably exceeding the period of light oscillations [106]. The luminescence of different materials has been known for ages, especially the one that is connected with biological processes (insects, mushrooms, fishes, microorganisms, etc.). Luminescence of many crystals, e.g. of ZnS, CdS, ZnO, etc., pure and doped, has been well known for a century as well. These crystals doped by proper activators were later applied in luminescent screens of TV sets, oscilloscopes and also for illumination purposes.

According to the excitation process, several types or classes of luminescence can be observed, e.g. as the cathodo-, iono-, electro-, thermo-, chemo- and mechano-luminescence. This review paper only discusses optically induced luminescence in Ch with focus on the luminescence of AGCh doped by rare -earth elements (RE).

Photoluminescence (PL) in solids is a process in which electrons are excited to higher energies by an external energy source and this energy, or its part, is later released as light. Luminescence can be excited by the UV, VIS, or IR photons, electrons, X-rays, ions and also by mechanical or chemical excitation. For a detailed discussion of this interesting area of physics and material science, see many books e.g. [106, 107].

The PL of solids is usually described as an intrinsic or extrinsic one. The intrinsic one means that there are no foreign atoms added on purpose to the matrix material as activators of the luminescence. The so- called activators create an electron level or several levels inside the forbidden gap of the solid and these levels take part in the creation or in the increasing intensity of luminescence bands in some spectral regions. The radiative electron transition in the intrinsic type of luminescence proceed either between the conduction and valence bands or between the conductive band and the levels of intrinsic defects, such as vacancies in "cationic" or "anionic" sub-lattices, and other possible intrinsic defects. The intensity of intrinsic PL in AGCh is relatively low. There is a strong Stokes shift and the empirical rule says that $E_{PL} \approx E_g/2$, where E_{PL} is energy of maximum PL band and E_g is energy band-gap. It is roughly valid not only for chalcogenides, but also for some oxide glasses ([19], p. 124).

The extrinsic luminescence of AGCh and thin films doped by RE is intensive and has many important applications. RE have many inner f electron levels; the electron transitions between them can radiate in the visible, NIR and MID IR regions; the RE-doped AGCh are frequently applied as luminophors for these regions. In a part of near infrared region (NIR) and in MID IR regions, oxidic matrices usually are nontransparent; Ch are applied as good materials for this region.

7.5.1.3 Luminescence of Chalcogenide Glasses Doped by RE

Luminescence of RE-doped AGCh, where RE are the activators, is much more intensive than the intrinsic one. Electron transitions between f-f levels can emit light in many different wavelengths (bands) depending on individual RE, on excitation wavelengths and on the composition and state of the matrix [107]. The strong luminescence of RE-doped AGCh can be applied e.g. in glasses for lasers, light up-convertors, fibre amplifiers, fibre lasers, etc.

Luminescent properties of chalcogenides doped by different RE can be determined experimentally. Many important parameters, such as luminescence intensity, electron transition probabilities, radiative life times, quantum efficiency of emission transitions, branching ratios for transitions from initial manifold electron levels, etc. can also be evaluated with a relatively good accuracy using Judd and Ofelt approach [108, 109]. According to Judd and Ofelt, the experimental values of oscillator strength f_{exp} depend on the intensity of individual absorption bands and can be calculated from (Eq. 7.19).

$$f_{\mathrm{exp}} = \frac{mc}{\pi e^2 N} \int \sigma(\nu)\, d\nu, \tag{7.19}$$

where m and e are electron mass and charge, respectively, $\sigma(\upsilon)$ is absorption cross-section, N is the density of RE^{3+} ions. Absorption cross section is given by $\sigma(\upsilon) = \alpha(\upsilon)/N$, where α is the absorption coefficient and N is the density of RE^{3+} ions (cm^{-3}).

A set of f_{exp} data can serve as a basis for the calculation of the Judd-Ofelt parameters [108, 109], Ω_t, using Eq. 7.20.

$$f_{\mathrm{exp}}\left(aJ, a'J'\right) \cong f_{calc}\left(aJ, a'J'\right) = \frac{8\pi^2 m\upsilon}{3h(2J+1)}\left[\frac{\left(n^2+2\right)^2}{9n}\sum_{t=2,4,6}\Omega_t\left|\left\langle aJ \left\| U^{(t)} \right\| a'J'\right\rangle\right|^2\right], \tag{7.20}$$

where f_{exp} and f_{calc} are experimental and calculated oscillator strengths, respectively, h is Planck's constant, m is electron mass, υ is the mean wavenumber of the absorption band, J is the ground-state total angular momentum. n is the refractive index of the material, Ω_t are the Judd-Ofelt phenomenological intensity parameters, and $\langle aJ \| U^{(t)} \| a'J' \rangle$ are reduced matrix elements of the tensor operator, $U^{(t)}$ of rank t, which can be evaluated by least-squares fitting methods, analogically to [108, 109] using Eq. 7.20.

Further details can be found in [110, 111].

The values of the Judd-Ofelt intensity parameters were calculated and measured for many glasses (see e.g. [112–114]). For Ge-Ga-Se glasses doped by Sm^{3+} ions, e.g. the following values were obtained [114]: $\Omega_2 = 7.42 \times 10^{-20}$ cm^2, $\Omega_4 = 14.43 \times 10^{-20}$ cm^2, $\Omega_6 = 6.43 \times 10^{-20}$ cm^2. These values are high; they are promising for the application of RE-doped glasses in light amplifiers and in NIR and MID IR lasers.

Luminescence in RE-doped AGCh proceeds among inner electron f-f transitions that can be excited either directly, or via the absorption of light in a glassy matrix with the following transport of the energy from the glassy matrix to the ions of RE. The 4f electrons of RE ions are placed below the bonding 5d and 6s electrons. These 5d and 6s electrons of RE screen the influence of the glassy matrix on the positions of energy levels of RE in chalcogenide glasses and amorphous films. The influence of glassy matrix is therefore not very significant.

The Ch matrixs have lower energy of phonons than that in oxides and halides; the number of phonons that should bridge the energy gap between the electron energy levels of RE ions for non-radiative electron transitions is therefore higher. The nonradiative decay rate, ω_p, which lowers the luminescence intensity due to multiphonon relaxation, depends on the energy gap between the energy levels of a corresponding RE transition, ΔE, and phonon energy, $\hbar\omega$. It is given by Miyakawa-Dexter equations (see e.g. [115]). At low temperatures we can say that

$$\omega_p = \omega_0 \exp\left(-\frac{\alpha \Delta E}{\hbar\omega}\right), \quad \alpha = \ln(p/g) - 1, \tag{7.21}$$

where $p = \Delta E/\hbar\omega$, g is the electron–phonon coupling strength, and ω_0 is the host-dependent constant. The probability of non-radiative electron transitions in AGCh is higher; therefore, much lower than in oxides with higher phonon energy. It results in a higher efficiency of NIR and MID IR luminescence in Ch. Due to low phonon energy in Ch, intensive luminescence can be observed even in technically very important NIR- and MID IR spectral regions.

Non-radiative transitions of electrons, which lower the intensity of emission, are more intensive in oxides than in Ch (Eq. 7.21), because the phonon energies in oxides are higher than those of Ch (sulphides ~340 cm^{-1}, selenides ~200 cm^{-1}, halides $\geq$ 500 cm^{-1}, oxides $\geq$1,000 cm^{-1}). Therefore, Ch form a relatively good matrix for RE doping [103, 112–117]. However, the solubility of RE in chalcogenides often is relatively low. It can be higher in Ga-, In- and Sb- containing glasses, or in chalcohalides (glasses containing some halogenides). According to Raman spectroscopy results [118], Ga in sulphides form ($GaS_{4/2}^{-}$) tetrahedra with one negative charge. Such tetrahedra were also supposed to be formed in GeS_2-Ga_2S_3 glasses by [119]. The negative charge of such centres can attract RE^{3+} ions and compensate their positive charge. At the same time, the RE^{3+} ions are connected via these $(GaS_{4/2})^{-}$ tetrahedra with a covalent chalcogenide matrix.

Similar effect may also be present in negatively charged halides of chalcohalide glasses (e.g. the AGCh with AgI) [120]. Chalcohalides are formed from

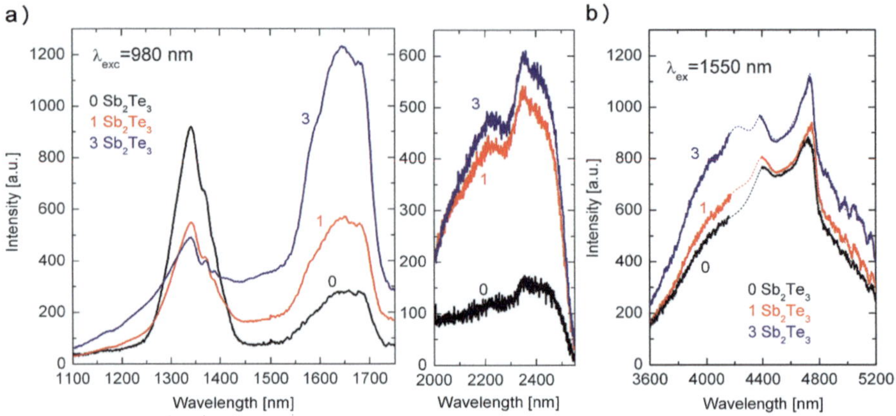

Fig. 7.39 Spectral dependence of intensity of luminescence of $80GeSe_2$ (20-x)Sb_2Se_3 xSb_2Te_3:$0.03Pr_2Se_3$ glasses; x = 0, 1,3 excited by 980 nm (**a**); dtto excited by 1550 nm (**b**). The luminescence in the spectral region 3,600–5,200 nm region is important for the detection of many chemical compounds [103], adapted

chalcogenides by doping with halides. Halide ions possess negative charge which can attract or surround RE^{3+} ions, compensate their positive charge and increase their solubility in AGCh. Chalcohalides have, unfortunately, higher phonon energy (<550 cm^{-1}) than chalcogenides and, therefore, higher probability of non-radiative electron transitions (Eq. 7.21), which negatively influences the intensity of luminescence. On the other side, if heavy-metal halides were used, the presence of heavy metals (Ag, Pb) and heavy halogens (e.g. iodides) increases the index of refraction; the intensity of luminescence transitions increases, too (Eq. 7.20). Non-radiative transitions have the opposite effect. The optimal composition shall be chosen by compromise.

Rare-earth elements have strong luminescence emission not only in the visible, but also in the NIR and MID IR parts of the spectrum (Figs. 7.39a, b [103]). While oxide glasses have limited transparency in the IR region, some chalcogenides are well transparent up to 25 μm (Fig. 7.37).

Chalcogenide glasses and thin films generally possess a higher index of refraction than oxide and halide glasses due to a higher polarizibility of their larger atoms. It supports higher values of spontaneous emission probability and causes larger emission cross-sections of radiative electron transitions between energy levels of RE (Eq. 7.19). All these effects are positive for a higher intensity of luminescent radiation and allow for the preparation of light amplifiers and lasers in a technically important IR region.

Intensive IR luminescence has been found not only in bulk glasses, but also in crystals and thin films doped by RE. Many chalcogenides doped by RE have been studied, e.g. As-S:Sm and As-In-S:Sm, Ge-Sb-Se:RE, Ge-Ga-Sb-Se:RE, Ge-Sb-S-PbX_2:RE, As-In-S:RE, Ge-Ga-Se:RE, Sb-S:RE, Ge-Sb-Se-Te:Pr, Ga-La-S:Re, Ga-La-Se:RE, where RE stands for rare-earth elements or ions, e.g. for Yb^{3+} or/and

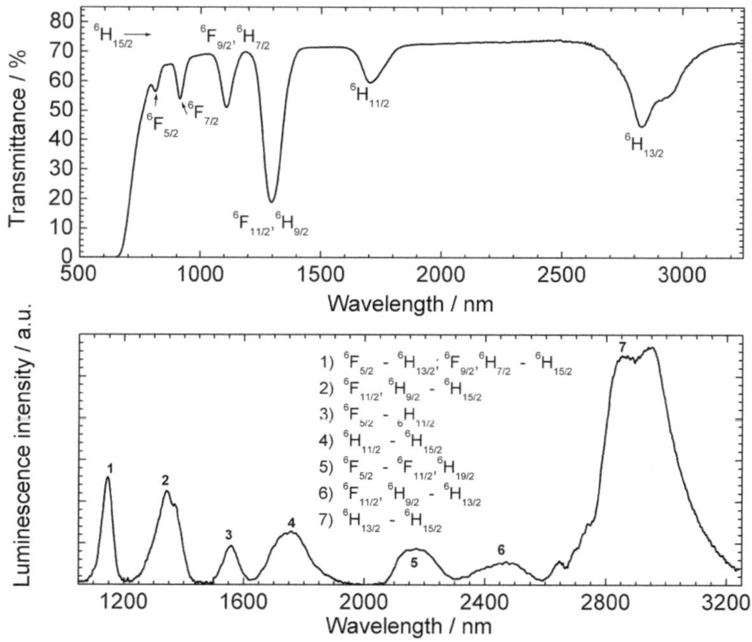

Fig. 7.40 The room temperature transmission and luminescence spectra of a $(Ge_{30}Ga_5Se_{65})_{99.8}$ $(Dy_2Se_3)_{0.2}$ glass. The luminescence was pumped by 808 nm light [121]

Tm^{3+}, Tb^{3+}, Dy^{3+}, Ho^{3+}, Pr^{3+}, Sm^{3+}, Er^{3+}, etc. Part of their content of S or Se can be substituted by Te atoms in order to shift the luminescence band position further in the IR region.

Examples of some absorption and emission bands in Dy or Sm doped chalcogenide glasses are given in (Figs. 7.40 and 7.41) [121, 122].

Doping of chalcogenide glasses by RE does not influence the IR and Raman spectra of the matrix glass as can be expected, because the density of RE ions dissolved in a glassy matrix is usually low, at the level <1 at.%. In spite of that, the IR absorption bands can be well pronounced (Figs. 7.40 and 7.41).

RE-doped AGCh can serve as IR light emitters in optics, optoelectronics, chemistry, in environment protection; as lasers in medicine (e.g. lasers for bloodless operations, drilling, and eye-safe lasers), IR amplifiers, fibres, night vision, defense facilities, etc. Thin films doped by RE play an important role in the development of planar optical integrated circuits and their components.

The PL of RE-doped Ch can also be used for up-conversion of light frequency, e.g. for increasing the efficiency of solar photovoltaic panels. Part of the sunlight is emitted in the NIR region, behind the sensitivity of Si solar cells ($\lambda \geq 1{,}100$ nm) and it is lost for photovoltaics (Fig. 7.42), [123]. This part of sunlight can be up-converted to shorter wavelengths, in which Si cells are sensitive. Such a process requires a two-IR-photon absorption. The efficiency, η, of such a process is not very

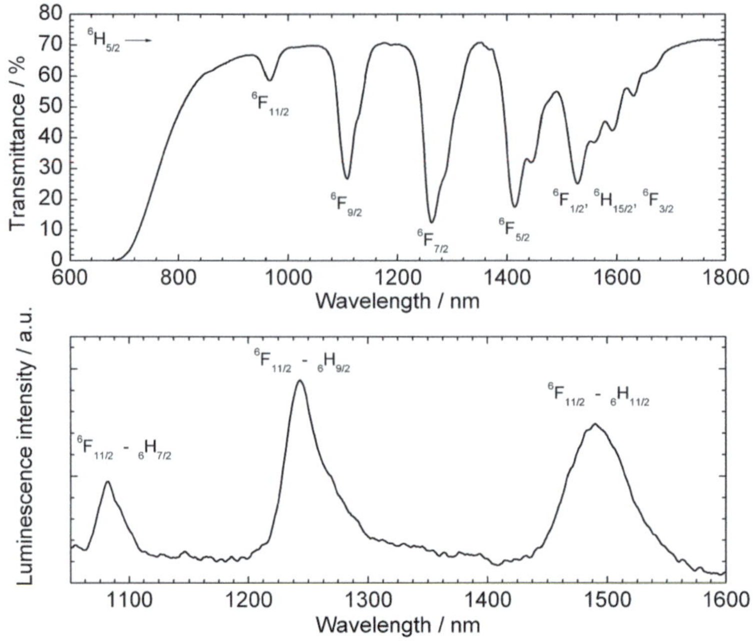

Fig. 7.41 The room temperature transmission and luminescence spectra of a $(Ge_{30}Ga_5Se_{65})_{99.8}(Sm_2Se_3)_{0.2}$ glass. The luminescence was pumped by 980 nm light [122]

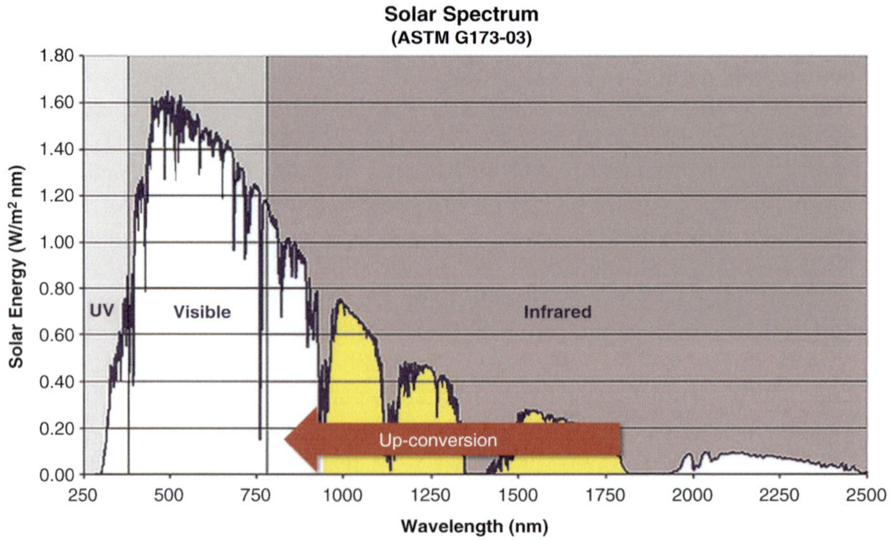

Fig. 7.42 Spectral dependence of Sun emission ([ASTM G173-03] [123], adapted). The *yellow* part is not absorbed in a crystalline silicon

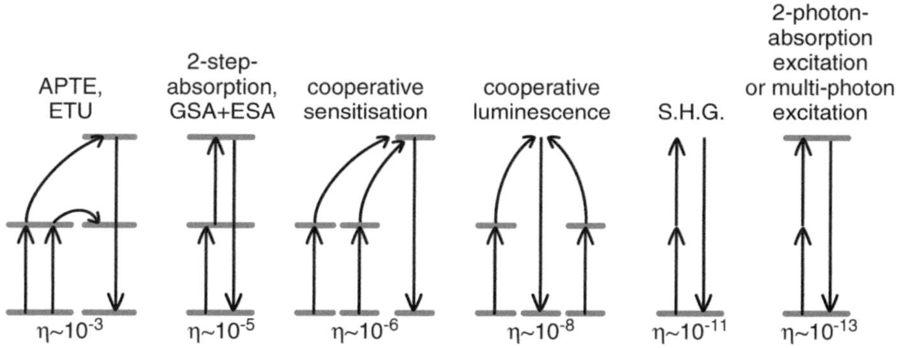

Fig. 7.43 A simplified scheme of up-conversion processes in solids. η represents maximum efficiency of individual processes. The most efficient one is a so-called energy transfer up-conversion (*ETU*) [124, 125]

high [124, 125] (Fig. 7.43); neither could be the increase in photovoltaic gain. As the Sun is a "green" source of energy, even relatively little increase in their efficiency can be important.

A detailed description and theoretical calculation of the different processes of up-conversion were given by Auzel [125]. According to him [125], the most efficient process is a so-called energy transfer up-conversion process (ETU) (Fig. 7.43). The next most efficient process is two-step absorption, described as ground- state absorption (GSA) followed by excited state absorption (ESA). Two-step absorption and cooperative sensitization take place via real, existing, energetic levels with higher efficiency as compared with processes that occur via virtual levels. The other processes in Fig. 7.43 have lower efficiency. The horizontal lines represent real existing energy levels. The vertical arrows indicate the excitation or de-excitation of the energy levels. In reality, many of the processes shown in Fig. 7.43 take place simultaneously in the up-convertor.

Most of "up-convertors" have similar structure; they contain an active ion, often an RE ion, which provides energy levels for absorption and de-excitation, and a host material, where the ion is embedded in AChF. AChF doped by RE usually are evaporated on the background of the Si photovoltaic cell.

An example of optical properties of a Ga-Ge-Sb-S glass doped by Er^{3+} pumped by 1,550 nm light, which shows several up-conversion bands, is given in (Fig. 7.44 [126]).

7.5.2 *Photoinduced Changes of Chalcogenides' Structure and Properties*

When a semiconducting Ch is exposed by absorbing light, X-rays, or electrons, its electrons can be excited to different higher levels. They can increase the

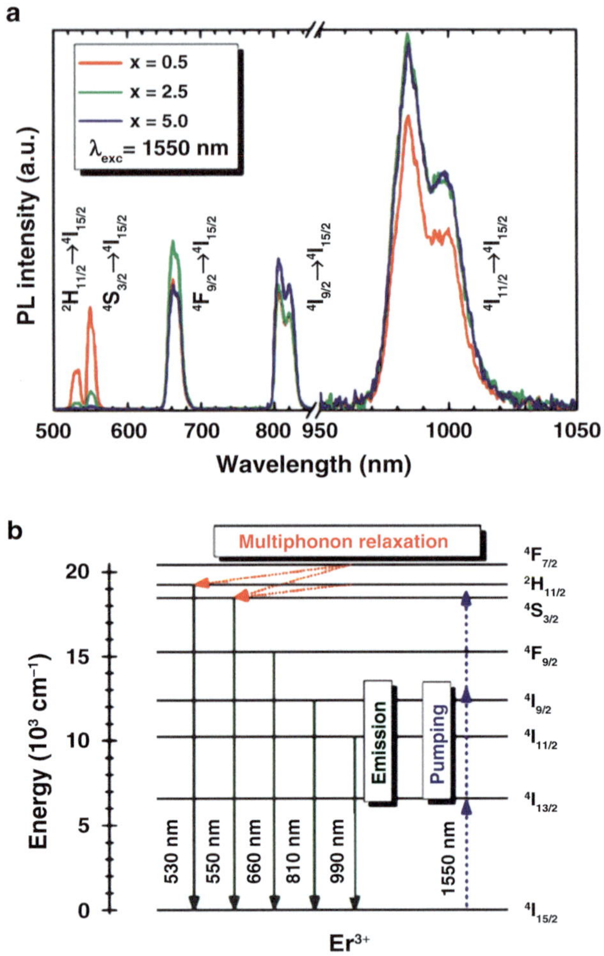

Fig. 7.44 Up-conversion bands induced by pumping with 1,550 nm light in a Ga-Ge-Sb-S glass doped by Er^{3+} (**a**), an energy diagram of individual luminescent transitions (**b**) [126]

photoconductivity or, when they are excited near a p-n junction, they can create local electrical fields (the photovoltaic effect). After a certain time period, the excited state relaxes by electron transitions among the localized energy levels in the forbidden gap, between conductivity and valence bands, or between local electron levels and the valence band. They can also relax between energy levels of a dopant. The difference in the energy of both states of these transitions (the excited and the lower one) can be transferred to heat by non-radiative transitions or irradiated as luminescence in different parts of the spectrum. Such a process can also induce light amplification and generation; it can change the index of refraction, create non-linear or plasmonic effects.

Excited free carriers can create or lower local electrical fields in the illuminated material. This effect has been used in Xerox facilities and in laser printers.

Electronic photoinduced effects have also been applied in saticon, vidicon, harpicon, X-ray and other sensors and panels, including solar ones (see also Sect. 7.6).

The excited state can also allow for changes of the atomic structure of crystals, of glasses, or of amorphous thin films (see e.g. [17–19, 21–24, 26, 29, 30, 32–35, 65, 69–71, 73, 74]). Excited electrons in the antibonding (conduction) band lower the chemical bond order, induce or support chemical reactions, or make structural changes easier. The energy of chemical bonds in chalcogenides is generally lower than in oxides or halides. The energy gap between valence and conduction bands, E_g, also is lower. As a consequence, the irradiation of AGCh in the visible and ultraviolet (UV) parts of the spectrum can interrupt some chemical bonds and induce or support chemical reactions (dissociation, decomposition, photolytic and photosynthetic reactions). It can also induce or support dissolution and diffusion of elements or ions (e.g. Ag, Cu) in ACh layers, or interdiffusion of layers [34]. Consequently, it can change not only the electronic, but also the atomic structure and change many physical and chemical properties of AGCh [17–19, 21–24, 29, 32–35, 69–71, 90, 92, 111, 127–132]. Such changes can be reversible, which means that the original state can be renewed either by the interruption of illumination, or by the annealing of the sample at higher temperatures (in AGCh it is usually around T_g or above, when atoms are partly movable). The original state of reversible processes can also be received by illumination by light of lower intensity or longer wavelengths. Such photoinduced effects in AGCh have been extensively studied in the last 40 years due to their present and potential applications in optics and optoelectronics, data storage, optomechanics, etc. They, or their parts, have been reviewed several times, see above.

Photostructural effects in solids are numerous, with different mechanism [17–19, 24, 29–35]. They are generally stronger in AGCh than in crystalline ones, as already mentioned. In amorphous solids, there is more space for atomic movement (higher free volume [13]); so changes of structure are easier. In crystalline compounds, optically induced changes of atomic structure are mostly confined to the surface of crystals or to its vicinity, or to the neighbourhood of crystal defects. Only rarely does the dense packing of atoms in crystals allow for greater changes in bond and atom distributions. In most cases, photoinduced changes in crystals are little or relax back after the exposure is finished, or they only proceed on the surface. A detailed description and discussion of these effects in CCh is beyond the scope of this paper; the *photoinduced changes of atomic structure of AGCh* are stressed. Disorder in glasses and especially in thin amorphous films of AGCh is high. Broader composition regions of glasses and a-thin films, as compared with crystals, are favourable for formation of many compositional and coordination defects, which are favourable for the creation of photoinduced effects.

Photoinduced atomic changes include the changes of short-range, medium-range and long-range orders, changes of surface properties, of chemical reactivity; they can induce chemical reactions such as photolysis or photosynthesis, photoinduced oxidation or photoinduced hydrolysis. Illumination can lead to photo-enhanced dissolution of metals (Ag, Cu), diffusion of elements or ions, and inter-diffusion between the layers (e.g. Bi + Se, Sb_2Te_3 + Bi_2Se_3, etc.), changes of viscosity [17–19, 22–24, 32–35]. Photoinduced atomic changes of structure and physico-chemical properties can lead to scalar and vectorial photoinduced effects depending on the polarization of exciting light. The mechanism of many of these photoinduced processes has not been fully clarified to date.

Photoinduced phenomena in AGCh have been found and described in many Ch, e.g. in binary and ternary systems As-S, As-Se, As-Se-I, As-Te, Sb-S, As-Ga-S, As-S-Te, Ge-Sb-S, Ge-Ga-S, Ge-Sb-Se, Ge-Sb-Te, Sb-Te, Ge-Te, Ge-Bi-Te, Ge-In-Te, Ag-As-S, Ag-As-Se, Ag-Ge-S, Ag-Ge-Se, Ge-S, Ge-Se, Bi-Se, and in many others, also in amorphous and crystalline Se.

Special and technically very important phenomena are photoinduced phase changes, especially the crystallization of ACh and amorphization of CCh, in PCMM (see e.g. [17–19, 21–24, 32–35, 131–141] and the papers quoted). They are discussed separately in Sect. 5.2 and Sect. 7.6.2.

A considerable effort has been given to the study of *photoinduced effects in thin films of As-S, As-Se*, pure and doped (see e.g. [17, 19, 21–24, 32, 34, 82]). Arsenicum sulphides and selenides have been selected by many researchers for several reasons: arsenic-containing AChF are relatively volatile and their thin films can be prepared easily by classical thermal evaporation without great changes of composition; evaporation does not need high temperatures; many photoinduced changes in these systems are strong enough; changes of structure can be followed relatively easily by Raman and IR spectroscopies; both systems can also serve as models for other AGCh.

Energies necessary for optically induced or optically enhanced dissociation of many chemical bonds or whole compounds are relatively low in AGCh, especially in As-S and As-Se systems, because the energy of the M-M and MY, where Y is for chalcogen, chemical bonds is not very different (Table 7.1, Eq. 7.7).

The exposure or annealing of an amorphous or glassy chalcogenide can shift the equilibrium of the reaction (Eq. 7.7) to the right- and to the left-hand side, respectively, depending on the system, its temperature, energy and intensity of light. For $h\nu_1 > h\nu_2$ and for $I_1 > I_2$, the reaction (7.7) in the As-S system is shifted to the right-hand side and the material is photo-darkened; by annealing ($T_2 > T_1$), or for $I_1 < I_2$, the reaction (7.7) is shifted to the left-hand side and photobleaching is observed. Illumination of some selenides or tellurides, which are bad glass formers, can cause crystallization rather than dissociation.

Photoinduced dissociation (photolysis) of a thin As_2S_3 film can form molecular fragments of lower molecular weight, e.g. according to a simplified reaction (Eq. 7.8), n >> 1. The long chains can be interrupted and shortened. The formation of the As_4S_4 molecular compound can be observed in Raman spectra (Fig. 7.45), [142].

Exposure of as-evaporated non-homogeneous films of a-chalcogenides, e.g. of As-S system films, with light of lower intensity and of longer wavelengths, can induce synthetic chemical reactions among the fragments. It results in chemical homogenization of as-evaporated films and also in their polymerization (Eq. 7.8,

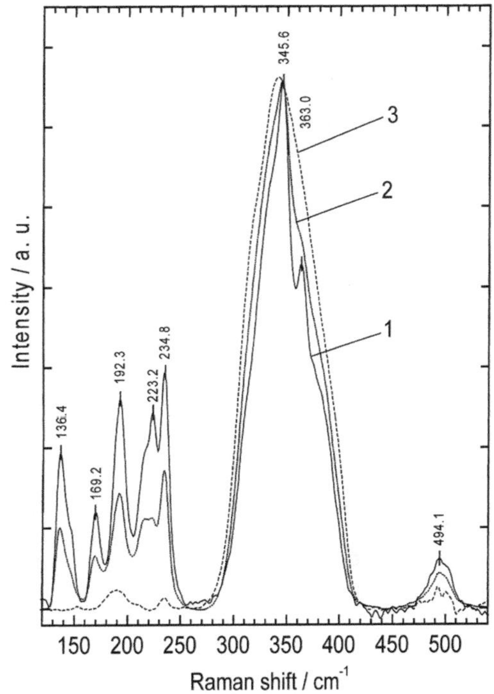

Fig. 7.45 Raman spectra of $As_{38}S_{62}$ films [142]. (*1*) the fresh evaporated film, (*2*) the exposed film and (*3*) the annealed film. The background was corrected, the intensity was normalised to the intensity of 345 cm^{-1} band

n >> 1, a backward chemical reaction). This process is accompanied by decrease of amplitudes of some bands of the Raman spectrum that correspond to As-As (235 cm^{-1}), S-S bonds (495 cm^{-1}), As_4S_4 vibrations (363 cm^{-1}, 169 cm^{-1}, 192 cm^{-1}, 228 cm^{-1}; Fig. 7.45, [142–144]). Similar photostructural and photochemical effects as in the As-S system have been found in As-Se, As-Ga-S, Ge-Sb-S, Sb-S systems [17, 34, 143]. Contrary to that, the exposure by light of short wavelengths and high intensity supports a forward reaction of Eq. 7.7.

Over-coordinated and under-coordinated atoms are very common in chalcogenide glasses. These defects can also be created, or their density enhanced, by irradiation [17, 128, 142].

Changes of optical transmission, photodarkening and photobleaching. Exposure of evaporated films shifts the absorption edge either towards longer wavelengths (the red shift, photo bleaching) or towards shorter wavelengths (the blue shift, photo darkening) (Fig. 7.46 [142]). The explanation of these shifts based on photosynthetic or photolytic reactions was proposed in [17, 34], see also above. It is supported by the fact that the absorption edge of As-S glasses is shifted with a change of stoichiometry analogically to photo-darkening. (Fig. 7.47 [128]). Tanaka connected the processes of photo-darkening with the creation of new defect levels inside the forbidden gap, or with broadening of the valence band but without any specification of the defect composition and structure [19, 22, 24]. The red or blue shift during

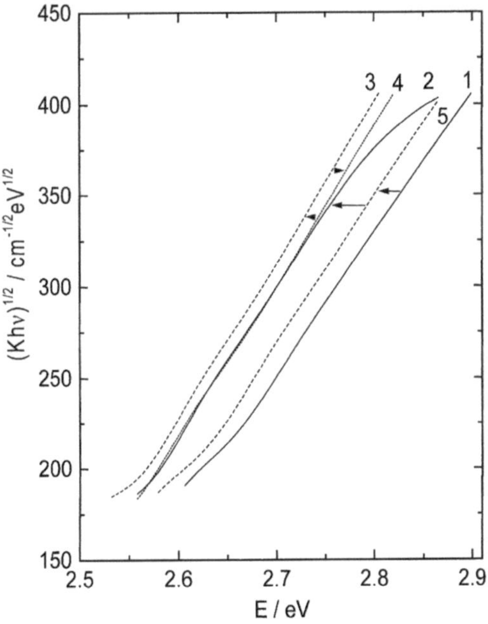

Fig. 7.46 Darkening – bleaching in As_2S_3 films. Dependence of $(Kh\upsilon)^{1/2}$ on photon energy [142]. *K* is the absorption coefficient (cm^{-1}), $h\upsilon$ is photon energy. *1* the fresh evaporated film, *2* the annealed film, *3* the film exposed after annealing, *4* the film was annealed, exposed and again annealed, *5* the exposed film

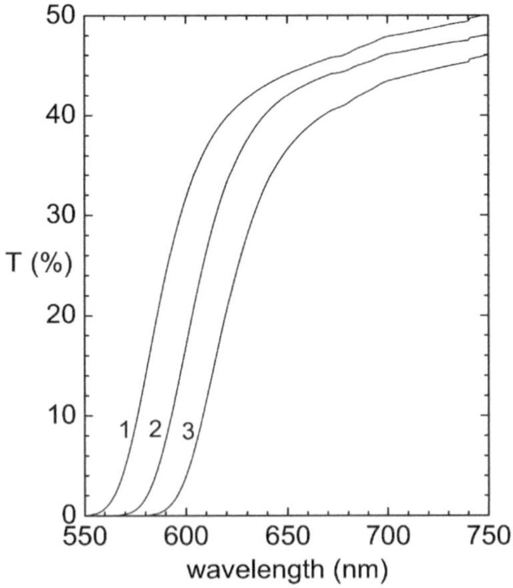

Fig. 7.47 Transmission spectra of As-S glasses: *1*. $As_{38}S_{62}$, *2*. $As_{40}S_{60}$, *3*. $As_{42}S_{58}$ [128]. The short-wavelength absorption edge is shifted not only by photodarkening (Fig. 7.48), but also by the change of composition (Fig. 7.49). The higher As content shifts the absorption edge towards longer wavelengths

photodarkening or photobleaching is not only composition, but also temperature-dependent, Fig. 7.48 [19]. The shift ΔE decreases with temperature. At temperatures close to T_g, the value of ΔE approaches zero. Such behaviour is similar in many different Ch (Fig. 7.48).

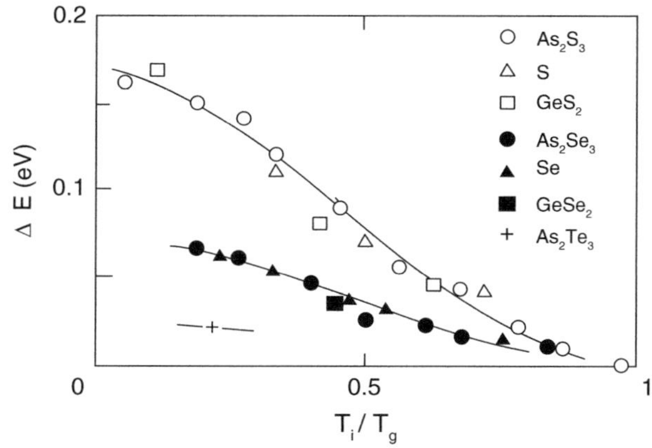

Fig. 7.48 Magnitudes of the red shift ΔE as functions of (a) temperature T_i normalized to T_g for several stoichiometric chalcogenides listed [19], p. 154. The photoinduced defect density decreases with increasing temperature and near T_g the density is close to zero. The mobility of atoms at this temperature is higher and the backward reaction (Eq. 7.7) is relatively fast

The short-wavelength absorption edge is shifted not only by photodarkening (Fig. 7.46), but also by change of composition (Fig. 7.47). The higher As content shifts the absorption edge towards longer wavelengths [142].

The mechanisms of isotropic irreversible photoinduced changes are relatively well understood. They can be explained by a simple chemical reaction described in Eqs. 7.7, 7.8, 7.22 by changes of a local bonding configuration due to photochemical reactions:

$$|\mathrm{As}-\mathrm{As}| + |\mathrm{S}-\mathrm{S}| = 2\,|\mathrm{As}-\mathrm{S}|\,. \tag{7.22}$$

The vertical lines in Eq. 7.22 denote the As-As, S-S and As-S chemical bonds. This reaction is temperature-dependent and the equilibrium constant K can be expressed by (Eq. 7.23)

$$K = \frac{[As-S]^2}{[As-As]\;[S-S]} = k\exp\left(\frac{\varepsilon}{kT}\right). \tag{7.23}$$

The values in square brackets (Eq. 7.23) represent the density (concentration) of individual chemical bonds. The energy ε is given by $\varepsilon = 2E_{\text{As-S}} - E_{\text{As-As}} - E_{\text{S-S}}$, where $E_{\text{X-Y}}$ are chemical bond energies between atoms X and Y.

This model is supported by Raman spectroscopy results and it is valid at least for As-Ch, Sb-S, Ge-Sb-S systems [34].

Changes of the index of refraction accompany the changes of structure and can also be induced by exposure. The index of refraction of a-chalcogenides can also be increased due to photodoping of AGCh by metals [17, 22, 24, 29, 32–35], e.g. by Ag, Cu, Sb and Al. Changes of the index of refraction cause the changes of the third-order non-linear susceptibility $\chi^{(3)}$ constant ([89, 90, 101], see also Sect. 5.1).

Exposure of fresh-evaporated and annealed AChF also changes their *reactivity and their solubility in chemical solvents,* see e.g. [17, 34, 69–71]. Generally, the differences in dissolution rates of exposed and unexposed parts are not significant.

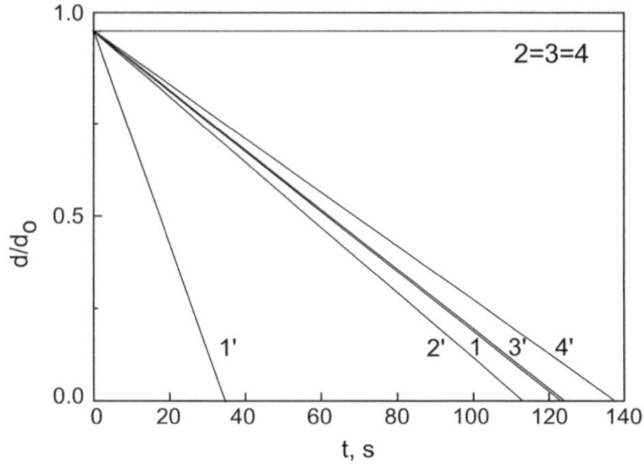

Fig. 7.49 Kinetics of the dissolution of an as-evaporated and exposed thin As_2S_3 film in an aqueous solution of Na_2CO_3 and Na_3PO_4 containing methol (pH = 12) [34]

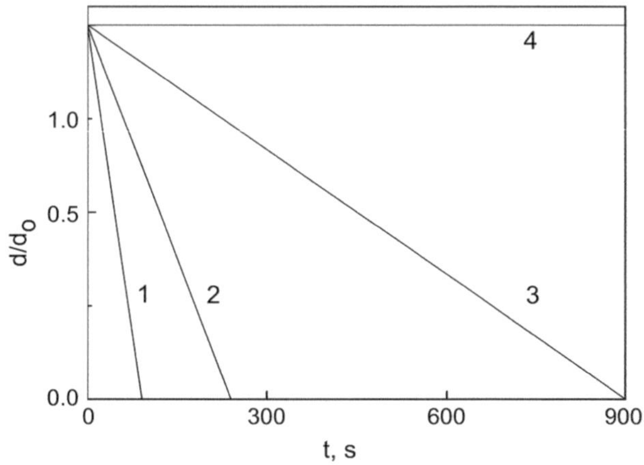

Fig. 7.50 Dissolution rates of a thin As_2S_3 film in CS_2 with I_2 [34]

These differences can be increased by using specific solvents and reactants [17, 34, 69–71, 127] (Figs. 7.49, 7.50, and 7.51). Dissolution rates can be influenced e.g. by presence of oxidizing (CS_2 +I_2, Fig. 7.50), or reducing agents (methol in the solution, Fig. 7.49, [34]) or by adsorption of quaternary alkyl-ammonium salts (Fig. 7.51).

Such a process can be illustrated by dissolution of arsenic-sulphur system films. As we have mentioned above, fresh-evaporated films of the As-S system contain a relatively large number of As_4S_4 molecules (Fig. 7.26, Raman bands near 362, 168, 189, 222 cm^{-1} [34, 128, 144]).

The dissolution of As_4S_4 in alkaline solutions is relatively slow, partly because of the elemental arsenic formed on the surface of the film, partly because As_4S_4 is relatively resistant to alkaline solutions:

$$3As_4S_4 + 24OH^- = 4AsS_3{}^{3-} + 4AsO_3{}^{3-} + 4As + 12H_2O. \tag{7.24}$$

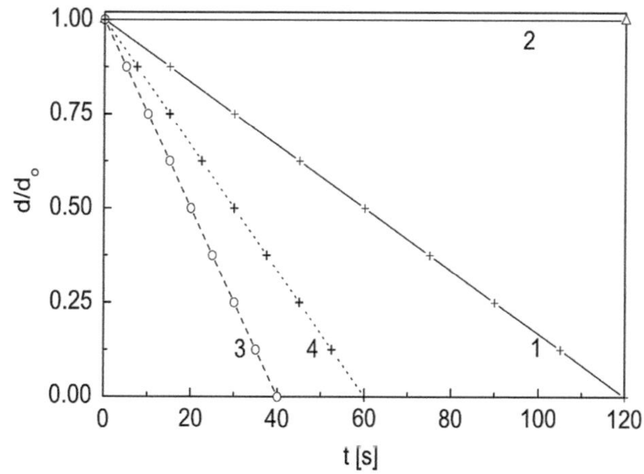

Fig. 7.51 A change of thickness of thin films of $As_{42}S_{58}$ during their dissolution in alkaline solutions [34]. *1* fresh evaporated film, *2* fresh evaporated film with chemisorbed cetyl trimethyl-ammonium bromide (CTAB), *3* exposed film, *4* exposed film with chemisorbed CTAB. *d* is the actual thickness of the film, d_0 is its original thickness

Dissolution of As_2S_3, which is the prevailing compound in the exposed parts of films, is much quicker

$$As_2S_3 + 6OH = AsO_3{}^{3-} + AsS_3{}^{3-} + 3H_2O. \tag{7.25}$$

Reducing agents added to the alkaline etchant can increase the content of As on the surface of a film and slow down the dissolution. The dissolution blocking effect of As deposited on the surface of a film was confirmed, when a thin film of As was evaporated on the surface of a chalcogenide film [142].

Oxidizing agents speed up the dissolution of unexposed films when oxidizing As_4S_4 probably according to a simplified reaction (7.26) (Fig. 7.50 [34]).

$$3As_4S_4 + 6I_2 + 48OH = 4AsS_3{}^{3-} + 12I + 8AsO_3{}^{3-} + 24H_2O \tag{7.26}$$

The dissolution rates of films with overstoichiometric excess of As (e.g. of $As_{42}S_{58}$) in alkaline solutions are lower than those of stoichiometric films. It supports the idea presented above (Eq. 7.24); films and bulk glasses with excess of arsenic contain a greater amount of As_4S_4 and of As-As bonds which are poorly soluble in alkaline solutions [34].

The solubility of fresh-evaporated and exposed films can be influenced drastically by absorption or chemisorption of some organic compounds, e.g. of aryl- or tetra alkyl-ammonium salts [34], in which at least one alkyl group is long enough (e.g. R = cetyl-, $C_{16}H_{33}$-). Cationic parts of these compounds are e.g. $(R(R')_3\,N)^+$, where R′ stands for other alkyl or aryl groups. This cationic part can be selectively chemisorbed on the unexposed parts of films; As-rich parts of a film are apparently preferred. By such sorption, dissolution of unexposed films can be stopped altogether (Fig. 7.51) [34]. The effect has not been elucidated so far. The nitrogen of $[NR_4]^+$, or $-[NR_3]^+$, $=[NR_2]^+$ and similar groups, where a horizontal line or two lines are for one or two chemical bonds, is positively charged and can be bonded to As-As structural units, which acts as Lewis' bases. The long alkyl (aryl) chain of alkyl-ammonium cations (e.g. cetyl) is hydrophobic; when the compound is chemisorbed, it protects the surface against the attack of an alkaline aqueous solution.

Positive and negative etching of ACh is possible [17, 33, 34, 128]. Etching of As-S thin films is positive in alkaline solvents, (the exposed part is dissolved more quickly), while the films with excess As, which were treated by CS_2 before etching, are etched negatively; the unexposed part of the film was dissolved more quickly. CS_2 can dissolve the free sulphur or sulphur-rich fragments of the film that are present in fresh evaporated films (Eqs. 7.22, 7.23, Fig. 7.50). The surface area of the unexposed part of the film increases considerably after such dissolution and can be dissolved in alkaline solvents more quickly [128].

The processes of dissolution of a-chalcogenides in reactive solutions are irreversible. They are important for the production of high-resolution photoresists, gratings and planar circuits [17, 34]. The commonly used polymethylmethacrylate polymer photoresists typically work for lines of the width >20 nm, while AChF can work even in smaller sizes and can have better contrast [69–71]. Very high resolution (≤20 nm, 5,000–10,000 lines/mm) can be obtained. ACh photoresists are more sensitive to X-rays and e-beam than those of organic polymers. Sensitivity can even be higher if pulsed laser light is applied [34]. The topic is also mentioned in Sect. 7.6.

The *photoinduced dissolution of silver and other metals* (e.g. Cu, Sb, Al) in chalcogenides also is very important for different applications; the reactivity of such films is also changed by doping. Ag-doped films are practically insoluble in alkaline solvents and can be used as photoresists, waveguides and planar optical circuits. The issue was reviewed in detail in Refs. [17, 33–35].

Structure and many physico-chemical properties of AGCh can also be changed by exposure reversibly (Fig. 7.52, [17, 34, 145]). *Reversible photoinduced changes of structure and properties* are found usually in bulk glasses, in well-annealed films or in films after their exposure by sub-band-gap light of lower intensity.

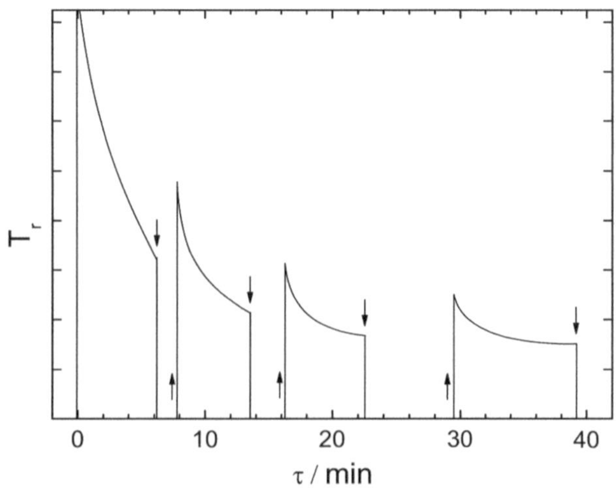

Fig. 7.52 A change in the transmissivity of annealed As_2S_3 thin films during the illumination at a constant intensity by 514.5-nm laser light. The *vertical arrows* correspond to the closing (↓) and opening (↑) of a light shutter. The measurements were made at room temperature [145]

The exposure of annealed a-chalcogenides also *changes reversibly the index of refraction*. The changes are much less significant than the irreversible ones. Part of the changes of the index of refraction, e.g. in arsenic chalcogenides, can be ascribed to changes of the volume (photoinduced density change).

The exposure of some amorphous chalcogenides can cause *photo-contraction and photo–expansion* [19]; after the exposure, the thickness (volume) is decreased or increased, respectively. Great photo-contraction has been seen in obliquely deposited films [146]. The angle between the vapour direction and the substrate was <90°, the most significant effects have been found for angle ≈40°. Obliquely deposited films possess a column-like structure with a large space between columns. After the exposure, the columns can collapse and interact among themselves; a more homogeneous film of higher density and lower thickness is then obtained. When an obliquely deposited film with a columnar structure was annealed before the exposure, or the vapours were deposited on a heated substrate, the photo contraction was lowered or it diminished altogether.

Photoinduced expansion has been found in many amorphous chalcogenides [17, 19, 22, 30, 147]. When the exposure was made by light of wavelengths corresponding to the Urbach tail of the absorption edge of ACh (the sub-band gap region, Fig. 7.35), the increase of thickness can be substantial and so-called "giant photo-expansion" was observed [19]. The a-film exposed through a small hole forms a dome due to photo expansion. This effect can be applied for micro-lense fabrication. The effect can be explained by changes in bond redistribution. In crystalline As_4S_n (n = 3, 4), and probably in the whole amorphous As-S system, As-As bonds are longer than As-S bonds (for example, in As_4S_4: As-As = 0.259 nm and As-S = 0.221 nm [145]). It is supposed [145, 147] that the band-gap illumination facilitates the breaking of a small fraction (≈4–8 %) of As-S bonds creating longer As-As bonds according to Eq. (7.22). The changes of bond lengths that are perpendicular to the substrate plane can contribute to the layer thickness change ΔX, $\Delta X = \Delta b.\sin\varphi$, [145], φ is the angle between a given bond direction and the substrate plane. The change in the bond length per bond can be given by $\Delta b = ½(b_{As\text{-}As} + b_{S\text{-}S}) - b_{As\text{-}S}$. The mean value of the layer thickness change $\overline{\Delta X}$ is given by [145]

$$\overline{\Delta X} = \int_0^{\pi} sin\varphi \; d\varphi = 2\Delta b/\pi. \tag{7.27}$$

For the bond lengths quoted above and for $b_{S\text{-}S} = 0.22$ nm, we obtain [145] $\overline{\Delta X/X_o} = 0.068$ for changes of all bonds. When compared with the experimentally found thickness increase, the fraction of As-S bonds changed by exposure = 6 % of the total amount can be evaluated. This value corresponds to other estimations [148]. Changes of the volume and surface profile by exposure can be used for the preparation of micro lenses or micro-lense arrays and can be applied in CCD cameras and in imaging machines.

Photoinduced optomechanical effect has been observed during the band gap illumination of amorphous chalcogenide films deposited on the top of a cantilever of an atomic force microscope [129, 130]. Exposure by polarized light can cause

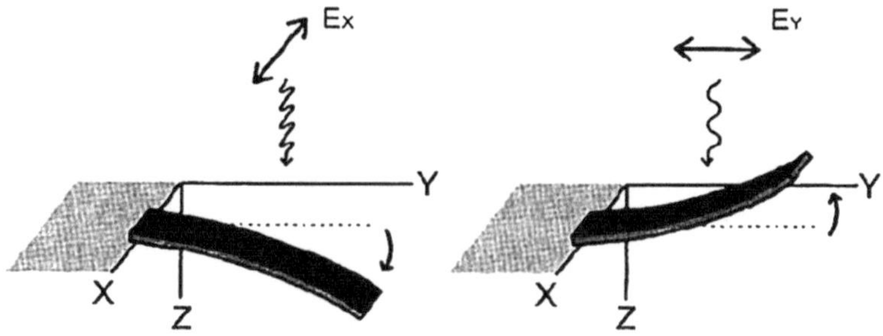

Fig. 7.53 Photoinduced photomechanical effect [129]. The cantilever of an atomic force microscope is covered by a thin film of Ch and illuminated by polarized light of a different polarization plane. The cantilever moves up and down by the change of polarization plane caused by a contraction or expansion of the Ch layer. The position of the cantilever can be controlled optically

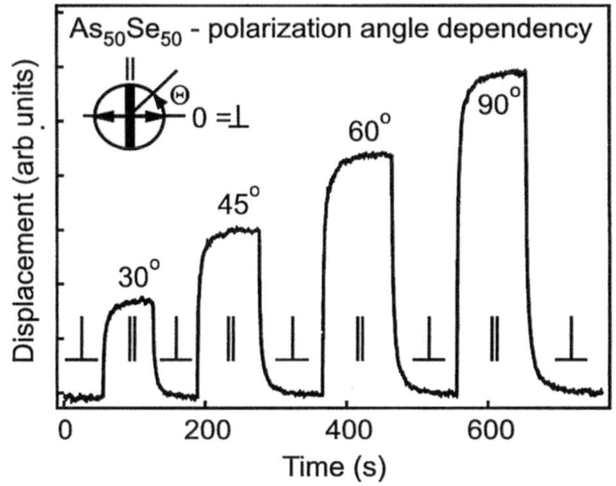

Fig. 7.54 The dependence of the deflection of a cantilever, coated with a-$As_{50}Se_{50}$, as a function of the angle Θ between the electric vector of the illuminating linearly polarized light and the normal one to the cantilever axis (see inset) [130]

either photo contraction or photo expansion depending on the polarization plane of light. The top of a cantilever can be moved up and down by changing this plane of polarization. The effect is highly reversible (Figs. 7.53 and 7.54) [129, 130]; it can be applied in mechanical devices controlled by light.

Optically induced changes of viscosity have been observed in a-Se and in several other Ch. In a-Se, they have been described as so-called photo melting [149], because the surface profile of the bulk Se was changed during the exposure even at low temperature (77 K), when the material was exposed. During the exposure, the sample of an Se film was immersed in a liquid N_2, so that thermal effects and heating to T_g could be eliminated. In an amorphous selenium, the decrease of viscosity and changes of the surface profile are explained by the interruption of weak van der

Waals bonds among the Se chains due to the exposure. Optically induced changes of viscosity have also been found in As_2S_3 [149]. The energy of excitation light used in [149] was lower than the energy gap, $\hbar\omega < E_g$, the light was not absorbed in a thin surface layer and penetrated deeper into the amorphous solid. It is supposed that light can excite and break van der Waals bonds and excite part (tail states) of valence band electrons.

In As_2S_3 bulk glasses, *photoinduced changes of hardness have been observed.* The films were hardened by annealing and softened by exposure.

Reversible photoinduced changes are generally dynamic, depending on the temperature and composition of the sample, e.g. photoinduced darkening can be lowered by heating to higher temperatures. Near T_g, photodarkening is of a low value, approaching zero (Fig. 7.48).

In some systems, photoinduced darkening diminishes (optical transmissivity increases) after the interruption of the excitation even at room temperatures (As-S system, Fig. 7.52, [145]).

The exposure of As_2Se_3 PLD films *increases the index of refraction.* The following annealing of these films decreases the index of refraction partly due to thermally induced expansion of amorphous layers.

When non-polarized light is used for the exposure of isotropic AGCh, the final state of the exposed material is optically isotropic. After the exposure made by polarized light, the resulting state of Ch can be optically anisotropic with preferential orientation of structural units (nano-volumes) in the amorphous solid. Changing the orientation of the polarizing plane of the exciting light can change the orientation of the structural "units". The annealing of films or glasses near the glass transition temperature, T_g, revert the changes back. When a material with optically induced anisotropy is illuminated by non-polarized light, the material becomes isotropic again. In anisotropic effects, only the orientation of some (not all) chemical bonds is probably changed. The bonds in anisotropic effects are reoriented without substantial changes in the number of "wrong" bonds and other defects. Exposure of ACh by polarized light can cause anisotropic crystallization of films. Optically isotropic ACh can become also optically anisotropic without any crystallization. A possible mechanism for the explanation of photoinduced anisotropy is based on *valence alternation pairs that are attributed to under- or over-coordinated atoms of chalcogen.* Such pairs are probably used as intermediate steps for photoinduced structural changes. There is a possibility that the lone electron pairs will be changed to ordinary chemical bonds of a new direction and the original bonding pairs would be converted into the lone ones.

Polarized light can also induce dichroism and change the relative transmission of AChF.

At very high light intensity, Ch can be locally evaporated, pyrolyzed or deteriorated altogether. The deteriorating effect is wavelength-sensitive.

More details on photoinduced effects can be found in [17–19, 21–24, 29, 30, 32–35, 69–71, 73, 82–84, 88, 117, 128–130, 133, 134, 137–140, 150–152] and in the papers quoted.

7.5.2.1 Phase Changes and Phase Change Memories

Special aspects of photoinduced effects in many AChF and in many crystalline compounds (alloys) are connected with photoinduced changes of the amorphous state to the crystalline one and vice versa. Such changes can be fast (several ns, linear speed of crystallization up to 80 m/s for films of the thickness of several nm). Such changes need low energy and can be repeated (cycled) many times (up to 10^5–10^{12}). The size of individual memory cells prepared from a proper phase change memory material (PCMM) can be very little and can be further scaled down. Phase changes can be induced not only by short optical pulses, but also by electrical ones. The underlying principle is the same. Optical reflectivity and electrical resistivity of the crystalline phase and of the amorphous phase of the same composition of PCMM are different in many Ch. Both states can be stable at room temperatures and also at temperatures below 90 °C; some can be stable at temperatures above 200 °C. All these properties are important for their application in non-volatile memories. The effect has attracted considerable attention and resulted in its application in CD, DVD, and Blu-ray discs as well as in commercial production of non-volatile electrical phase change random access memories (PC RAM) and their arrays (see e.g.[18, 19, 21–24, 29, 32, 73, 83, 91, 133–140] and the cited papers). They are very perspective both scientifically and commercially; below we mention some historical remarks on the subject.

The study of these effects and materials started in 1964 by a description of the threshold switching effect (see below). S.R. Ovshinsky was one of the pioneers in this area and he recognized the application potential of these effects and patented several devices for phase change random access memories (PC RAM) production. In 1968, he described repeatable threshold and memory switching in the Te-As-Si-Ge system [133]. The prototype of a 1,024-bit chip was presented in 1978; however, it was not produced. N. Yamada et al. [134] started to use fast switching alloys from the GeTe-Sb_2Te_3 system for optical data storage. These alloys can have great optical contrast, $C = (R_c - R_a)/R_c$ where R_c and R_a is optical reflectivity of crystalline and amorphous films, respectively. The alloys also have good cyclability (rewriting capability). Similar modified materials were successfully applied later in optical storage discs (CD 500 Mb 1990), then in DVD discs (1998) and Blu-ray disc in 2003 (up to 100 Gb/disc with triple layers, (Fig. 7.55) [152]). The tremendous success of optical recording (several billions of DVDs are sold per year) renewed interest in electrical recording, and resulted in commercial production of phase change random access memories (PC RAM) and in memory arrays controlled by short ns electrical pulses (they have already been produced by Micron and Samsung Co. and possibly others).

Prior to discussing the principle of optical and electrical storage of the data in PCM, we present few remarks on memory and threshold switching.

Electrical conductivity of most of AGCh is for lower electrical fields $<10^4$ V/cm relatively low and ohmic. With growing electrical field, non-ohmic conductivity occurs (Fig. 7.56a) [137]. At the threshold voltage (U_T, about 10^6 V/cm), conductivity switches very fast to a state with low resistance and the electrical

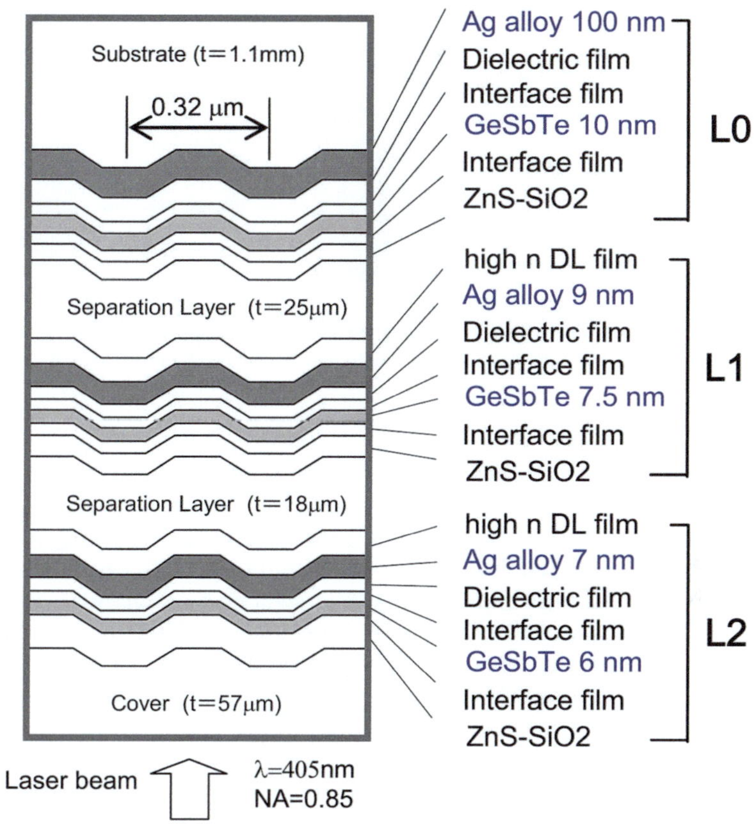

Fig. 7.55 A schematic view of a triple layer Blu-Ray disc [152]. The light (405 nm) proceeds from the *bottom*

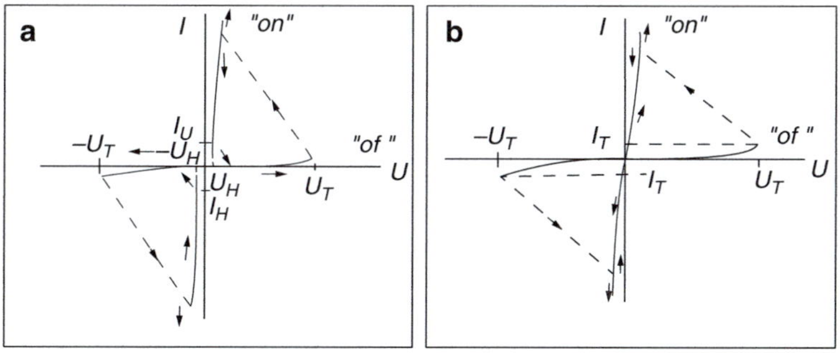

Fig. 7.56 I-U characteristics of threshold switching (**a**) and of memory switching (**b**) [137]

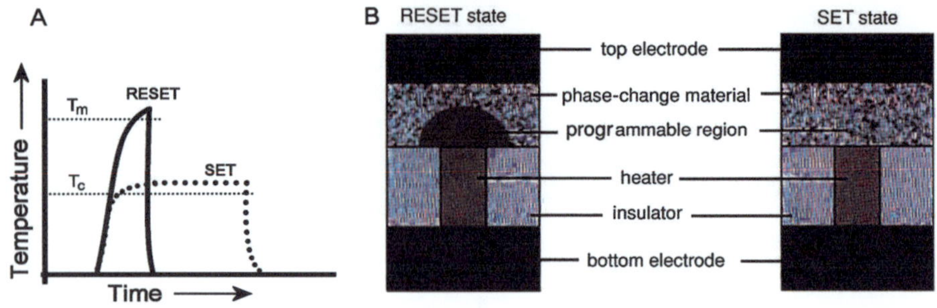

Fig. 7.57 (**a**) Schematic size of the RESET and SET pulses, (**b**) mushroom in a PCM cell (the RESET state, *left*; the SET state, *right*). The size of a mushroom can be programmed [23], p. 268, adapted

current is limited by a serial resistor. The lowering of the applied voltage lowers the current to the value I_H, and then the resistivity switches again to its original high value. Contrary to many CCh, the repeatability of the switching process in AGCh can be high, the durability and reproducibility in ACh can be reasonable for many applications. Materials for threshold switching are glasses with a poor crystallization ability. Models of threshold switching are several, both thermal and electronic mechanisms play important roles [17–19, 22–24, 29–32, 133, 135–138].

Memory switching (Fig. 7.56b) [137] has the original part (ohmic and non-ohmic) very similar to the one of threshold switching. At the voltage U_T, the conductivity switches fast to a high level, but the lowering or removal of the applied voltage preserves the high conductivity state of the material (the ON or SET state, Figs. 7.56b and 7.57) [23]. Contrary to the threshold process, part of the material (a filament) becomes crystalline with high conductivity and high stability even without any applied voltage.

Amorphization of PCMM (writing, RESET, Fig. 7.57) needs a more intensive pulse which melts a thin crystallized film. The melted part of the film is then cooled very quickly (10^9–10^{11} K/s) and the amorphous (glassy) state is formed. For information erasing (Fig. 7.57), the pulse increases the temperature of the amorphous (glassy) material above its crystallization temperature (T_c), ($T_c < T_m$), and the film crystallizes (SET state). The erasing pulse is longer, the crystallization is slower than melting. The whole process can be repeated many times (up to 10^{12}) for the time being; the effort to increase it is considerable.

This principle has been applied for construction of optical and electrical PCM cells. Its application in the very successful optical memories has already been mentioned (CD, DVD, Blu-ray discs); electrical memories are also produced commercially; they are used in mobile phones and other mobile facilities. Their size can be further scaled down and it is supposed that in the near future they will replace the present-day "flash" memories, which cannot be scaled down too much, as they are approaching their physical limits. When they are further scaled down, their neighbour cells could interact and change their charge (content). An

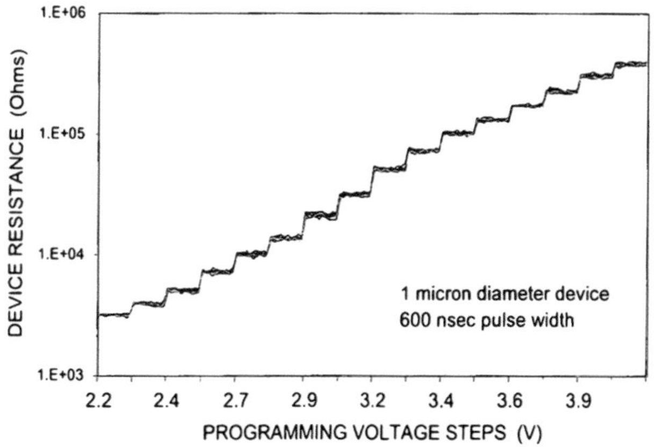

Fig. 7.58 Resistance steps experimentally found when a 600-μs pulse were used for the amorphization of PCMM [138]. Many steps can be formed, but in practice their number is much lower because the resistivity of the amorphous phase is temperature-dependent and there is also some shift of resistivity due to ageing processes. The overlapping of neighbour levels due to such processes shall be avoided

announcement has been made recently that this limit can be broken by producing 3D flash memories that can have greater data storage capacity. This idea is being actively researched, and the realization can be fast.

The amorphous "mushroom cap" (Fig. 7.57b, RESET state, the upper part of the figure) can be of different size. It can be partly or fully crystallized and the whole resistivity is different. Several resistance levels can be then obtained and used for a so-called multilevel recording (Fig. 7.58, [138]). Not only 0 (zero) and 1 (one) states will be obtained, as it is necessary for von Neumann's logic; but also 2, 3, and possibly more. The different resistivity levels can be recognized by their resistivity. The "memory cells" with more than two memory (resistivity) states are probably used by animal and human brains for sorting and elimination of unimportant signals of lower intensity [138].

Optical pulses of several ns, used for writing and erasing, have very sharp edges and the decrease of temperature of the PCMM is very sharp too. A fast cooling rate ($\approx 10^9$–10^{11} K/s) of a very thin film (thickness $\approx 6 \leq \times \leq 70$ nm) preserves the amorphous structure of the melted spot; it becomes glassy and has different optical reflectivity or electrical resistivity than the crystalline one. The crystallization temperature of many PCMM is above 150 °C, so data are not erased bellow 80 °C even after a long time. The size of data points decreases with the capacity of the discs (Fig. 7.59, [139]). The information retention time in commercial DVD and Blu-ray discs is usually $\geq$10 years, for special, e.g. archival purposes, it can be $>10^3$ years (obtained by extrapolation, Fig. 7.60 [150]). The retention time can also be increased by keeping memories at lower temperatures.

For the purpose of data storage, many potential PCMM have been studied. All of them are bad glass formers; they need very fast cooling for the preservation of glassy

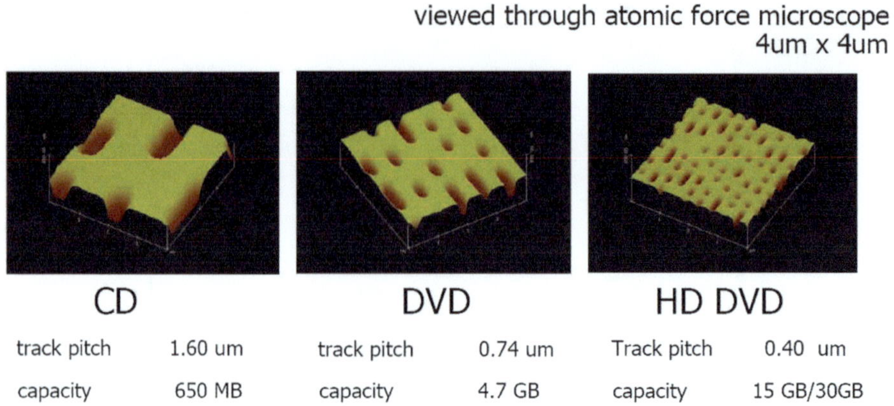

Fig. 7.59 The data points in CD, DVD and in Blu ray discs are of different size and density. The holes are viewed through an atomic force microscope, 4×4 μm [139]

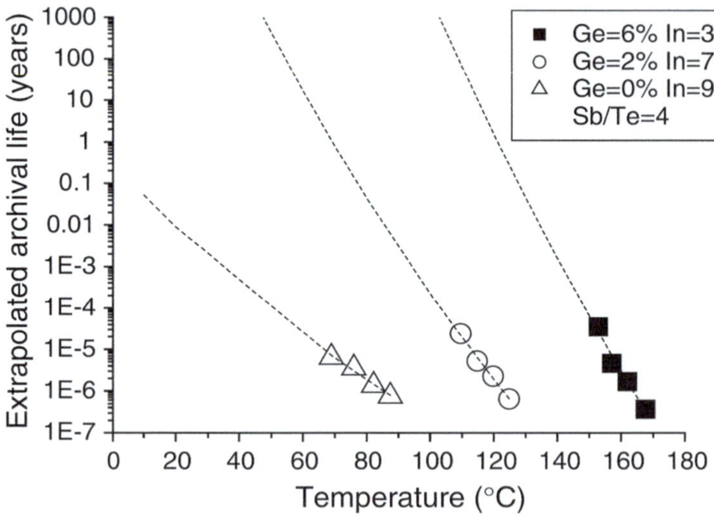

Fig. 7.60 Archival life stability of a written amorphous mark for doped Sb-Te phase-change materials. In inclusion various dopants are given. The *dotted line* shows the extrapolation to room temperature [150]

state. Some PCMM have already been applied: tellurides, selenides, intermetallic compounds, alloys, binary, ternary, stoichiometric and non-stoichiometric; e.g. from the Ge-Sb-Te, Sb-Te, ($Sb_{70}Te_{30}$+Sb, Ge, Ag), Sb-Se, Sn-Te-Se, Ge-Te-Se, Sb-Se-Te, In-Se, Ge-Te, In-Sb-Te, Ge-Sn-Te, Au-Ge-Sn-Te, Ag-In-Sb-Te systems, etc.; many substitutions in both sublattices of the crystalline structure are possible. In the crystalline state, they form narrow-gap semiconductors, intermetallic compounds; in the amorphous form, they are semiconducting. The example of the crystallization

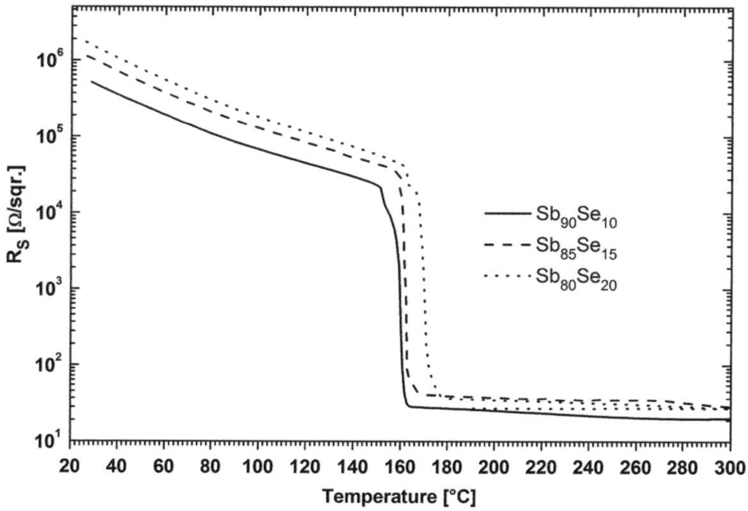

Fig. 7.61 The temperature dependence of electrical resistivity of some samples from the Sb-rich Sb-Se system [151]

of a Sb-rich Sb-Se PCMM due to increased temperature, as measured by van der Pauw method, is given in Fig. 7.61.

Not only tellurides, but also materials without any antimony or germanium, or without any chalcogen, such as in systems InSb, InGaSb, GeSbSnIn, GeSbIn, GeSb, GaSb, GeTe etc. have been applied for PCMM preparation. Commercial systems are still based on Ch. Some of these PCMM contain large amount of antimony, e.g. the Sb-rich Sb-Te and Sb-Se alloys, $Ag_{3.4}In_{3.7}Sb_{76.4}Te_{16.5}$ (AIST), and Sb-rich Sb-Ge, GaSb, and InSb systems (mostly doped), (see e.g. Fig. 7.61). The composition of Sb-rich alloys is often close to Sb-rich eutectic, e.g. it is close to $Ge_{10}Sb_{90}$, $Ga_{10}Sb_{90}$; they crystallize quickly, too. High-speed optical recording can be obtained using e.g. AIST, InSb, $In_{12}Ga_{38}Sb_{50}$, $In_{25}Ga_{25}Sb_{50}$ based alloys. Their structure is again close to Sb crystals (A7, rhombohedral, a slightly deformed cubic cell), all atoms randomly occupy 6(c) sites of the R3m space group (Fig. 7.18).

Exclusive positions in the group of PCMMs have those based on the Ge-Sb-Te system and used in DVD ($Ge_2Sb_2Te_5$) and "Blu-ray" discs ($Ge_8Sb_2Te_{11}$). The alloys based on $Ge_8Bi_2Te_{11}$ have similar properties as $Ge_8Sb_2Te_{11}$. To make the writing spot smaller and exclude optical diffraction, blue light (405 nm) is used for reading, writing and erasing in "Blu-ray" discs.

The statistical occupation of all or part of crystal positions in some PCMM is apparently caused by: (a) chemical similarities of constituting atoms (atomic radii, bond energy, bond symmetry, electronegativity, covalent or semimetallic character of chemical bonds, formation of intermediary compounds, alloys); (b) very fast phase changes, all processes are very far from the equilibrium [131, 140]; metastable solid solutions are also formed in the isotropic crystalline phase, crystallization proceeds into a single phase.

It is generally accepted that in the structure of the Ge-Sb-Te-based PCMM, the positions taken by Te atoms are in the "anionic" sublattice; the positions taken by (Ge and Sb) atoms are in the "cationic" sublattice (Fig. 7.17, [73]). In another already applied PCMM, namely in AIST, it was proposed that all cell positions are occupied statistically (randomly) by all atoms [74]. The reason for such a difference is not clear. It cannot be excluded that some of the proposed models may contain inaccuracies. The Sb and Te atoms are neighbours in the periodic table of elements with very similar electronic structure, and they can be distinguished from the X-ray diffraction or absorption only with difficulties. The problem is still investigated.

Glassy thin films from the Ge-Sb-Te system are crystallized by short laser pulses in a metastable fcc cubic structure, (Fig. 7.17) [73]. This fact can be caused by fast crystallization from the isotropic glassy system to the isotropic crystal lattice without necessity for transport of individual atoms for a long distance. However, the fcc structure is also obtained by the annealing of Ge-Sb-Te films at temperatures near 230 °C, when crystallization is not fast. The transport factors are, therefore, not the most important factors. The energy barrier between a metastable fcc structure and a stable hexagonal one is probably higher than the one between the amorphous and fcc structures. The transition to stable hexagonal modification (Fig. 7.19) needs higher temperature.

The composition of individual PCMM is not confined to chemical compounds, even in crystals; it can be changed in broad regions. Crystalline PCMM behave partly as alloys, their properties can be smoothly tuned by composition, see e.g. (Fig. 7.62, [152]).

The formation of polycrystals with a composition far from a stoichiometric one can also be caused by *kinetic reasons*. An explanation is illustrated in Fig. 7.63

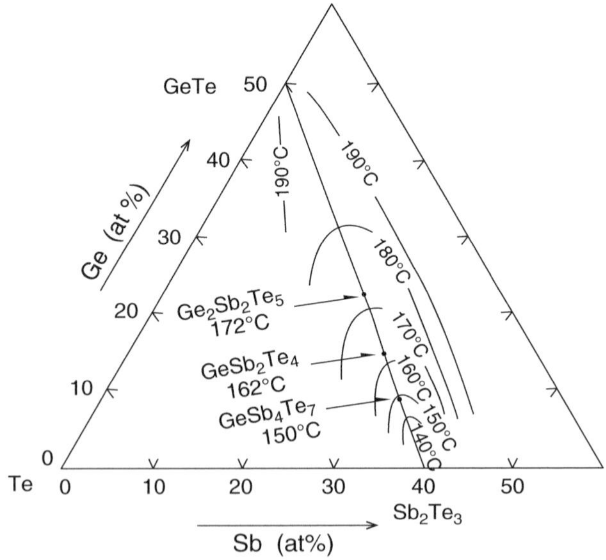

Fig. 7.62 Optically measured crystallization temperatures of PCMM of the Ge-Sb-Te system. The heating rate was 100 °C/min [152]

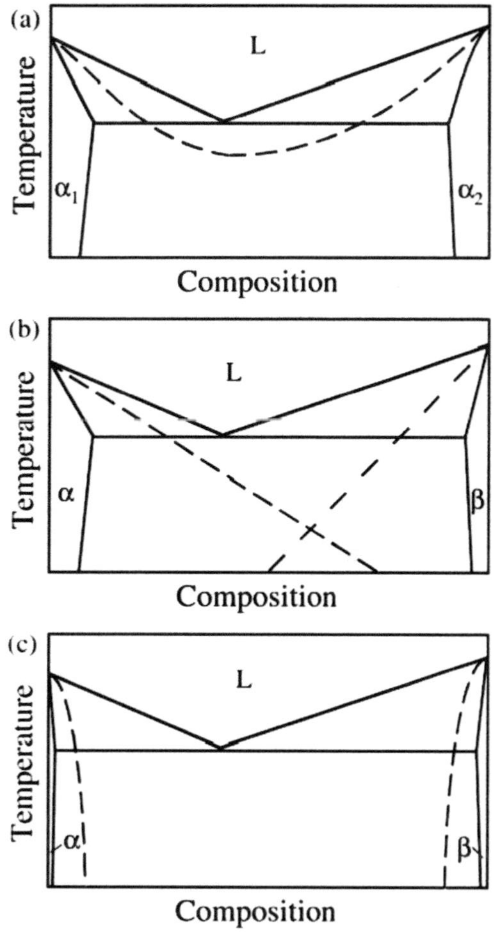

Fig. 7.63 A binary system in which two solid alloys α and β of limited miscibility (the *solid line*) are formed in equilibrium [153]. When the rate of the cooling of a melt is increasing, the solubility of both alloys increases from (**c**) to (**a**) (the *dashed lines*). The part (**a**) denotes a full miscibility of both solid phases

[153]. Rapid (several ns) non-equilibrium crystallization takes place in all PCMM. The crystals formed can be very far from the equilibrium. Therefore, the mutual solubility of phases α and β in the phase diagram given in (Fig. 7.63) can be much higher than in the equilibrium state. The solid phase could even have the same composition as the parent liquid without any segregation (Fig. 7.63a). A broad range of homogeneity is then received. The crystallization rate (Fig. 7.63) increases from the bottom (Fig. 7.63c) to the upper part of the phase diagram (Fig. 7.63a). The composition of obtained crystals is given below the dashed lines.

The properties of PCMM are changed smoothly with changes of composition without any abrupt changes in the vicinity of individual compounds (Fig. 7.62). Such properties are evidently enabled by the formation of metastable structures not only in amorphous but also in quickly crystallized films - many crystal positions can be probably occupied statistically or partly statistically by atomic substitution, similarly to intermediate compounds. Due to that, very high crystallization rates are possible and there is no necessity for long-distance diffusion of individual atoms and formations of separated two- or poly- phase systems. However, a comprehensive and clear explanation of the structure of heavily doped (alloyed) PCMM which has been approved by experiment is missing. We speculate on this topic in Refs. [131, 140].

The structure of *single crystals* of many present-day PCMMs, e.g. of $Ge_2Sb_2Te_5$, Sb_2Te_3 pure and doped, $GeSb_2Te_4$ and $GeSb_4Te_7$, $PbBi_2Te_4$ and others, is well-known; their crystals are of a hexagonal symmetry (see e.g. Figs. 7.19 and 7.20). Their single crystals were prepared and studied by our research group in 1970s, e.g. in papers [132, 141, 154–156] and in many others. High density of defects was found even in their single crystals; their structure is close to the equilibrium state, as they were grown very slowly. It was supposed that part of such defects is of the antisite type, e.g. antimony at the Te position (Sb_{Te}) and Te at the Sb position (Te_{Sb}). In Sb_2Te_3 single crystals, antisite defects were found, too [132]. It can be then supposed that much higher density of such defects is formed in fast crystallized PCMM than in single crystals that are in a state close to the thermodynamic equilibrium. The formation of high density of defects in PCMM can speed up crystallization and eliminate long-distance transport of individual atoms. The atoms can just jump to a nearby position. This effect can play an important role in the fast speed in PCM cells.

7.5.2.2 Structure of Amorphous PCMM

The structure of individual amorphous phases of PCMM can differ significantly; it depends on prehistory (temperature, composition, thermal history and on other external forces). The evaluated cooling rates of the amorphous phase of PCMM in PCM cells are very high (10^9–10^{11} K/s), as we have already mentioned. However, such evaluation is of approximate nature. Similar evaluation is also valid for the melting temperature and crystallization temperature of PCM. Experimentally, both are difficult to determine precisely, at least for very thin ($\approx$10 nm) films which are covered by other layers (heating, isolating, Fig. 7.55). Synchrotron experiments, e.g. Extended X-rays Absorption Fine Structure (EXAFS) studies and X-rays Absorption Near Edge Structure (XANES) revealed a lot of important results [23]. The structure of amorphous PCMM is isotropic, close to the crystalline phase – no long-distance diffusion is necessary for fast crystallization into cubic modification. It is especially valid for Sb-rich alloys (compounds), e.g. for AIST, which have statistically occupied all crystal positions. This structure corresponds to a slightly deformed cubic lattice.

Bonding angles of atoms in amorphous solids are not very far from those in the crystalline state. Coordination numbers of individual atoms are different, several values for every atom were found in [157] (Fig. 7.64). Chemical bonds of a-Ge-Sb-Te films are mostly covalent with a small charge that is slightly negative for Te atoms, slightly positive for Sb or Ge; the bonds are then slightly polar [157]. The partial distribution functions of chemical bonds (Sb-Ag, Sb-In, Sb-Sb and Sb-Te) has several maxima in AIST [157].

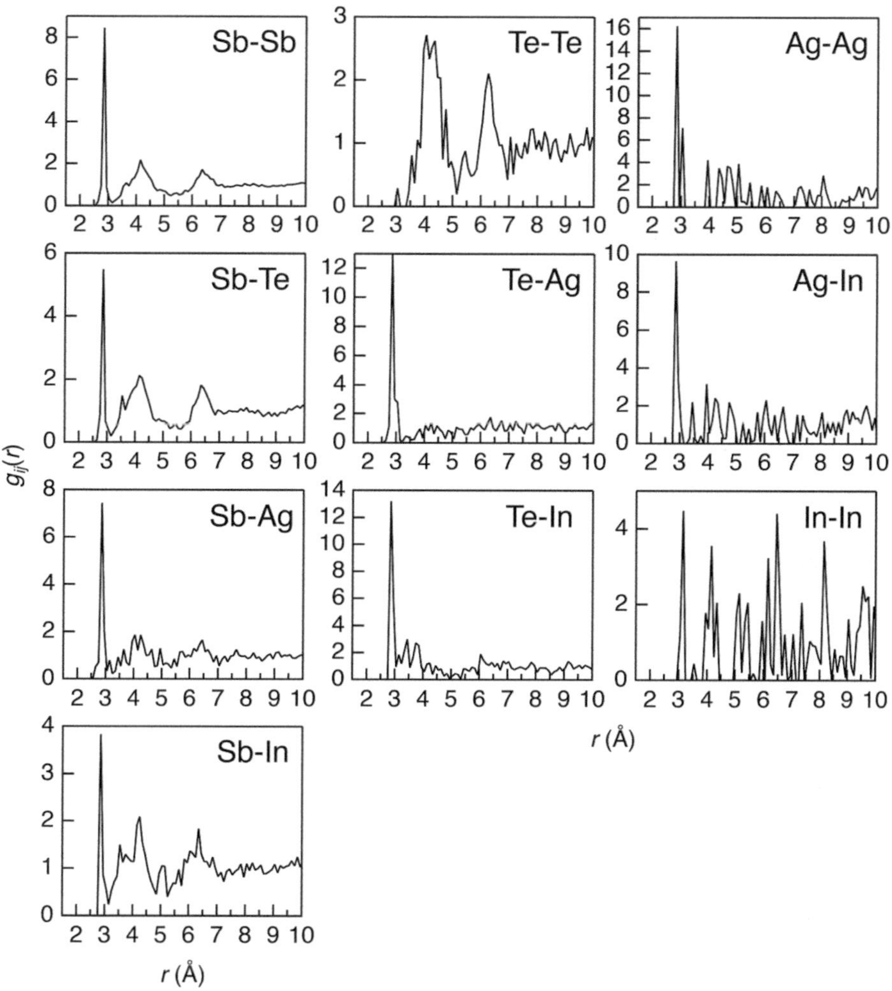

Fig. 7.64 Partial radial distribution $g_{ij}(r)$ structure functions of a- AIST of the density function-reverse Monte Carlo (DF-RMC) [157]

According to Raman spectroscopy results, pyramids of SbY_3, tetrahedra of GeY_4 and some Sb-Sb, Te-Te, and Ge-Ge bond vibrations have been found in the amorphous AIST, where Y stands for chalcogen. The excess Sb or Ge in amorphous films of Sb-rich $Ge_2Sb_2Te_5$ can form Sb-Sb, Ge-Ge or Sb-Ge bonds, e.g. in the form of $SbTe_{3-x}Ge_x$ or $SbTe_{3-x}Sb_x$ pyramids [158]. Similarly, Sb-Sb bonds in Sb-rich Sb-Se and Sb-Te PCMM systems can be formed there. Sb and Te atoms cannot be distinguished properly by X-rays methods, as they are neighbours in the periodic table and have similar electronic structure. There are further experimental difficulties for an exact determination of structural parameters of thin PCMM films, as we have already mentioned. The thickness of the films is several nm, the active PCMM in real cells are surrounded by several different layers (Fig. 7.55); the processes are several ns long. Calculation and modelling of the structure and its changes are therefore very promising sources of information, which attract many research groups who aim to either approve or disapprove it. Due to the above-mentioned difficulties, experimental confirmation of calculation results is very often still missing.

A profound understanding of the structure of PCMM and of the fast crystallization process is still not available, but the effort made to understand it has been very intensive. A detailed review of intermediate results can be found in [18, 23, 157, 159]. A proper understanding can lead to preparation of new PCMM and to further improvement of their properties.

PCM are very prospective in terms of future computers, potentially as full solid-state non-volatile memory (SSM) or Ovonic Universal Memory (OUM) without any moving parts. Prospective is their application in cognitive memories, too. However, numerous issues will have to be solved before the commercialization of these ideas has taken place (see also Sect. 7.6).

7.6 Applications of Crystalline and Amorphous Chalcogenides and Their Thin Films

7.6.1 Crystalline Chalcogenides

CCh are materials with numerous important applications. Several of them have already been mentioned. Some CCh are dielectrics, some are metallic, or semimetallic, most of them are semiconductors with the energy gap between 0.1 and 3 eV. Crystals can be nonstoichiometric containing excess of metal or chalcogen. For their application, CCh are very often doped by other elements.

Semiconducting Ch can form a p-n junction easily by doping with excess intrinsic element(s) or by foreign dopants. The p-n junctions are used in thermoelectric or Peltier elements, as well as in optical sensors, photovoltaic elements for harvesting of sunlight, etc. The research in photovoltaic elements has aroused considerable interest; the active ChF can be thinner and less expensive than the crystalline Si, but their efficiency is still lower than the one of the crystalline Si and their long-time stability is to a certain extent limited because of photolysis. Some of them contain relatively rare (Ga, In) elements or Cd, which is an element with toxic properties. Many CCh are applied as optical sensors, mainly for the IR region (PbS, PbSe, CdTe, HgTe, etc.).

The application of CCh as luminescent materials has already been mentioned. They can be used in the visible part of the spectrum (ZnS, CdS, etc.); however, a more important characteristic is their possibility to emit light in the IR region. Some RE-doped Ch are used as lasers also in the NIR and MID IR regions. Some Ch have a long luminescence decay time (tens of minutes) and are used for illumination during electrical break-down or blackout in important areas, in subways, nuclear power stations, in defense facilities, etc. Some crystalline materials are used as lasers (e.g. ZnSe:Bi, etc.); they have been recently made not only from uniform single crystals, but also as ceramic or glass-ceramic materials made by vacuum sintering of nanoparticles at higher mechanical pressure. The size of such nanoparticles has to be much smaller that the wavelength of light. They can then be fully transparent.

Many CCh are used as important materials in electronics and in optics (e.g. crystals of CaS, PbS, PbSe, PbTe, HgTe, CdTe, Sb_2Te_3, Sb_2Se_3, Bi_2S_3, Bi_2Se_3, Bi_2Te_3, $Ge_2Sb_2Te_5$, $Ge_8Sb_2Te_{11}$ and many others). They are applied in different areas; CCh are used for their good transmittance in the infrared (IR) region as optical elements, and for their luminescence as lasers and light amplifiers and light generators. Many CCh have been applied as light, temperature and composition sensors (e.g. PbS, PbSe, HgS, HgSe, HgTe, PbTe, CdTe, Sb_2S_3) for night vision, for remote temperature detection, for security and defense applications, etc.

Some CCh or chalcohalides are important ferroelectric materials with ultra-high permittivity (e.g. SbSI, BiSeI, SbSeI).

A substantial group of selenides and tellurides (e.g. Sb_2Te_3, Bi_2Te_3, Sb_2Se_3, Bi_2Se_3 and their solid solutions) were applied as important thermoelectric materials for room and not very high temperatures [3, 4], some were exploited as Peltier cells for cooling, e.g. in portable automotive fridges, for cooling of sensors or for cooling of bio-medical samples in microscopy, etc. At the moment, some of these compounds are important PCMM and promising plasmonic materials with many present and future applications, e.g. in high resolution microscopy, specific detection of very low concentrations of biological molecules, enzymes, drugs, etc.; in surface plasmon-based sensors. Many of them can be realized with chalcogenide films. Further increase of data density can be realized by plasmonics applications, in which the light diffraction limit can be broken.

Some chalcogenides are important catalysts in chemical industry, e.g. MoS_2, $MoSe_2$, WSe_2 and many others (see, e.g. [5]). The last mentioned group of compounds have a layered structure similar to graphite, and have recently been intensively studied as 2D graphen-like materials. The p-n junction in WSe_2 crystals was proposed as a material for light-emitting diodes [160].

7.6.2 Amorphous and Glassy Chalcogenides

The considerable effort invested into the application of AGCh has been very successful in many areas; some applications started more than 100 years ago, some of them are very recent and are going to be commercialized in the future. Some of the applications have already been mentioned. AGCh from the As-S, As-Se, As-S-Se, Ge-S, Ge-As-S, As-Se-Te, Ge-As-Se, Ge-As-Se-Te, Ge-La-S, Ge-(Ga,Sb)-S, Ge-Sb-Se, Ge-Bi-S, Ge-Bi-Se, Se, As_2S_3, As_2Se_3, Ge-Sb-Se and Ge-Sb-S systems have already been applied due to *good transmittance in the IR part of spectra*

(Fig. 7.37) [9, 10, 19, 24, 25, 27, 29, 32, 37, 49, 50, 87] etc.; as IR optical elements, lenses, micro lenses, filters, antireflection layers, gratings, fibres, polarizers, optical waveguides, in integrated optical circuits, as fibres for the IR region and for all optical switching.

Thermal imaging with optics made from GCh was developed for defence and police applications, for hunters and, recently, also for drivers and firemen to detect "warm" living bodies at night, in smog, fog, etc. It has also been applied for remote temperature sensing of non-living objects. It should be added that the active part of such facilities also contains an IR sensor, often made from CCh such as PbS, PbSe, CdTe, HgTe, HgSe, etc., or from amorphous ones. For long- distance sensing in the Earth atmosphere, measurements should be done in atmospheric transmission windows of light between 3 and 5 μm and between 8 and 12 μm. The latter-mentioned transmission window can also be used for living-body detection. The optics made from a single-crystal Ge can be used too, instead of GCh; however, it is much more expensive and it also is difficult to prepare large single Ge crystals. For optical elements, the often-used glasses are those from the Ge-Sb-Se, Ge-Sb-S, Ge-As-S, As-S, As-Se, GeSe-GaSe systems. The advantage of glasses is a possibility to make large optical elements and to form them by glass moulding, which is less expensive and more effective. High-pressure hot pressing is also used for element shaping. Some AGCh have low glass-transition temperature and a high index of refraction. They have been used as immersion media in optical elements for the IR region (Fig. 7.65).

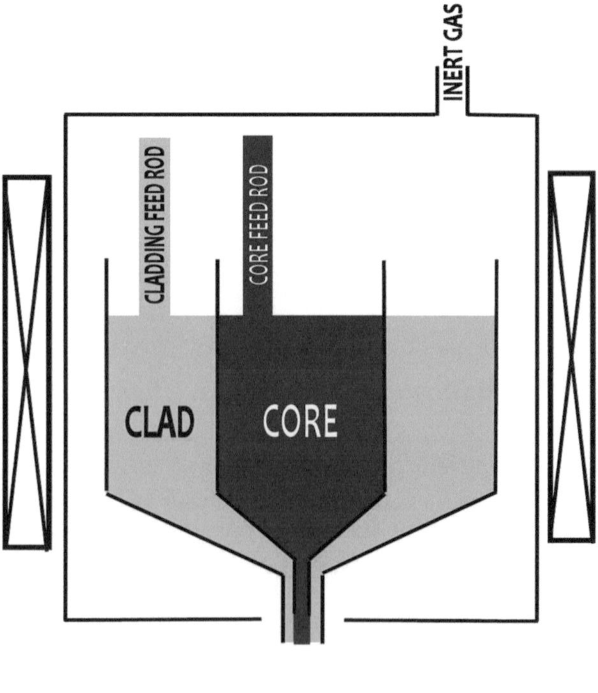

Fig. 7.65 A scheme of drawing an optical fibre from two glassy melts. The core and cladding glasses have a different value of the index of refraction. Due to that, the light is confined inside the fibre even when it is bent

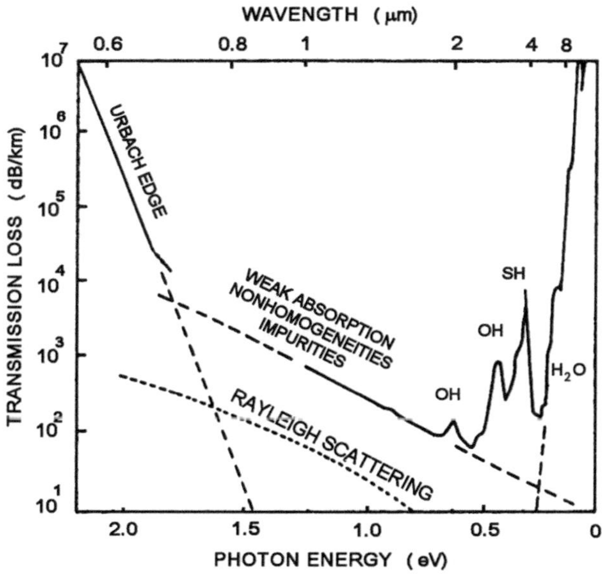

Fig. 7.66 Scattering and absorption losses of ChG [161]

Chalcogenide films also are promising as materials for optical limiters, which can be used for protection of eyes and sensitive detectors against intense radiation (see e.g. [89], p. 151). Due to great non-linear effects and two-photon absorption, the limiting factor of such devices is relatively substantial ([89], p. 153).

Very important applications can also be found in IR fibres made from GCh. They are made from two glasses of a different index of refraction (Fig. 7.65).

Preparation of precursors for chalcogenide fibre drawing needs very pure materials free of S-H, Ge-O, Ge-OH, Se-H groups, which can lower their transparency in the IR region. For illustration, Fig. 7.66 presents absorption and scattering losses of AGCh [161].

The minimum absorption coefficient in chalcogenide fibres is approximately two or more orders higher than in SiO_2 optical fibres. It is because the energies of chemical bonds in AGCh are lower than in silica, and such bonds can be broken (dissociated) more easily, and form e.g. As-As and S-S or As-As and Se-Se bonds with higher absorption in the IR region. In arsenic chalcogenides, also As_4S_4 or As_4Se_4 molecules, which can scatter light, can be formed at not very high temperatures. Unfortunately, some impurities can be removed from the GCh with a great difficulty (see, e.g. Table 7.2) [49].

The short-wavelength-as well as long-wavelength absorption edges of AGCh are dependent on composition and can be tuned in relatively broad ranges (Fig. 7.37).

Optical losses in chalcogenide fibres are greater than in silica optical fibres as we have mentioned; therefore, chalcogenide fibres are not good for long-distance data or for high-intensity light transport; they can be used for IR detection (IR

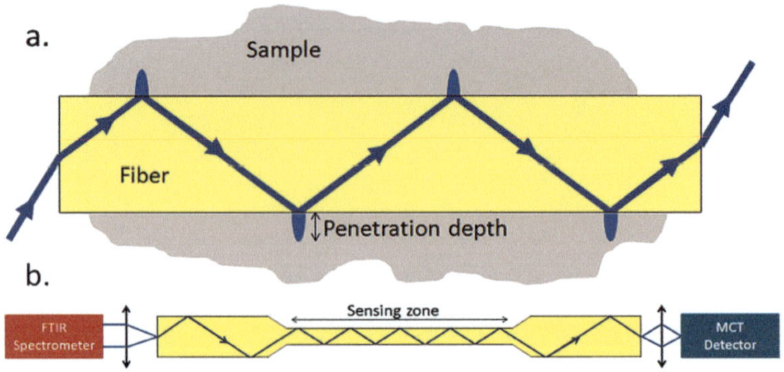

Fig. 7.67 A scheme of unclad part of fibre for evanescent spectroscopy. A part of the fibre is unclad and part of light penetrates from the fibre into the sample that is usually in a liquid or gaseous state (**a**). The light is specifically absorbed (**b**) in the sample around the fibre and its lowered intensity is measured by a detector [87]

Table 7.2 Impurity groups in vitreous arsenic chalcogenides [49]

Impurity group	Impurity in the group	Typical impurity content (ppm at.)
Light elements (gas-forming impurities)	Hydrogen, oxygen, carbon, nitrogen	10–100
Metals	Transition and other metals, silicon	0.1–1.0
Analogs of elements-macro components	Phosphorus, antimony, sulphur, selenium	1–100
Embedded in the glass network	Hydrogen, oxygen, nitrogen, halogens (OH, SH, SeH, NH, AsH groups)	0.1–10
Dissolved compounds	CO_2, COS, H_2O, N_2	0.01–10
Heterogeneous inclusions	Carbon, silicon dioxide	10^6–10^9 particles, cm^{-3}

fingerprints of chemical compounds including CO_2, NO, NO_2, poisonous and explosive ones; they are very useful for medical applications, sensors, in vivo sensors, bloodless operations, teeth drilling, etc.). The detection sensitivity of such sensors can be very high. For analytical purposes, evanescent spectroscopy can also be used (Fig. 7.67). Part of an IR Ch fibre for evanescent spectroscopy is without any cladding (a sensing zone, Fig. 7.67). Part of the light penetrates out of glass and can be used for determination of chemical compounds in liquids and gasses, e.g. of content of alcohol in wine, CO_2 content in gases, for environmental applications (water pollution, food safety measurements, for biomedical application, for in vivo studies, etc.).

A special attention has been paid to searching of possible remote civilizations on planets in the space within the European Darwin mission framework, where the presence of H_2O, O_3 and CO_2 as very probable attributes of life has been searched with Ch-based IR spectroscopy [87, 162].

One of the important applications of Ch is also connected with Ch-based glass ceramics and with optical elements obtained from nanocrystalline particles by hot sintering. In both cases, the hardness and resistance to fracture propagation is higher compared to glass, which is very important for defense and field applications. A controlled crystallization of Ch for glass-ceramic applications is more difficult than a controlled crystallization of silicate glasses [163]. For good transparency of such ceramics, the size of nanocrystals should be much smaller than the wavelength of the light. The refractive index of nanocrystals in glass ceramics should not differ significantly from the one of the glass in order to lower the Rayleigh scattering. Both these requirements can only be satisfied with difficulties, especially in chalcohalide glasses which have frequently been studied for glass ceramics [163].

Nano-and micro-crystallized glasses possess enhanced non-linearity. By using the surface plasmon resonance in noble metal nanoparticles, strong enhancement of the third-order optical non-linear coefficient was achieved [164]. Nanocrystals of CuTe have shown strong plasmonic properties and were used in surface-enhanced Raman spectroscopy (SERS) as probes [165].

An important and interesting aspect of optical fibre application is connected with holey fibres. They were originally proposed for obtaining low attenuation of light in silica optical fibres, but they have also been prepared and used from GCh. Light is transported in the air inside a holey fibre and the losses are caused just by the roughness of the inner walls of the hollow core of fibres [166]. The improvement of optical transmission of holey fibres can be demonstrated on silica fibres, which were studied more intensively. In spite of the fact that light in holey fibres (in silica, 20 μm core, $\lambda = 1{,}550$ nm) made 2.8 million reflections per km, the losses obtained were only 1 db/km, which corresponds to unexpected light reflectivity of 0.99999992! [167].

A set of holey fibres has some properties of photonic crystals, e.g. compound hollow core GeAsSeTe glasses have a non-linear index of refraction n_2 (m^2/W) of three orders higher than the one in the silica hollow fibres. Such sets of fibres are also called chalcogenide micro structured optical fibres (MOF). For IR applications, they are made from a bunch of Ch hollow fibres (Fig. 7.36). Optical losses in such fibres are orders of magnitude lower than in corresponding bulk glasses. They contain many hollow fibres that are melted together and then clad (Fig. 7.36). The glassy material and the air possess a different index of refraction and MOF work like photonic crystals. MOF can be used not only for light transport, but also for generation of supercontinuum, for spectroscopic sensors, for intense non-linear acoustic modulation driven by light, for nanowire plasmonics, and probably for many other applications. Ch possess a high index of non-linearity, which is important for the optoelectronic application. For example the value of $\chi^{(3)}$ in As_2Se_3 is 600 times higher than that in pure silica and 100 times higher than that in tellurite (TeO_2-based) glasses. Attenuation in different chalcogenide MOF is from <0.05 db/m (e.g. for an $As_{38}Se_{42}$ glass to more than 50 db/m for a Te-As-Se glass [166]).

Ch in the form of fibres, membranes or thin films have also been used as ionic or optical sensors. Thin fibre sensors can be placed in solutions, living cells, blood vessels, or other parts of living bodies, or used for many biomedical applications. They can e.g. measure the content of heavy metals in blood, in urine or other body liquids or parts, etc. (see e.g. [15]).

Crystallization of AGCh can be slow, or even very slow, and it can be applied for "glass-crystalline materials" preparation. Crystallization of AGCh can also be very fast and applied in "phase change memories". ACh are in this case extremely important as active materials for non-volatile PCMM (DVD, Blu-Ray discs, PC RAM, multilevel memories, cognitive devices, etc. [18, 23]).

The PCMM for electrical PCM are in great demand for portable electrical non-volatile memories. They should replace present-day "flash" memories which formed $\approx$34 % of the semiconductor memory market ($\approx$20.3 billion USD/year) [168]. PCM need less energy, are quicker, and their size can be scaled down for higher density of data storage. The range of electrical non-volatile memories which have already been commercialized comprises also multi-level memories (see Sect. 7.5). Considerable effort has been made in the study of so-called Ovonic universal memories (OUM), which will have no moving parts and potentially can replace also the computer hard disc. If successful, OUM could reduce the size and energy consumption and increase mechanical resistance of present-day computers or movable electronic devices.

An important effort in the study and future application of PCMM is connected with *cognitive abilities of computers*. Signals connected with partial crystallization of PCMM can be added and after several additions ($\approx$ memorizing), the cell will be fully crystallized, which can be equivalent to learning; the data or the programme will then be changed.

Promising are also so-called interfacial PCM (iPCM) which are of a different type. They could be based on a $GeTe/Sb_2Te_3$ superlattice formed from many layers, several nm thick, in which Ge atoms can be supposedly switched between octahedral and tetrahedral positions of both compound layers. It results in a reduction of necessary switching energy, improves the rewriting capability and increases the switching speed. Interestingly, a similar effect was obtained by inserting a thin HfO_2 layer between GST and the heating electrode [169]. The mechanism is yet to be elucidated.

In order to increase the storage density of optical discs, the application of "super-resolution near field structure" (Fig. 7.68) [170] was proposed. It uses a thin film of Sb that is illuminated by an optical pulse and becomes transparent only in the middle of the beam due to non-linear effects (Fig. 7.68). The active PCMM, which is below the Sb film, is illuminated only in the middle part of the original beam, which allows for a further reduction of the size of data points [170].

Further effort in the application of phase change memories is connected with lower sensitivity of AGCh to X-rays and γ-rays as compared with Si-based devices and circuits. In this connection, PCMM are studied for application in space and for defense purposes. Very promising (also for space and other research) is the application of PCM in *reconfigurable electronics* which can further increase computer speed up to 10^4 times [171].

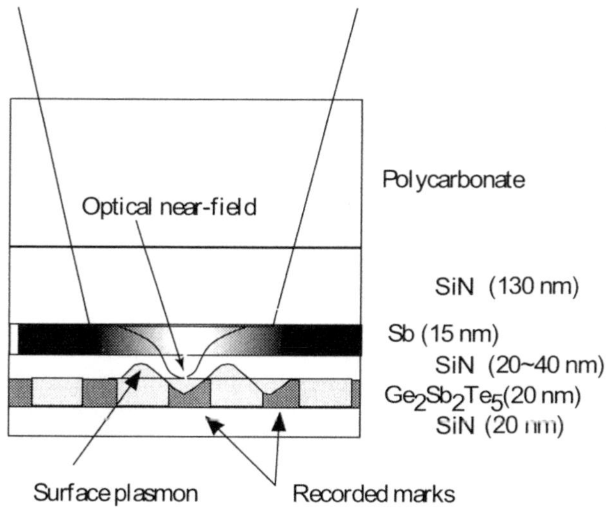

Fig. 7.68 Super-resolution near-field structure (super-RENS) using Sb and GeSbTe (GST) layers [170]

Active research has been done in the area of new materials for PCM, with higher T_c, longer retention time, which is important for archivation; with quicker transformation, higher data density, higher cyclability, with a possibility of multilevel storage and with lower energy consumption. Considerable effort has been made to prepare materials and circuits that would work closer to human brain functions and can help to understand how the brain works. Such memories can be eventually implemented to the brain, or replace some of its functions. It could help people with brain defects or brain injuries.

Some PCMM issues are also discussed in Sect. 7.5.

A relatively new area of potential application of AGCh is *electrochemical metallization memories (ECM) or conductive bridge random access memories* (ECB RAM) [172–174]. Thin films of Ch doped by Ag or Cu can form a solid-state electrolyte. It can be placed between two electrodes (Ag and an inert one). In this electrolyte, a conductive filament (bridge) from Ag/Cu atoms can be formed or interrupted in an electrochemical process which is reversible (Fig. 7.69, Eq. 7.28)

$$Ag^{+} + e = Ag^{\circ}. \tag{7.28}$$

The write/erase cycles are induced by a change of polarity of the external electrical field. An advantage of such non-volatile memories is low (room) working temperature; their size can be scaled down relatively easily. At the moment, such memories are slower ($\approx 30\ \mu s$) than the PCM ones, they generally have shorter retention time and lower cyclability when compared with PCM. They are, however, in an early state of research. Their parameters would be improved. In any case, resistive

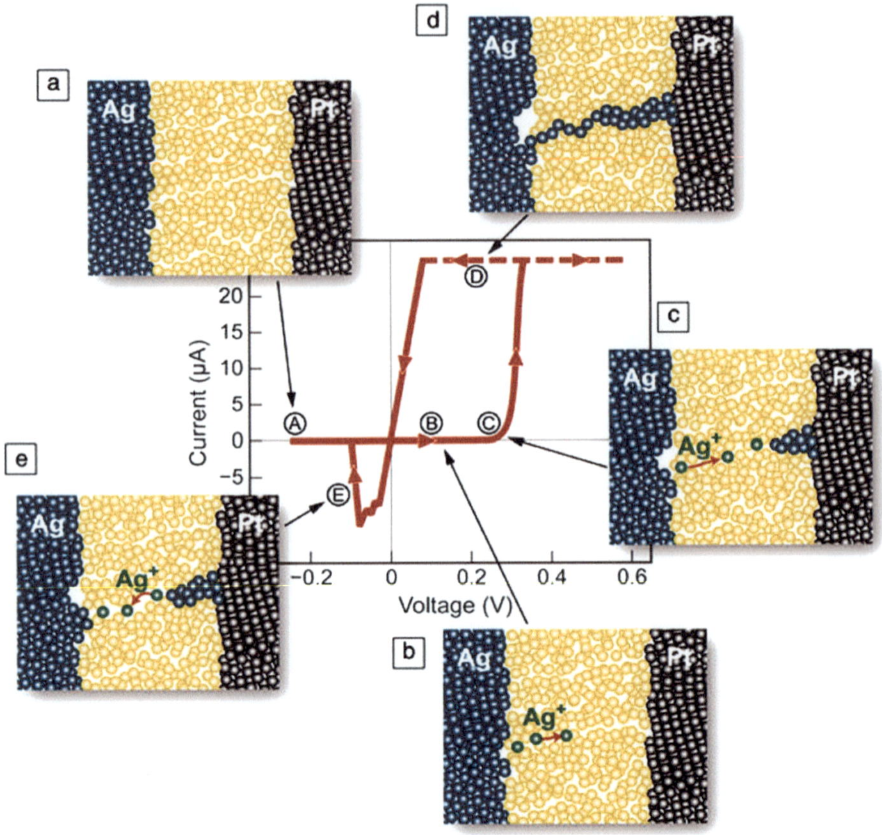

Fig. 7.69 A scheme of an electrochemical metallization memory cell [173]. (**a**) the cell before the writing process, (**b**) the voltage applied starts the electrolysis, (**c**) the Ag "filament" starts to be formed, (**d**) the Ag filament connects both electrodes and the conductivity of the cell increases (on state), (**e**) if the polarity of applied voltage is changed, the filament starts to be dissolved and the conductivity drops (off state). The process can be repeated many times

memories of the type ECM or ECB RAM based on electrochemical cells with Ag^+ ionic conductivity are an alternative solution of future non-volatile memories. ECM cells bear much resemblance with neural synapses – the interconnections between neurons in animals or human beings. This fact can help to build brain-inspired computing systems [174].

Another important application based on AChF is in *solid state batteries.* Great free volume of a-chalcogenides enables an easy doping of AGCh by Ag^+, Li^+ or other ions. The mobility of Ag^+ and Li^+ ions in AGCh is high and Ag-doped AGCh can form a thin solid electrolyte in such batteries (fully solid-state batteries). The batteries can be very thin (one or several μm thick) and they are easily incorporated into paper or plastics cards. One electrode is formed by an Ag layer; the second one is made from a chalcogen-rich Ch. Mobile charge carriers are often Ag^+ or Li^+ ions. The battery works due to electrochemical reactions (Eqs. 7.28, 7.27, and 7.30)

$$e^- + Y = Y^- \tag{7.29}$$

or

$$2e^- + Y = Y^{2-}, \tag{7.30}$$

on one electrode, where Y stands for chalcogen, mostly for selen or sulphur. On other electrode proceeds the oxidation of Ag according

$$Ag = Ag^+ + e^- \tag{7.31}$$

AGCh can be applied in *luminescence and up-conversion of light*, especially for Si photovoltaic cells, as mentioned in Sect. 5.1. Ch films evaporated on the rear part of an Si cell are used; they can have a good optical contact with the back part of Si cells. A theoretical prediction of maximum efficiency of Si cells with up-conversion is 37.4 % for an ideal up-converter on the rear part of the cell [124], and 30 % without any up-converter under non-concentrated sunlight. By concentrated sunlight, the up-conversion efficiency could be $\leq$55 % [124]. The most efficient process is ETU (Fig. 7.43), or two-step absorption via the ground state absorption (GSA) and the following excited step absorption (ESA). The other mechanisms (Fig. 7.43) have low efficiency. In real up-convertors, several processes can take place simultaneously. This issue has also been intensively studied. We discussed part of it in Sect. 7.5.

The up-conversion process has also been used in IR-light indication cards which convert the invisible IR radiation to the visible one.

Important from the viewpoint of application is the viscosity of glasses and its temperature dependence. Viscosity spans a broad range of values depending on composition and temperature. A proper value of viscosity allows for preparation of glassy fibres, hollow fibres and microstructured optical fibres. Lenses and other optical elements can be produced by hot pressing. It also allows for preparation of optical waveguides by embossing.

The excellent optical sensitivity (photoconductivity) of Se and of many Ch has been used for very long time in different photocells (see e.g. [175]). Photoconductivity of an amorphous Se has been also used in the *Xerox process*, as we have already mentioned. Some authors consider xerography as a "big invention of the twentieth century" [9]. Great density of electron and hole traps in ACh and AChF reduce the free path of charge carriers and lower their mobility (Figs. 7.32 and 7.70). This electrical property allows for the existence of high-resolution xerography and laser printers as well as some high-resolution television (TV) pickup tubes (see e.g. [9, 175]). The Xerox copiers and laser printers which work on basis of the same principle substantially changed administrative work in all areas of everyday life.

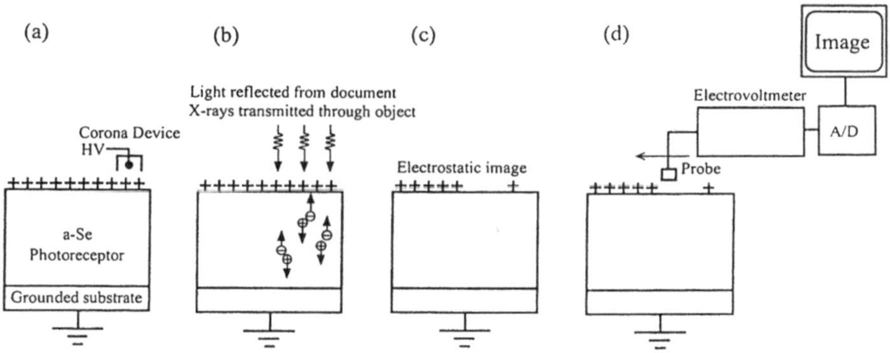

Fig. 7.70 A flat Panel for X-Ray Image. The *a-Se:0.3 %* As +20 ppm Cl, stabilized, coated onto an active detector for digital radiography [176]. For the explanation of the principle see Sect. 5.1

Due to illumination of ACh by light or by X-rays, the conductivity of the illuminated parts, which were previously charged, increases and their charge decreases. The voltage profile of the illuminated (exposed) and unexposed surface can be detected by electron probe reading (Fig. 7.70, [175, 176]) or by charged fine toner particles that adhere to the sensitive cylinder and can be electrostatically transferred to paper (transparencies, etc.) and then fixed by heating, which melts the thermoplastic part of the toner particles. The advantages of Xerox copying and laser printing are high resolution, speed, and high contrast. Resolution of the X-ray imaging systems is limited by pixel size, which can be small, (e.g. 50 μm), which is fully acceptable for most applications [176]. The contrast of X-ray panel pictures can be controlled, so also soft body tissues can be envisaged. The panel figures can be viewed immediately, computer-processed, sent, or stored. Due to higher sensitivity of X-ray radiography based on ACh films, the necessary X-ray irradiation (dose) is lower. The price of X-ray panel equipment is, however, higher than the price of a classical facility with photographic films.

The a-Se and other ACh have been replaced in xerography and in laser printers by an amorphous Si (higher hardness for heavy-duty printers), or by organic photoconductors in common office printers. Organic polymer photoconductors are less expensive and do not contain any arsenic or Se, which are not very environment-friendly. For X-ray radiography, organic polymer photoconductors have too low absorption of X-rays and their sensitivity is low, while heavier elements such as Se and selenides absorb much more. An amorphous Se in such devices shall be stabilized by small amounts of As and Cl. Stabilized Se has high room-temperature resistivity and, therefore, it can be corona charged to higher voltage.

Free holes (electrons) in an amorphous Se in sufficiently high electrical fields exhibit avalanche multiplication (up to 800 times; [176]), which can be used for the construction of television or other pickup tubes of very high sensitivity (Fig. 7.71). Highly sensitive ACh have been used, for example, in Hitachi Saticon and Harpicon pickup tubes (see e.g. Fig. 7.71). A commercially available Saticon uses Se-As-Te alloys (Fig. 7.72, [176]). Due to the Te content, sensitivity increases in the red region

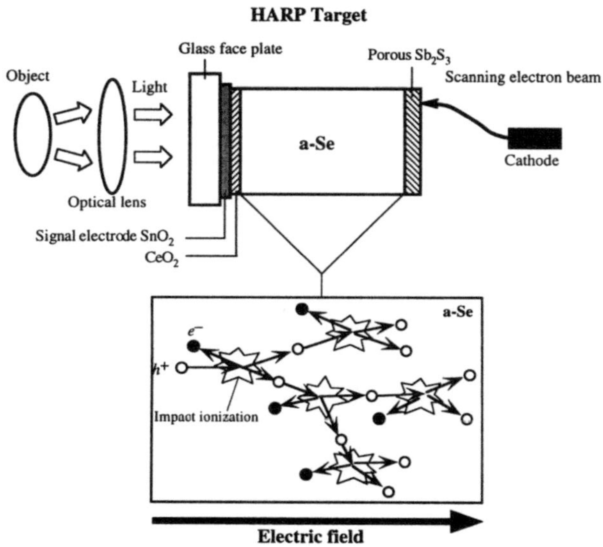

Fig. 7.71 A scheme of a Harpicon tube of very high sensitivity. One photo-generated hole which is the higher mobility charge carrier in most Ch creates many e-hole pairs due to the avalanche breakdown [176]

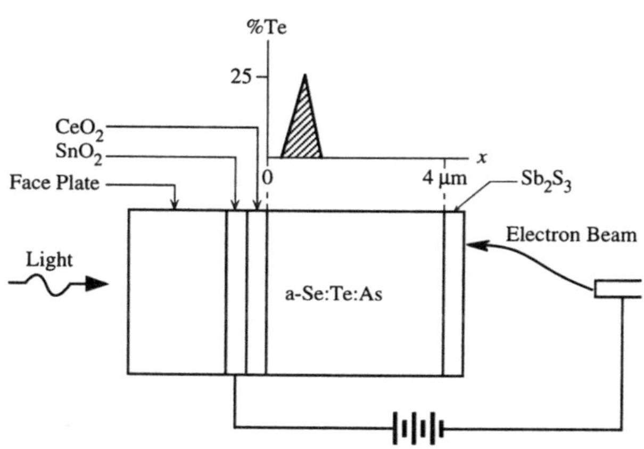

Fig. 7.72 A scheme of a Saticon tube [176]. Tellurium is concentrated in a narrow region. The sensitivity in the red spectral region is substantially increased

and the tubes are panchromatic. Super-sensitive Harpicon (high-gain avalanche rushing vidicon devices - HARP) of NHK Japan have been used for high-definition television cameras. They possess high sensitivity (100–1,000) times higher than that of CCDs. NHK have constructed a very sensitive coloured TV camera (the famous midnight photograph of Mt. Fuji). They are able to capture even starlight images. Their high-gain results from the multiplication effect: one photo-generated hole can result in several hundred of electron–hole pairs (Fig. 7.71). The details of such applications are given in [176]. Chalcogenide-based camera tubes were also used in an automatic lunar vehicle (Lunochod) of the former Soviet Union.

Very promising are also the applications of Ch that belong to the plasmonics (see e.g. [177–179]), which is an important part of nanophotonics. Plasmonics is based on the interaction processes between electromagnetic radiation and free charge carriers of higher density of metals or degenerated (narrow-gap) semiconductors. Many narrow-gap tellurides and selenides (e.g. Sb_2Te_3, Bi_2Te_3, Bi_2Se_3) possess very high density of free carriers and have plasmonic properties. The electromagnetic field can be confined in dimensions smaller than the wavelengths of light, which promises that the optics could break the optical diffraction limit and go to sub-wavelengths dimensions. Plasmonic devices can also enhance the intensity of the Raman and IR spectra.

Acousto-optic and acoustoelectric elements made from AGCh have been used, e.g. *for the delay of electrical signal.* Their application is based on the fact that the acoustic signal proceeds through GCh much more slowly than the electronic signal, and the signal can be delayed depending on material and its thickness.

Good transmittance in the IR region together with strong luminescence of AGCh has been used in light amplifiers and lasers. A very important application of Ch is in surgical IR lasers, IR emitting eye-safe lasers (e.g. for optical radar; eyes cannot be damaged by light at wavelength $\geq 2\ \mu m$). Two-dimensional CCh like MoS_2, Bi_2Se_3, Bi_2Te_3, etc. have become promising materials for photonic and optoelectronic applications ([6] and the papers cited). Among others, they have shown saturable absorption properties.

Topological insulators such as Bi_2Se_3, Bi_2Te_3 are materials with the insulating bulk state and gap-less surface states [179]. Such topological insulators are an immerging class of materials for broadband optoelectronic applications. They have ultrafast non-linear responses and have huge potential in the development of nanophotonic devices, such as optical limiters, etc.

AGCh are also promising in the area of *all optical integrated circuits;* they allow for various passive and active functionalities and *optical signal processing (*light up-conversion, signal couplers, frequency mixing, amplifying, modulation, signal generation, etc.). They can be easily patterned on a submicron scale for preparation of waveguides, gratings, and holographic elements. When doped with RE, they have good laser glass properties with high non-linearities (see also Sect. 7.5). They can be used for preparation of integrated pulsed laser sources [89]. Non-linear properties can be enhanced by presence of metallic nanoparticles or semiconductors (plasmonic effects). An enhancement by a factor 10^3, when compared to the bulk value, has been announced [89]. The addition of Ag, e.g. to As_2Se_3, has led to increase of the non-linear refractive index ranging for $Ag_{20}(As_{0.4}Se_{0.6})_{80}$ from 2,000 to 27,000 times of that of fused silica [89]. It is believed that this material from Ag-As-Se system is the most promising for all optical switching at 1.55 μm [89].

Transition-metal-doped chalcogenide lasers, e.g. a Cr^{2+} doped ZnSe laser [89], are of great interest because of their versatility, broad room temperature wavelength tunability, high optical efficiency and their potential to be scaled to high powers via direct diode or fibre laser pumping.

7.7 Conclusion

CCh and AGCh are important materials with many interesting physico-chemical properties and important and perspective applications in many areas of technology. Ch are still in the centre of intensive fundamental and application research; very promising new avenues for research are multilevel memories, cognitive computing, plasmonics, reconfigurable electronics, space applications and metamaterials which can be used by light in dimensions below the diffraction limits of light. Further promising areas are also fully optical circuits and, possibly, fully optical computers that can increase their "computing speed" and capabilities in several orders of magnitude. In this short review we were not able to describe and discus all areas and all applications. The reader can find further information not only in scientific journals and books but also in the references provided in this paper.

According to web of Sci. more than 10,000 papers have been published in the period 2004 –Feb. 2014, including 260 review papers (Web of Sci, topic: "(chalcogenide*)", 2004 – February 2014).

Because of a limited scope of this review, we were not able to discuss many interesting and important effects; we had to be concise. For those who want or need to learn more about a particular topic, a lot of original and review papers and books are available; this review paper only mentioned few of them. The science and various applications of chalcogenides are still very attractive, only in the last 5 years, more than nine books have been published [9, 18, 19, 21, 23–25, 180, 181], many review papers (see, e.g. [17, 54, 69, 166]) and some are in preparation. This fact provides evidence that the area is still in the centre of scientific and application interest.

Considerable progress has also been made in the understanding of the disordered state of materials and their changes. However, only few of the studies were mentioned here.

We hope that this review will be useful as an introduction to this important area and also will help M.Sc., Ph.D. students and other young researchers, scientists and engineers who work in solid state chemistry, physics and material science, electronics and optics and general audience interested in progress in science and technology. We hope that our young colleagues can find here a lot of interesting and important information on unknown areas for their future research and investigation into possible application.

Acknowledgement The authors would like to thank the Czech Science Foundation for grants 203/09/0827 and project CZ.1.07/2.3.00/20.0254 ReAdMat and for the institutional support RVO:61389013 of IMC AS CR.

References

1. Korbel P, Novak M (1999) Encyclopedia of minerals. Rebo International, Lisse, in Czech
2. Turjanica ID et al (1968) Some optical photoelectric and ferroelectric properties of mixed crystals SbS_xSe_{1-x}. Czech J Phys 18:1465

3. Nolas GS, Sharp J, Goldsmid HJ (2001) Thermoelectrics, basic principles and new materials development. Springer, Berlin
4. Sherrer H, Sherrer S (2006) Chap. 27. In: Rowe DM (ed) Thermoelectrics handbook, macro to nano. CRC, Boca Raton, 2006 and papers cited
5. Cheetham AK, Day P (1992) Solid state chemistry, compounds. Clarendon Press, Oxford
6. Bonaccorso F, Sun Z (2014) Solution processing of graphene, topological insulators and other 2d crystals for ultrafast photonics. Opt Mat Express 4:63
7. Tang P et al (2013) Topological insulator: Bi_2Te_3 saturable absorber for the passive Q-switching operation of an in-band pumped 1645-nm Er:YAG ceramic laser. IEEE Photonics J 5:1500707
8. Bao W et al (2013) High mobility ambipolar MoS_2 field-effect transistors: substrate and dielectric effects. Appl Phys Lett 102:042104
9. Mikla VI, Mikla VV (2012) Amorphous chalcogenides, the past, present and future. Elsevier, London/Waltham
10. Kokorina VF (1996) Glasses for infrared optics. CRC, Boca Raton
11. Kasap SO (1991) Photoreceptors: chalcogenides. In: Diamond A (ed) The handbook of imaging materials. Dekker, New York
12. Chizhikov DM, Schastlivij VP (1964) Selene and selenides. Nauka, Moscow, in Russian
13. Zallen R (1983) The physics of amorphous solids. Wiley, New York
14. Li X, Wang M et al (2011) Inorganic Sn-X-complex-induced 1D, 2D, and 3D copper sulfide superstructures from anisotropic hexagonal nanoplate building blocks. Chem-Eur J 17:10357
15. Yamane M, Asahara Y (2000) Glasses for photonics. Cambridge University Press, Cambridge
16. Fuxi G, Lei X (eds) (2006) Photonics glassses. World Scientific, London
17. Frumar M et al (2011) Amorphous and glassy semiconducting chalcogenides. In: Bhattacharya P, Fornari R, Kamimura H (eds) Comprehensive semiconductor science and technology, vol 4, Materials preparation and properties. Elsevier, Amsterdam, pp 206–261
18. Raoux S, Wuttig M (eds) (2009) Phase change materials, science and applications. Springer, New York
19. Tanaka K, Shimakawa K (2011) Amorphous chalcogenide semiconductors and related materials. Springer, New York
20. Bleckan DI (2004) Crystalline and glassforming chalcogenides Si, Ge, Sn and alloys on their bases. Vidavnictvo "Zakarpatia", in Russian
21. Wang RP (ed) (2013) Amorphous chalcogenides, advances and applications. Pan Stanford Publication, Singapore
22. Kolobov AV (ed) (2003) Photo-induced metastability in amorphous semiconductors. Wiley-VCH GmbH, Weinheim
23. Kolobov AV, Tominaga J (2012) Chalcogenides metastability and phase change phenomena. Springer, Berlin
24. Adam J-L, Zhang X (eds) (2014) Chalcogenide glasses, preparation, properties and applications. Woodhead Publishing, Oxford
25. Hilton AR (2010) Chalcogenide glasses for infrared optics. McGraw-Hill, New York
26. Feltz A (1983) Amorphe und glasartige anorganische Festkörper. Akademie-Verlag, Berlin, in German
27. Borisova ZU (1983) Chalcogenide semiconducting glasses. Izd Leningrad University, Leningrad, in Russian
28. Borisova ZU (1972) Chemistry of glassforming semiconductors. Izd Leningrad University, in Russian
29. Popescu MA (2000) Non-crystalline chalcogenides. Kluwer, Dordrecht
30. Boolchand P (ed) (2000) Insulating and semiconducting glasses. World Scientific Publishing, Singapore
31. Tauc J (ed) (1974) Amorphous and liquid semiconductors. Plenum Press, London
32. Elliott SR (1990) Physics of amorphous materials, 2nd edn. Longmann, Essex
33. Frumar M, Wagner T (2003) Ag doped chalcogenide glasses and their applications. Curr Opin Solid State Mater Sci 7:117–126

34. Frumar M et al Photoinduced phenomena in amorphous and glassy chalcogenides. In [22], pp 23–44
35. Wagner T, Frumar M Optically induced diffusion and dissolution of metals in amorphous chalcogenides. In [22], pp 160–181
36. Chizhikov DM, Schastlivij VP (1966) Tellur and tellurides. Nauka, Moscow, in Russian
37. Shelimova LE et al (2000) Homologous series of layered tetradymite-like compounds in the Sb-Te and GeTe-Sb_2Te_3 systems. Inorg Mater 36:768
38. Kifune K et al (2005) Extremely long period-stacking structure in the Sb-Te binary system. Acta Crystallogr B61:492
39. Shelimova LE et al (1965) Ge-Te system in the region of GeTe compound. Zh Neorg Chimii 10:1200
40. Kozyukhin SA et al (2011) Phase-change-memory materials based on system chalcogenides and their application in phase-change random-access memory. Nanotechnol Russ 6:227
41. Kroger FA (1964) The chemistry of imperfect crystals. North Holland Publishing, Amsterdam
42. Ipser H et al (1982) The germanium-selenium phase-diagram. Monatsh Chem 113:389
43. Shelimova LE et al (2000) Structural and electrical properties of layered tetradymite-like compounds in the GeTe-Bi_2Te_3 and GeTe-Sb_2Te_3 systems. Inorg Mater 36:235
44. Okamoto H (1998) C-Ti (carbon-titanium). J Phase Equilib 19:89
45. Hansen M, Anderko K (1953) Constitution of binary alloys. McGraw-Hill, New York
46. Gordon P (1968) Principles of phase diagrams in material systems. McGraw-Hill, New York
47. Giessen B, Willens R (1970) Rapidly quenched (splat-cooled) metastable alloy phases; their phase-diagram representation, preparation methods, occurrence, and properties. In: Alper AM (ed) Phase diagrams, material science and technology, vol 3. Academic, New York, p 104
48. Feldmann C (2013) Metastable solids – terra incognita awaiting discovery. Angew Chem Int Ed 52:7610
49. Churbanov M, Plotnichenko V. In [50], p 209
50. Fairman R, Ushkov B (2004) Semiconducting chalcogenide glass III, applications of chalcogenide glasses. Elsevier, Amsterdam
51. Shaefer H (1961) Chemische transport reactionen. Chemie Verl, Berlin, in German
52. Frumar M et al (1972) Preparation and some physical properties of single-crystals of semiconducting compound $PbSb_2Se_4$. Collect Czech Chem Commun 37:2317
53. Horak J et al (1972) Reflectivity of iodine-doped Sb_2Te_3 crystals. Phys Status Solidi (a) 14:289
54. Orava J et al Deposition techniques for chalcogenide thin films. In [24], p 290
55. Reso D et al (2011) Hot wire chemical vapor deposition of $Ge_2Sb_2Te_5$ thin films. J Electrochem Soc 158:D187–D190
56. Frumar M et al Pulsed laser ablation and deposition of chalcogenide thin films. In [57], p 81
57. Popescu M (ed) (2005) Pulsed laser deposition of optoelectronic films. INOUE, Bucharest
58. Focsa C et al (2009) Laser ablation of As_xSe_{100-x} chalcogenide glasses: plume investigations. Appl Surf Sci 255:5307
59. Nemec P et al (2004) Structure, thermally and optically induced effects in amorphous As_2Se_3 films prepared by pulsed laser deposition. J Phys Chem Solids 65:1253
60. Houska J et al (2014) Laser desorption time-of-flight mass spectrometry of atomic switch memory $Ge_2Sb_2Te_5$ bulk materials and its thin films. Rapid Commun Mass Spectrom 28:699
61. Zha Y et al (2013) A review on solution processing of chalcogenide glasses for optical components. Opt Mat Express B, 1259 and papers cited
62. Kohoutek T et al (2013) Large-area inverse opal structures in a bulk chalcogenide glass by spin-coating and thin-film transfer. Opt Mat 36:390
63. Yitai Q Chemical preparation and characterization of nano crystalline materials. In Nalwa S (ed) In [66], p 423
64. Meister S et al (2006) Synthesis and characterization of phase-change nanowires. Nano-Lett 6:1514

65. Caldwell MA, Raoux S et al (2010) Synthesis and size-dependent crystallization of colloidal germanium telluride nanoparticles. J Mat Chem 20:1285
66. Nalwa HS (ed) (2000) Handbook of nanostructured materials and nanotechnology, vol 1, Synthesis and processing. Academic, London
67. Barber DJ, Freestone IC (1990) An investigation of the origin of the color of the Lycurgus cap by analytical transmission electron-microscopy. Archaeometry 32:33–45
68. Duval D, Risbut SH, Semiconductor quantum dots: progress in processing. In [66], p 481
69. Jain H, Vlcek M Chalcogenide glass resists for litography. In [24], p 562
70. Vlcek M et al (2003) Selective etching of chalcogenides and its application for fabrication of diffractive optical elements. J Non-Cryst Solids 326–327:515
71. Jain H, Vlcek M (2008) Glasses for lithography. J Non-Cryst Solids 354:1401
72. Hughes MA et al (2013) On the analogy between photoluminescence and carrier-type reversal in Bi- and Pb-doped glasses. Opt Express 21:8101
73. Yamada N Development of materials for third generation optical storage media. In [18], p 214
74. Matsunaga T, Yamada N (2002) A study of highly symmetrical crystal structures, commonly seen in high-speed phase-change materials, using synchrotron radiatio. Jpn J Appl Phys 41:1674
75. Petrov II et al (1968) The determination of structure of $Ge_2Sb_2Te_5$ and $GeSb_4Te_7$ by difraction of electrons. Kristallografiya 13:417, in Russian
76. Talybov AG, Vainstein BK (1962) The determination of the superstructure of $PbBi_4Te_7$ alloy by difraction of electrons. Kristallografya 7:43, in Russian
77. Talybov AG (1964) The determination of super structure III of $PbBi_4Te_7$ alloy by difraction of electrons. Kristallografa 9:57, in Russian
78. Bokij GB, Romanova EM (1961) Structural elements of complex thioarsenides. Kristallografiya 6:869, in Russian
79. Born L, Hellner E (1960) A structural proposal for boulangerite. Am Mineral 45:1266
80. Frumar M (1979) Semiconducting chalcogenides of the 4th and 5th groups of periodic law. DSc theses, Charles University, Prague, in Czech
81. Gan F, Xu L (eds) (2006) Photonic glasses. World Scientific, Singapore, p 56
82. Frumar M et al (1997) Photoinduced effects in amorphous chalcogenides. Chem Pap 51:310
83. Yamada N et al (1991) Rapid-phase transitions of GeTe-Sb_2Te_3 pseudobinary amorphous thin-films for an optical disk memory. J Appl Phys 69:2849
84. Frumar M et al (2009) On the atomic structure of thin amorphous Ge-Sb-Te films. Phys Status Solidi B 246:1871
85. Fritzsche H Electronic properties of amorphous semiconductors. In [31], p 297
86. Tauc J Optical properties of amorphous semiconductors. In [31]
87. Cui S et al (2013) From selenium-to tellurium-based glass optical fibers for infrared spectroscopies. Molecules 18:5373
88. Shimakawa K et al (2013) Terahertz and direct current losses and the origin of non-drude terahertz conductivity in the crystalline states of phase change materials. J Appl Phys 114:233105
89. Zakery A, Elliott SR (2007) Optical nonlinearities in chalcogenide glasses and their applications. Springer, Berlin
90. Frumar M et al (2003) Optically and thermally induced changes of structure, linear and non-linear optical properties of chalcogenides thin films. J Non-Cryst Solids 326 & 327:399, and papers quoted
91. Ticha H, Tichy L (2002) Semiempirical relation between non-linear susceptibility (refractive index), linear refractive index and optical gap and its application to amorphous chalcogenides. J Optoelectron Adv Mat 4:381
92. Vogel EM (1991) Nonlinear optical phenomena in glass. Phys Chem Glasses 32:231
93. Petit L et al (2006) Effect of the substitution of S for Se on the structure and non-linear optical properties of the glasses in the system $Ge_{0.18}Ga_{0.05}Sb_{0.07}S_{0.70-x}Se_x$. J Non-Cryst Solids 352:5413

94. Petit L et al (2007) Correlation between the nonlinear refractive index and structure of germanium-based chalcogenide glasses. Mater Res Bul 42:2107
95. Shen X et al (2011) Preparation and third-order optical nonlinearity of glass ceramics based on GeS_2-Ga_2S_3-CsCl pseudo-ternary system. J Non-Cryst Solids 357:2316
96. Singh J (ed) (2006) Optical properties of condensed matter. Wiley, Chester
97. Tanaka K (2005) Optical nonlinearity in photonic glasses. J Mater Sci Mater Electron 16:633
98. Cherukulappurath S et al (2004) 4f coherent imager system and its application to nonlinear optical measurements. J Opt Soc Am B21:273
99. Boudebs G, Cherukulappurath S (2005) Nonlinear refraction measurements in presence of nonlinear absorption using phase object in a 4f system. Opt Commun 250:416
100. Yang J et al (2011) Measurement of nonlinear refraction using a reversal 4f coherent imaging system with phase objects. Opt Commun 284:4223
101. Boyd RW (2003) Non-linear optics, 2nd edn. Academic, San Diego
102. Wemple SH, Di Domenico M (1971) Behavior of electronic dielectric constant in covalent and ionic materials. Phys Rev B3:1338
103. Frumarova B et al (2013) Preparation and physical properties of luminescent $80GeSe_2$ (20-x)Sb_2Se_3 xSb_2Tey:Pr_2Se_3 glasses; x=0, 1, 3, 10; y=2, 3, 4. J Lumin 134:558
104. Nasu H et al (1989) Optical 3rd-harmonic generation from some high-index glasses. J Non-Cryst Solids 110:229
105. Stolen RH, Tom WK (1987) Self-organized phase-matched harmonic-generation in optical fibers. Opt Lett 12:585
106. Vij DR (ed) (1998) Luminescence of solids. Plenum Press, New York, and references quoted
107. Liu G, Jacquier B (eds) (2010) Spectroscopic properties of rare earths in optical materials. Springer, Berlin
108. Judd BR (1962) Optical absorption intensities of rare-earth ions. Phys Rev 127:750
109. Ofelt GS (1962) Intensities of crystal spectra of rare-earth ions. J Chem Phys 37:511
110. Edgar A Optical properties of glasses. In [96], p 159
111. Digonnet MJE (1993) Rare-earth-doped fibre lasers and amplifiers. Marcel Dekker, New York
112. Nemec P et al (1999) Optical properties of germanium-gallium-selenide glasses doped by samarium. J Optoelectron Adv Mat 1:33
113. Sourkova P et al (2009) Spectroscopy of infrared transitions of Pr^{3+} ions in Ga-Ge-Sb-Se glasses. J Lumin 129:1148
114. Nemec P, Frumar M (2008) Compositional dependence of spectroscopic parameters of Dy^{3+} ions in Ge-Ga-Se glasses. Mater Lett 62:2799
115. Deol RS et al (1993) Improved fluoride glasses for 1.3 μm optical amplifiers. J Non-Cryst Solids 161:257
116. Frumarova B et al (2013) Preparation of As-In-S:Sm thin films and a study of their properties. Thin Solid Films 548:429
117. Sanghera JS et al Rare Earth doped infrared-transmitting glass fibers. In [111], p 449
118. Barnier S et al (1992) Raman and infrared studies of the structure of gallium sulfide based glasses. Mater Sci Eng B14:413
119. Heo J et al (1998) Raman spectroscopic analysis on the solubility mechanism of La^{3+} in GeS_2-Ca_2S_3 glasses. J Non-Cryst Solids 238:115
120. Ren J et al (2011) Intense near-infrared and midinfrared luminescence from the Dy^{3+}-doped $GeSe_2$-Ca_2Se_3-MI (M=K, Cs, Ag) chalcohalide glasses at 1.32, 1.73, and 2.67 μm. J Appl Phys 109:033105
121. Nemec P et al (2000) Optical properties of low-phonon-energy $Ge_{30}Ga_5Se_{65}$: Dy_2Se_3 chalcogenide glasses. J Phys Chem Solids 61:1583
122. Nemec P et al (1999) Optical properties of germanium-gallium-selenide glasses doped by samarium. J Optoelectron Adv Mat 1:33
123. ASTM G173-03, Solar Spectrum; from: Kim S, Heo J, Joint Symp. MSE for the 21th century, poster

124. Strümpel C et al (2007) Modifying the solar spectrum to enhance silicon solar cell efficiency – an overview of available materials. Sol Energy Mater Sol Cells 91:238
125. Auzel F (2004) Upconversion and anti-stokes processes with f and d ions in solids. Chem Rev 104:139, In [133], p 243
126. Strizik L et al (2014) Green, red and near-infrared photon up-conversion in Ga-Ge-Sb-S:Er^{3+} amorphous chalcogenides. J Lumin 147:209
127. Frumar M et al (1988) Photoinduced changes of structure and properties of amorphous chalcogenides. React Solids 5:341
128. Frumar M et al (2001) Structure and optically induced changes of reactivity and optical properties of amorphous chalcogenides. In: Thorpe MF, Tichy L (eds) Properties and application of amorphous materials. Kluwer, Dordrecht, pp 321–328
129. Krecmer P et al (1997) Reversible nanocontraction and dilatation in a solid induced by polarized light. Science 277:1799–1802
130. Stuchlik M et al The optomechanical effect in amorphous chalcogenide films. In [22], p 112
131. Frumar M et al (2007) Phase change memory materials-composition, structure, and properties. J Mater Sci Mater Electron 18:169
132. Horak J et al (2005) Defect structure of $Sb_{2-x}MnxTe_3$ single crystals. J Solid State Chem 178:2907
133. Ovshinsky SR (1968) Reversible electrical switching phenomena in disordered structures. Phys Rev Lett 21:1450
134. Yamada N (1987) High-speed overwritable phase-change optical disk material. Jpn J Appl Phys 26(26–4):61
135. Fritzsche H Switching and memory in amorphous semiconductors. In [31], p 313
136. Fritzsche H Electronic properties of amorphous semiconductors. In [31], p 221
137. Adler D, Moss SC (1972) Amorphous memories and bistable switches. J Vac Sci Technol 9:1182
138. Ovshinsky SR (2004) The ovonic cognitive computer – a new paradigm. 3rd E\PCOS 04. Principality of Liechtenstein, Sept 2004. http://www.epcos.org/library/papers/pdf_2004/01paper_ovshinsky.pdf
139. Yamada H *HD DVD promising design*. 3rd E\PCOS 04. Principality of Liechtenstein, 4–7 Sept 2004. http://www.epcos.org/library/papers/pdf_2004/02pp_yamada.pdf
140. Frumar M et al (2005) Non-volatile phase change memory materials and their induced changes. E\PCOS 05. Cambridge, UK. 3–6 Sept 2005. http://www.epcos.org/library/papers/pdf_2005/Frumar.pdf
141. Frumar M et al (1972) Preparation and some physical properties of semiconducting $GeSb_2Te_4$ crystals. Mater Res Bul 7:1075
142. Frumar M et al (1999) Raman spectra and photostructural changes in the short-range order of amorphous As-S chalcogenides. J Non-Cryst Solids 256:105
143. Frumar M et al (2001) Photoinduced changes of structure and properties of amorphous binary and ternary chalcogenides. J Optoelectron Adv Mat 3:177
144. Frumar M et al (1997) Photoinduced changes of the reactivity of amorphous chalcogenide layers. J Non-Cryst Solids 221:27
145. Frumar M et al (1984) Reversible photodarkening and structural-changes in As_2S_3 thin-films. Philos Mag B50:483
146. Singh B et al (1979) Photo-contraction effect in amorphous $Se_{1-x}Ge_x$ films. Solid State Commun 29:167
147. Frumar M et al (1983) A model for photostructural changes in the amorphous As-S system. J Non-Cryst Solids 59 & 60:921
148. Hamanaka H et al (1977) Reversible photostructural change in melt-quenched As_2S_3 glass. Solid State Commun 23:63
149. Tanaka K Sub-gap photoinduced phenomena in chalcogenide glasses. In [22], p 69
150. Lankhorst MHR (2003) Prospects of doped Sb-Te phase-change materials for high-speed recording. Jpn J Appl Phys 42:863

151. Hromadko L et al (2014) Physico-chemical properties of the thin films of the Sb_xSe_{100-x} system (x=90, 85, 80). Thin Solid Films 569:17
152. Yamada N et al (1987) High-speed overwritable phase-change optical disk material. Jpn J Appl Phys 26:61
153. Bramberger M (2001)) Rapid solidification and microstructure, Chapt 8.2. In: Ready JF (ed) LIA handbook of laser materials processing. Magnolia Publishing and Springer, Laser Institute of America, Orlando, p 268
154. Fritzsche H Light-induced structural changes in glasses. In [30], p 653
155. Frumar M, Horak J (1971) Preparation and some physical properties of semiconducting $PbBi_4Te_7$ crystals. Phys Status Solidi (a) 6:K133
156. Frumar M et al (1974) Some physical-properties of semiconducting $GeSb_4Te_7$ crystals. Phys Status Solidi (a) 22:535
157. Matsunaga T et al (2011) From local structure to nanosecond recrystallization dynamics in AgInSbTe phase-change materials. Nat Mater 10:129, supplementary Fig. 2
158. Kozyukhin S (2013) Structural changes in doped $Ge_2Sb_2Te_5$ thin films studied by Raman spectroscopy. Phys Procedia 44:82
159. Hyot B Chalcogenide for phase optical and electrical memories. In [24], p 597
160. Ross JS et al (2014) Electrically tunable excitonic light-emitting diodes based on monolayer WSe_2 p-n junctions. Nat Nanotechnol 9:268
161. Kanamori T et al (1985) Transmission loss characteristics of $As_{40}S_{60}$ and $As_{38}Ge_5Se_{57}$ glass unclad fibers. J Non-Cryst Solids 69:231
162. Conseil C et al (2012) Te-based chalcohalide glasses for far-infrared optical fiber. Opt Mat Express 2:1470
163. Calvez L Transparent chalcogenide glass-ceramics. In [24], p 310
164. Tao H et al (2013) Optical non-linearity in nano- and micro-crystallized glasses. J Non-Cryst Solids 377:146
165. Li W et al (2013) CuTe nanocrystals: shape and size control, plasmonic properties, and use as SERS probes and photothermal agents. J Am Chem Soc 135:7098
166. Troles J Chalcogenide microstructured optical fibres for infrared applications. In [24], p 411
167. Roberts PJ (2005) Ultimate low loss of hollow-core photonic crystal fibres. Opt Express 13:236
168. Hudgens SJ (2008) The future of phase-change semiconductor memory devices. J Non-Cryst Solids 354:2748
169. Hubert Q et al (2013) Detailed analysis of the role of thin-HfO_2 interfacial layer in $Ge_2Sb_2Te_5$-based PCM. IEEE Trans Electron Devices 60:2267
170. Tominaga J Application of Ge-Sb-Te glasses for ultrahigh-density optical storage. In [22], p 327
171. Bolda C (2007) Introduction to reconfigurable computing architectures. Springer Science and Business Media, Berlin
172. Kolar J et al (2014) Down-scaling of resistive switching to nanoscale using porous anodic alumina membranes. J Mater Chem C 2:349
173. Valov T (2011) Electrochemical metallization memories-fundamentals, applications, prospects. Nanotechnology 22:254003
174. Lu W et al (2012) Electrochemical metallization cells-blending nanoionics into nanoelectronics? MRS Bulletin 37:124
175. Kasap S, Capper P (eds) (2006) Springer handbook of electronic and photonic materials. Springer Science and Business Media, New York
176. Kasap S, Rowlands J (2000) Amorphous chalcogenide photoconductors and imaging technologies. In [30]
177. Samson ZL et al (2010) Chalcogenide glasses in active plasmonics. Phys Status Solidi RRL 4:234
178. Gu Q (2010) Plasmonic metallic nanostructures for efficient absorption enhancement in ultrathin CdTe-based photovoltaic cells. J Phys D Appl Phys 43:465101

179. Hasan M, Kane C (2010) Topological insulators. Rev Mod Phys 82:3045
180. Massobrio C, Du J, Bernasconi M, Salmon PS (eds) (2015) Molecular dynamics simulations of disordered materials, network glasses to phase-change memory alloys. Springer Science and Business Media, Inc, New York
181. Nishi Y (ed) (2014) Advances in non-volatile memory and storage technology. Chapts. 1, 5–7. Woodhed, Amsterdam

Chapter 8
Plasmonic Gas and Chemical Sensing

Andreas Tittl, Harald Giessen, and Na Liu

Abstract Sensitive and robust detection of gases and chemical reactions constitutes a cornerstone of scientific research and industrial applications. In an effort to reach progressively smaller reagent concentrations and sensing volumes, optical sensor technology has experienced a paradigm shift from extended thin-film systems towards engineered nanoscale devices. In this size regime, plasmonic particles and nanostructures provide an ideal toolkit for the realization of novel sensing concepts. This is due to their unique ability to simultaneously focus light into subwavelength hotspots of the electromagnetic field and to transmit minute changes of the local environment back into the farfield as a modulation of their optical response. Since the basic building blocks of a plasmonic system are commonly noble metal nanoparticles or nanostructures, plasmonics can easily be integrated with a plethora of chemically or catalytically active materials and compounds to detect processes ranging from hydrogen absorption in palladium to the detection of trinitrotoluene (TNT). In this review, we will discuss a multitude of plasmonic sensing strategies, spanning the technological scale from simple plasmonic particles embedded in extended films to highly engineered complex plasmonic nanostructures. Due to their flexibility and excellent sensing performance, plasmonic structures may open an exciting pathway towards the detection of chemical and catalytic events down to the single molecule level.

A. Tittl • H. Giessen
4th Physics Institute and Research Center Scope, University of Stuttgart, D-70569 Stuttgart, Germany
e-mail: h.giessen@pi4.uni-stuttgart.de

N. Liu (✉)
Max Planck Institute for Intelligent Systems, Heisenbergstrasse 3, D-70569 Stuttgart, Germany
e-mail: laura.liu@is.mpg.de

M. Bardosova, T. Wagner (eds.), *Nanomaterials and Nanoarchitectures*, NATO Science for Peace and Security Series C: Environmental Security, DOI 10.1007/978-94-017-9921-8_8

8.1 Introduction

In the last decade, plasmonics has developed into an expansive and vibrant field, with applications as diverse as negative refraction, enhanced photovoltaics, and targeted drug delivery [1–3]. This progress has been driven by major advances in the fabrication of nanostructured systems, including both "top-down" approaches such as electron-beam or nanoimprint lithography as well as "bottom-up" techniques like nanosphere lithography [4–6]. The unprecedented ability to structure metals and dielectrics on the nanoscale has allowed plasmonics to move from giga- and terahertz into the near-infrared and visible spectral ranges. Since the response of plasmonic structures in the latter range allows for remote and non-invasive read-out using standard optical spectroscopy, this move has been a crucial driving factor for the field of plasmonic sensing [7–11].

Most plasmonic sensing concepts rely on the ability of small metal particles to concentrate incident light into deep subwavelength volumes. This behavior is due to the collective oscillations of the conduction electrons against the restoring force of the positively charged nuclei, which can be resonantly driven by an external electromagnetic field [12]. The spectral behavior of such plasmonic resonances can be calculated analytically for geometrically simple systems, such as a subwavelength gold sphere in vacuum, yielding a Lorentzian lineshape. However, for more complex systems, the plasmon resonance position and lineshape are highly dependent on the composition, geometry, and dielectric environment of the nanostructure [13].

The greatly enhanced electromagnetic fields at the surface of a resonant plasmonic structure allow it to probe minute changes in the surrounding environment. Due to the evanescent nature of the fields, the sensing volume of plasmonic nanostructures is only marginally larger than the structures themselves, making them ideal local probes for neighboring changes in a medium. Consequently, plasmonic probes are especially relevant for monitoring of chemical or catalytic reactions, where the reaction efficiency can crucially depend on localized changes in reagent concentration or catalyst morphology.

In order to quantify the sensing performance of a plasmonic system, we can consider its refractive index sensitivity S, defined as the peak shift in nanometers of the plasmonic resonance per refractive index unit (RIU). However, since the final accuracy of the peak tracking depends both on the magnitude of the peak shift and on the resonance linewidth, a more suitable choice for plasmonic systems is the figure of merit (FOM), namely, dividing the refractive index sensitivity by the resonance linewidth.

Thus there are two key approaches for optimizing the figure of merit in a plasmonic sensing geometry: reducing the resonance linewidth, and increasing the peak shift by boosting the local field-enhancement. In general, plasmon oscillations in nanoparticles experience damping arising from two contributions: intrinsic damping due electron-electron collisions and impurity scattering in the metal, and radiation damping due to the direct decay of the collective oscillations into photons. Utilizing noble metals for the resonant nanoparticles helps to reduce intrinsic damping. However, radiation damping, which constitutes the dominating effect for particles larger than a few tens of nanometers, cannot easily be overcome in simple plasmonic systems [14, 15].

By arranging multiple plasmonic particles into a defined geometry, coupling effects may happen and can significantly reduce the linewidth of the plasmonic resonance [16]. One way to achieve this is to couple a broad dipolar plasmonic resonance to a spectrally narrow quadrupolar mode, which cannot be excited from the farfield. The coupling of two such modes can lead to a narrow spectral feature in the otherwise broad dipolar lineshape, which exhibits a greatly increased FOM compared to simple plasmonic structures in the same spectral range, a feature known as the plasmonic analogue of electromagnetically induced transparency (EIT) [17–19].

In order to increase the electromagnetic field-enhancement, plasmonics has taken cues from classical antenna theory. In triangular "bowtie" nanoparticle geometries, the electromagnetic fields can be enhanced and concentrated at the tips of the structure. By utilizing complex Yagi-Uda antenna designs to the nanoscale, efficient and directed energy transfer between the far-field and the near-field can be realized [20, 21]. Using the coupling between, e.g., two rod-type nanoantennas can further enhance the near-fields compared to single antenna elements. When decreasing the distance between the two rods, the electromagnetic fields can be confined to a progressively smaller volume, leading to extremely high enhancement factors, only limited by fabrication tolerances and quantum mechanical tunneling effects below sub 1 nm gaps [22–24].

When such highly sensitive plasmonic structures are adapted for sensing applications, two main concepts for the detection of gases and chemical reagents emerge: in *direct* sensing configurations, the reagents actively influences the plasmonic entity, changing its material properties and in turn its optical response, whereas in *indirect* sensing schemes, the plasmonic structure is located adjacent to some reagent-sensitive material and simply probes the dielectric changes in that system [25].

One experimental issue connecting both sensing schemes is the correlation of gas or reagent concentration changes with changes in the optical response. To facilitate this analysis, one commonly extracts a single "sensor signal" value from the full spectral response. Depending on the system under consideration, observables such as the intensity at a given wavelength, the resonance position, or the resonance peak's full width at half maximum (FWHM) can be tracked in relation to environmental changes. This allows for a straightforward characterization and performance comparison between plasmonic sensor designs with different structural and configurational parameters.

In the following, we will review both direct and indirect sensing schemes, covering a wide technological range from plasmonically functionalized thin films through highly engineered chemically grown particles composed of multiple materials all the way up to complex nanostructured geometries. The presented sensor devices are selected to showcase the wide array of reagents that can be investigated using plasmonic systems as well as the most versatile and high-performance complex plasmonic nanoprobes. By combining such highly sensitive nanodevices with novel reagent-specific materials, plasmonic sensors can be used to investigate a plethora of chemically and biologically significant reagents, with nanometer spatial resolution and possible down to the single molecule level.

8.2 Thin Films with Embedded Plasmonic Particles

One straightforward approach to optical sensing is the use of thin films of reagent-sensitive material deposited on a substrate. Such films combine robust fabrication via standard evaporation or sputtering techniques with simple optical transmission or reflection measurements for read-out [26, 27].

In order to enhance the sensor performance and tailor the optical response of such thin films, small plasmonic particles can be added using co-deposition or multi-step processes. When considering thin films with embedded plasmonic particles, two main concepts can be distinguished. In *plasmonically enhanced* systems, small noble metal particles in the thin film act simply as light concentrators and probes to enhance the dielectric changes in the surrounding material. In *catalytically enhanced* systems, gold nanoparticles in direct contact with, e.g., transition metal oxide films can substantially increase the catalytic activity of such material systems. Here, the optical response stems largely from changes in the transition metal oxide. In the following, we will first focus on two examples of catalytically enhanced thin-film systems.

8.2.1 Carbon Monoxide Sensing with Plasmonically Functionalized Thin Films

Due to its high toxicity, sensitive and reliable detection of carbon monoxide is vital for industrial process and environmental monitoring applications. Ando et al. have demonstrated a plasmonically functionalized Au-CuO nanocomposite film for CO sensing, fabricated by first depositing Au particles on a substrate via DC sputtering and then covering them with a CuO film via pyrolysis [28]. Figure 8.1a shows a transmission electron microscope (TEM) image of the nanocomposite film, where small Au particles of average diameter 24 nm surrounded by a CuO matrix can be clearly identified. The authors performed absorbance measurements at a temperature of 300 °C and observed a plasmonic absorbance peak at a wavelength of around 800 nm. The magnitude of this peak increased substantially upon exposure of the film to 1,000 ppm and 1 % CO in dry air (Fig. 8.1b). In addition, when calibrating their sensor device with various concentrations of CO in dry air, the authors found a linear relation between the absorbance change and the logarithm of the CO concentration over a wide range from 50 to 10,000 ppm CO in dry air (Fig. 8.1c), allowing for straightforward quantitative sensing.

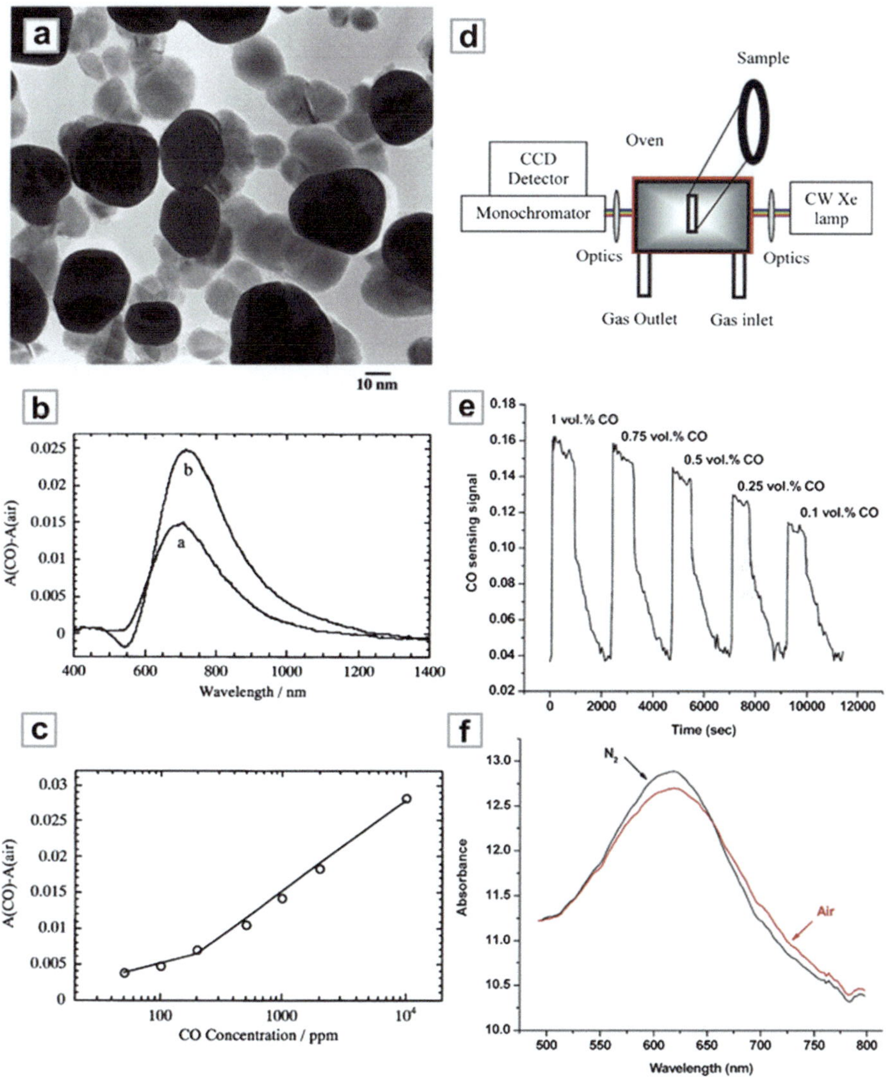

Fig. 8.1 Carbon monoxide sensing with plasmonically functionalized thin films. (**a**) Transmission electron microscope image of a Au-CuO nanocomposite film. (**b**) Relative CO-induced absorbance change of the Au-CuO composite film. (**c**) Correlation between CO concentration and relative absorbance change (**a–c** from [28]). (**d**) Schematic of a setup used for high temperature CO sensing experiments. (**e**) Sensing signal response curve of the Au-YSZ nanocomposite film at 500 °C. (**f**) Absorbance of the Au-YSZ nanocomposite film in N_2 and air atmosphere at 500 °C (**d–f** from [29])

A similar approach for the detection of CO was presented by Sirinakis et al. who used gold nanoparticles embedded in an yttria-stabilized zirconia (YSZ) matrix [29]. Their device was fabricated using radio frequency magnetron co-sputtering followed by an annealing process and optical characterization was performed in a quartz transmission cell surrounded by a tube furnace (Fig. 8.1d). The authors observed a pronounced Lorentz-like absorbance peak at a wavelength of around 600 nm, which spectrally narrowed and blueshifted upon CO exposure. This change allowed the authors to detect CO concentrations ranging from 0.1 to 1 vol.% in dry air at a temperature of 500 °C (Fig. 8.1e). However, they observed no change in the optical response when exposing their sample to CO in nitrogen carrier gas. The authors attributed this behavior to the reactivity of the YSZ film to oxygen, since switching from nitrogen carrier gas to dry air already introduced a broadening and redshift of the plasmon resonance in their system (Fig. 8.1f). This allowed them to explain the blueshift and narrowing of the plasmon resonance through charge transfer into the Au nanoparticles caused by the oxidation of CO and the corresponding reduction of the YSZ matrix at high temperatures. This behavior makes the system by Sirinakis et al. a prime example of *catalytically enhanced* sensing with plasmonically functionalized thin films.

8.2.2 Hydrogen Sensing with Plasmonically Functionalized Thin Films

Other than carbon monoxide, hydrogen detection is an additional safety factor in large-scale industrial chemical processing [30]. Hydrogen forms a combustible mixture with air at concentrations ranging from 4 to 76 vol.%, which can be ignited easily with sparks or at a sufficiently high temperature. Therefore, sensitive hydrogen detection is a cornerstone of alternative energy concepts such as water splitting or, in more general terms, solar-to-chemical energy conversion [31].

In this framework, Buso et al. have utilized a SiO_2 matrix containing Au and NiO nanoparticles for sensitive detection of hydrogen, where the SiO_2 matrix was chosen due to its extremely high specific surface area [32]. High-resolution TEM images in Fig. 8.2a show the evenly distributed Au particles embedded in the SiO_2 host material, surrounded by finely dispersed crystallite NiO particles. In optical measurements, the authors found a plasmon resonance peak at a wavelength of 563 nm, which underwent a drastic blueshift of 22 nm when exposing the sensor device to 1 vol.% H_2 in dry air. This pronounced change in the optical behavior (Fig. 8.2b) could not be sufficiently explained in terms of catalytic charge transfer into the Au particle, which led the authors to relate their system to *plasmonically enhanced* detection of hydrogen. There was, however, a second sensing pathway when exposing the sensor device to CO with a weaker but still measurable response (Fig. 8.2b). The authors assumed an injection of electrons from the NiO crystallites into the Au nanoparticles during CO exposure and subsequent oxidation, which led to the observed changes in the plasmonic response. Thus, their system combined both plasmonically and catalytically enhanced approaches for detection of hydrogen and carbon monoxide.

In a further study on this system, Dharmalingam et al. investigated Au-YSZ nanocomposite films for varying sizes of the embedded Au particles for detection

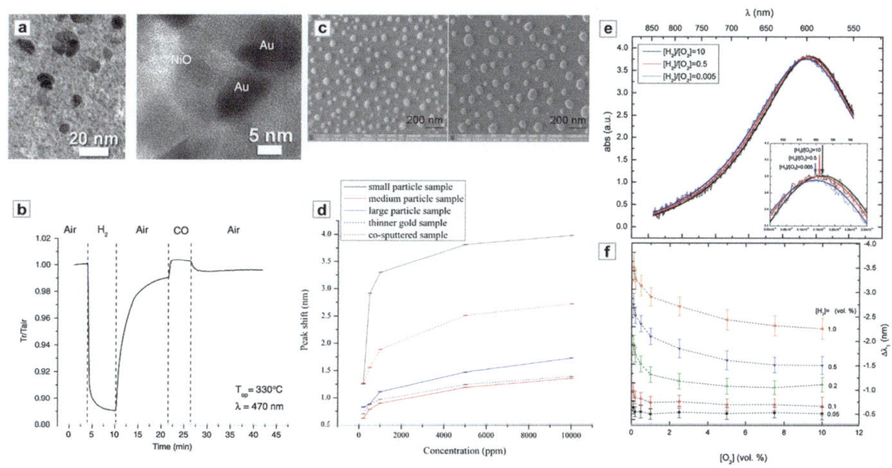

Fig. 8.2 Hydrogen sensing with plasmonically functionalized thin films. (**a**) Cross-sectional and high-resolution TEM images of a SiO_2 matrix containing Au and NiO nanoparticles. (**b**) Dynamic optical response of a film containing 6 % Au exposed to cycles of 1 vol.% H_2 in air and 1 vol.% CO in air (**a**, **b** from [32]). (**c**) SEM images of small-particle and medium-particle Au-YSZ nanocomposite films. (**d**) Hydrogen sensing calibration curves for nanocomposite samples with different particle sizes and preparation methods (**c**, **d** from [33]). (**e**) Absorbance spectra and Lorentzian fits for a Au-YSZ film under various redox conditions at 500 °C. (**f**) Change in plasmon resonance position as a function of O_2 concentration upon exposure to various concentrations of H_2 (**e**, **f** from [34])

of CO and H_2 [33]. For the parameter variation, the authors fabricated their samples using layer-by-layer physical vapor deposition (PVD), where Au was deposited first to form small islands on the substrate and then covered with YSZ (Fig. 8.2c). They then exposed the film to different concentrations of H_2 and CO at a temperature of 500 °C and monitored the spectral changes of the film's plasmon resonance peak at a wavelength of around 580 nm. Calibration curves for several Au nanoparticle configurations are shown in Fig. 8.2d. The authors found that the film with the smallest Au particles (mean crystallite size of approx. 50 nm) yielded the largest response to hydrogen exposure (Fig. 8.2d). When comparing this response to a co-deposited nanocomposite film, where the Au particles were embedded in an YSZ matrix, they obtained an enhancement of the sensing performance by a factor of 1.5. However, this enhancement is accompanied by a decrease of the sensitivity to CO. Still, the study by Dharmalingam et al. outlines how sensitivities and selectivity in Au-YSZ nanocomposite films can be tuned by a straightforward variation of one of the structure parameters, accelerating the development of integrated sensor chips for multiple reagents.

Au-YSZ can also serve as a broader model system for the investigation of electrochemical reactions in nanocomposite geometries, as demonstrated by the work of Rogers et al. [34]. They monitored the optical response of Au nanoparticles embedded in an YSZ matrix during H_2 exposure. However, by simultaneously tuning the concentration of H_2 and O_2, they were able to achieve much better control over the electrochemical reactions occurring in the nanocomposite film. They studied hydrogen to oxygen ratios in nitrogen carrier gas ranging from 10 (1 vol.% H_2, 0.1 vol.% O_2) to 0.005 (0.05 vol.% H_2, 10 vol.% O_2) and found a clear influence of the mixing ratio on the optical response (Fig. 8.2e). When comparing the spectral shifts in their nanocomposite system for different H_2 and O_2 concentrations, the authors made two key observations: an increase of the H_2 concentration generally led to a blueshift and a narrowing of the plasmon resonance, whereas an increase of the O_2 concentration led to a redshift and a broadening (Fig. 8.2f). This behavior could be explained by reaction-induced charge transfer between the oxygen ions in the YSZ matrix and the Au nanoparticles, where the blue- and redshifts were associated with charge injection and removal, respectively. Taking a further step, the authors were able to correlate the plasmon resonance position (and thus the square root of the charge density in the Au nanoparticles) with the gas mixture ratio through an electrochemical model, paving the way towards the all-optical interrogation of different oxygen ion transport and charge transfer mechanisms.

8.2.3 Chemical Sensing in Harsh and Corrosive Environments

In recent years, industrial energy production has experienced an increasing shift from traditional fossil-based technologies towards renewable sources. However, large-scale petrochemical processing is still a vital step in the production of a multitude of products ranging from plastics to pharmaceuticals. Many important petrochemical reactions take place at high temperatures and otherwise harsh or corrosive environments, which requires constant and reliable process monitoring. Due to their durability, sensitivity and all-optical readout, plasmonically functionalized thin films are well positioned to detect a variety of reagents in such challenging environments.

Towards this goal, Ohodnicki et al. have utilized an Au/TiO_2 nanocomposite film for the detection of hydrogen at temperatures as high as 850 °C [35]. The authors prepared their nanocomposite through sol-gel deposition techniques and obtained a TiO_2 matrix with a well-controlled amount of dispersed Au crystallites (Fig. 8.3a). TiO_2 was chosen due to its reactivity to a variety of gaseous reagents combined with excellent thermal stability. They then examined the sensing performance of their device by exposing it to 4 vol.% H_2 in N_2 carrier gas at temperatures of 600 °C and 850 °C and recording the transmittance changes close to the plasmon resonance position at a wavelength of 625 nm (Fig. 8.3b). The authors observed pronounced, reversible and reproducible transmittance changes for both temperatures and found that the response time of their system could be substantially reduced from 33 to only 3 min by increasing the temperature from 600 °C to 850 °C. This rapid and reproducible response makes the Au/TiO_2 system suitable for chemical process monitoring, especially at high temperatures associated with advanced petrochemical processing or reactions in solid oxide fuel cells.

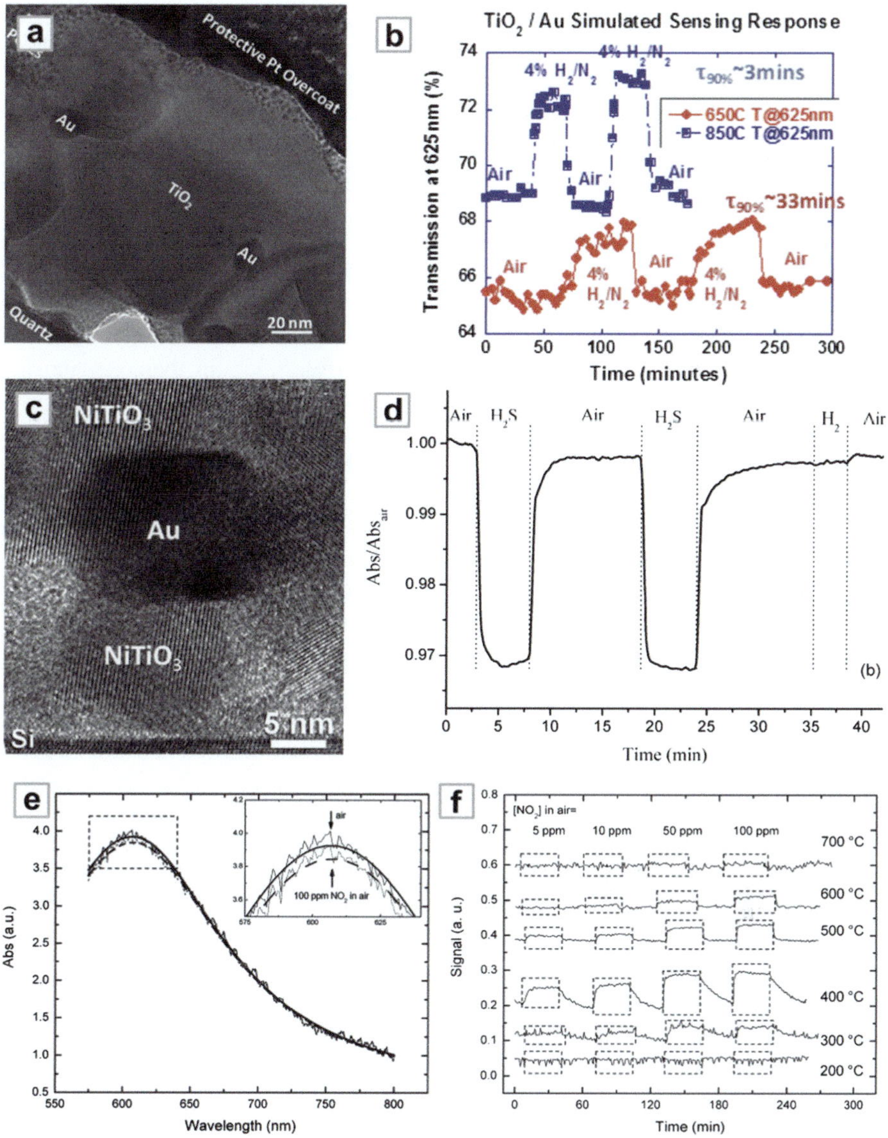

Fig. 8.3 Sensing in harsh and corrosive environments. (**a**) Cross-sectional TEM images for an Au/TiO_2 nanocomposite thin film. (**b**) Sensing response of an Au/TiO_2 film to 4 vol.% H_2 in different environmental conditions (**a**, **b** from [35]). (**c**) Bright-field TEM images of an Au-70TiO_2-30NiO nanocomposite film. (**d**) Dynamic response of the Au-TiO_2-NiO nanocomposite film to gas concentrations of 0.01 vol.% H_2S and 1 vol.% H_2 in air (**c**, **d** from [36]). (**e**) Absorbance spectra and Lorentzian fits for an Au-YSZ film exposed to air and 100 ppm NO_2 in air. (**f**) Sensor signal intensity over time for a Au-YSZ film exposed to air (signal decrease) and 5, 10, 50, and 100 ppm NO_2 in air (signal increase) at 200, 300, 400, 500, 600, and 700 °C (**e**, **f** from [37])

Besides high-temperature chemical monitoring devices, sensors for the detection of reagents in harsh or corrosive environments also play a pivotal role in controlling large-scale chemical processing plants. Hydrogen sulfide (H_2S) is a prominent example for a toxic and corrosive gas that is produced in chemical reactions ranging from natural gas manufacturing to food production. Della Gaspera et al. utilized a TiO_2-NiO film with embedded Au nanoparticles for the detection of H_2S down to the parts per million (ppm) range [36]. The authors produced the films using a sol-gel method with a subsequent annealing step at 500–600 °C and examined three composition ratios of the TiO_2-NiO system: 70TiO_2-30NiO, 50TiO_2-50NiO, and 30TiO_2-70NiO. A bright-field TEM image of an Au-70TiO_2-NiO nanocomposite film also showed that annealing at 600 °C could result in the formation of $NiTiO_3$ islands in addition to the NiO and TiO_2 phases (Fig. 8.3c). In their gas sensing experiments, the authors exposed the three nanocomposite films to 100 ppm H_2S and 1 vol.% H_2 in air and monitored the absorbance change at resonance over time (Fig. 8.3d). They found that the Au-70TiO_2-NiO film yielded the best dynamic behavior with large absorbance changes as well as quick response times. The sample also showed only a negligible response to H_2 versus H_2S, even at a concentration ratio of 100:1, demonstrating the excellent selectivity of the device.

So far, we have mainly covered sensing applications for chemical process engineering. However, the monitoring of exhausts released into the environment is equally important. Many combustion processes release large amounts of nitrous oxides (NO_x), which contribute to the formation of acid rain and degrade water quality on a global scale. Thus, the detection of NO_x close to the source of formation in, e.g., a turbine jet engine is crucial for the development of low-emission machinery. Here, Rogers et al. extended their previous work on Au/YSZ nanocomposites towards the detection of NO_x in small amounts and at temperatures between 500 °C and 800 °C [37]. They prepared their nanocomposite using co-magnetron PVD with a subsequent annealing step and obtained films with an average grain size of approx. 28 nm for both Au and YSZ crystallites. The film was then exposed to different ppm range concentrations of NO_2 in air and the optical absorbance change was recorded for different temperatures. When comparing the absorbance spectra for 100 ppm NO_2 in air with air at a temperature of 500 °C, a redshift and a broadening of the plasmon resonance could be observed (Fig. 8.3e). The authors attributed this behavior to the removal of electrons from the Au nanoparticles, caused by oxygen ion production during the dissociation of NO_2 on the Au-YSZ nanocomposite. To further characterize the temperature dependence of the sensor response, they exposed their sensor device to concentrations of 5, 10, 50, and 100 ppm NO_2 in air at different temperatures ranging from 200 °C to 700 °C. To simplify analysis, the authors subtracted the optical spectra during gas exposure from the "baseline" spectrum in air and defined their sensor signal as the difference between maximum and minimum of the resulting curve (Fig. 8.3e). They found the highest sensor signal change for a temperature of 400 °C whereas the best trade-off between magnitude of the signal and response time was observed at a temperature of 500 °C, with a sensitivity down to 5 ppm NO_2 in air. Furthermore, they were able to resolve concentrations of 50 ppm NO_2 in air in a wide temperature range from 300 °C to 700 °C, which makes their design ideally suited for industrial NO_2 monitoring.

8.3 Shell-Isolated Nanoparticle-Enhanced Raman Spectroscopy

One important application of plasmonic enhancement is the amplification of Raman scattering signals from molecules adsorbed on roughened or nanostructured metallic surfaces [38]. In general, Raman spectroscopy is used to investigate rotational and vibrational transitions in molecules through the measurement of inelastically scattered photons from the sample. However, inelastic scattering signals in such systems are generally much weaker than the elastic Rayleigh scattering and thus are difficult to detect. Here, deposition of molecules onto, e.g., a roughened or nanopatterned metal film can greatly enhance the inelastic response. This signal increase can be understood in terms of surface plasmon propagation at the metal-molecule interface, which leads to strongly enhanced electromagnetic fields as well as to a chemical effect due to charge transfer complexes. Depending on the material and morphology of the substrate, surface-enhanced Raman scattering (SERS) geometries can reach enhancement factors in the 10^{10} range, which makes them suitable for the detection of single molecules [39].

In order to investigate the nanoscale properties of novel catalytic or reactive materials, current approaches need to be extended towards localized optical probes that can be freely dispersed on any surface or morphology, while retaining the high enhancement factors crucial for Raman scattering measurements. Li et al. have utilized small silica-coated gold nanoparticles to perform shell-isolated nanoparticle-enhanced Raman spectroscopy (SHINERS) [40]. For fabrication, the authors chemically synthesized spherical gold particles with a diameter of around 55 nm, which were subsequently coated with a thin, uniform, and optically transparent silica shell (Fig. 8.4a). The purpose of the thin shell is twofold; it offers controllable separation of the gold core from the surface to be studied as well as protection of the core from the chemical or biological environment. In addition, the amount of field-enhancement around the nanoprobes can be tailored by varying the shell thickness. The authors first applied their technique to hydrogen adsorption on platinum. They spread the shell-isolated nanoprobes on a Pt(111) single-crystal electrode covered by a 0.1 M $NaClO_4$ electrolyte and started the hydrogen generation reaction by applying a suitable electrode potential. Through the use of particles with an ultrathin SiO_2 shell of only 2 nm, they were able to detect the Raman signal of the adsorbed hydrogen. To further demonstrate the versatility of their approach for environmental monitoring and food safety applications, the authors studied pesticide residues on fruits. They dispersed their nanoprobes on orange skin with and without methyl parathion contamination and recorded the Raman scattering signal (Fig. 8.4c). They were able to clearly detect the presence

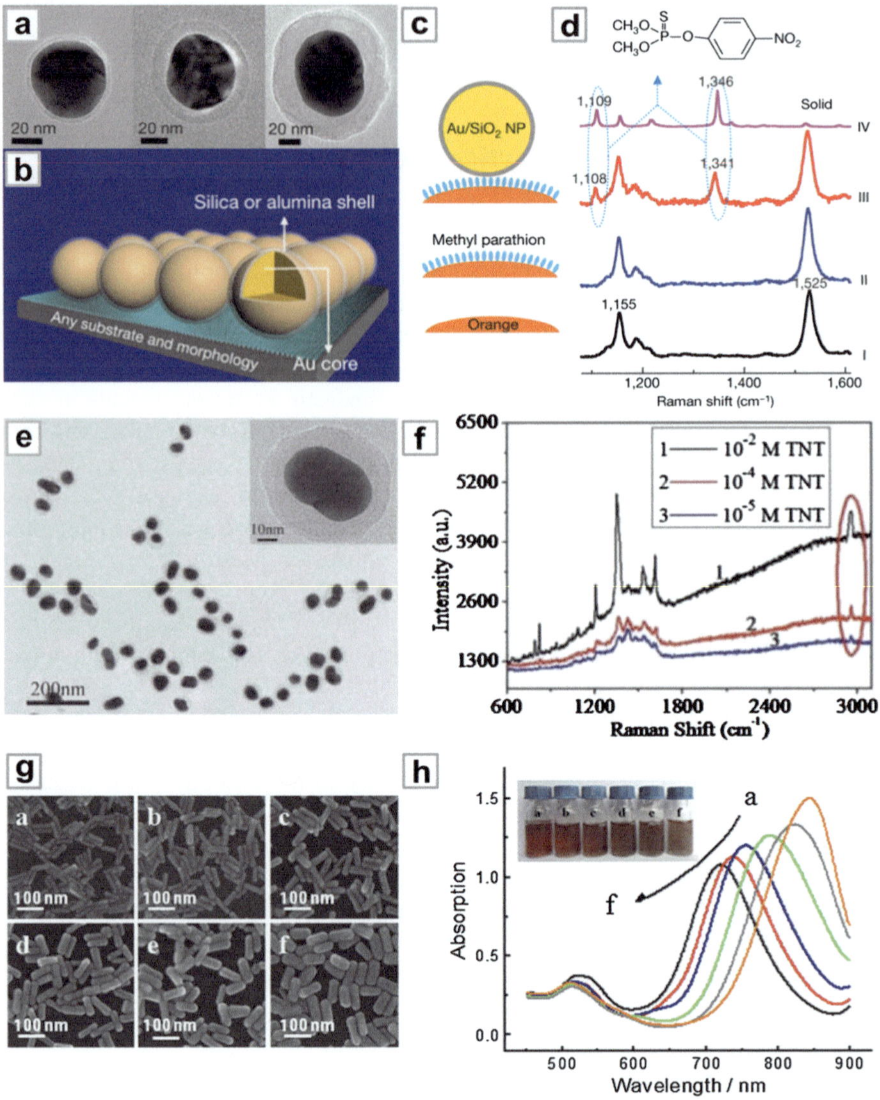

Fig. 8.4 Shell-isolated nanoparticle-enhanced Raman spectroscopy. (**a**) TEM images of Au/SiO_2 core – shell nanoparticles with different shell thicknesses. (**b**) Schematic of the sensing principle. Shell-isolated nanoparticles can be used to probe any surface or morphology. (**c**) Schematic of a SHINERS experiment for the detection of pesticides on citrus fruits. (**d**) Normal Raman and SHINERS spectra for oranges with and without pesticide contamination by parathion (**a**–**d** from [40]). (**e**) TEM images of Au/PAT core – shell nanoparticles with different shell thicknesses (**f**) SERS spectra of Au@2 nm PAT NPs + TNT (10^{-2} M to 10^{-5} M) (**e**, **f** from [41]). (**g**, **h**) Extinction spectra and SEM images of Au NRs with different aspect ratios prepared by tuning the volumes of the Au seed solution (**g**, **h** from [42])

of parathion on the orange skin samples in their nanoparticle-enhanced Raman spectra, whereas measurements without the nanoprobes yielded no signal changes (Fig. 8.4d). Thus, SHINERS offers a clear advantage in sensitivity and flexibility over standard SERS approaches and constitutes an important milestone on the pathway from plasmonically functionalized thin films to more complex sensing devices.

The reliable monitoring of environmental safety is a major area of chemical sensing. One crucial aspect for land and air traveling is the detection of explosive materials in low amounts and on diverse surfaces and morphologies. To achieve this, Qian et al. demonstrated the use of functionalized shell-isolated nanoprobes for the selective detection of trinitrotoluene (TNT) [41]. They first synthesized gold nanoparticles with a diameter of 50 nm, which were then coated with an ultrathin 2 nm shell of poly- (2-aminothiophenol) (PAT). In a final processing step, the PAT was functionalized with amino groups to facilitate the binding of TNT. This resulted in uniform nanoprobes with high yield and controllable shell thickness (Fig. 8.4e). The authors then performed SHINERS measurements on their nanoprobes in the presence of different amounts of TNT ranging from 10^{-2} to 10^{-5} M and observed several TNT-specific Raman peaks in their spectra for all concentrations (Fig. 8.4f). To elucidate the mechanism for the TNT-PAT binding, the authors focused on the Raman peak around 2,955 cm^{-1}, which they attributed to the formation of a Meisenheimer complex between the TNT molecules and the amino groups at the PAT surface. By tracking signal changes only at the wavenumber associated with the highly specific Meisenheimer complex, the authors were able to demonstrate excellent selectivity of their sensor geometry even when exposed to very similar reagents such as dinitrotoluene (DNT).

To optimize the sensing performance of SHINERS-based devices even further, more complex particle geometries with tunable properties need to be synthesized. In this context, Zhang et al. investigated gold nanorods covered with an ultrathin shell of silica [42]. By varying the amount of gold seed solution in their synthesis process, they were able to reproducibly grow nanorods with different lengths and aspect ratios ranging from 2.2 to 3.7 (Fig. 8.4g). In addition, the shell thickness could be varied from 3 to 7 nm. To optically characterize their shell-isolated nanoparticles, the authors performed absorbance measurements and found that the plasmon resonance wavelength of their nanoprobes could be tuned from around 700 nm well beyond 800 nm (Fig. 8.4h). With such flexibility, SHINERS-based sensor geometries can easily be tuned to the wavelength of an exciting laser beam, maximizing the Raman enhancement and allowing for reagent detection with greatly improved sensitivity.

8.4 Label-Free Refractive Index Gas Sensing

Many of the sensing approaches presented so far have relied on a combination of passive plasmonic antennas, often composed of noble-metal nanoparticles, with a chemically active surrounding material. This material allows the composite system to exhibit selectivity to specific reagents and provides a reactor for chemical surface or bulk processes. Label-free sensing devices, on the other hand, focus on the reagent-induced refractive index changes in the vicinity of plasmonic structures. Since the adsorption of biological molecules or the generation of new compounds in chemical processes causes fairly large refractive index changes, label-free methods are well suited for the investigation of a plethora of liquid-phase systems [43]. However, the label-free detection of chemical reagents and processes in the gaseous phase is significantly more challenging, mainly due to the minute refractive index changes associated with gas concentration variations.

Bingham et al. utilized high-resolution localized surface plasmon resonance (HR-LSPR) spectroscopy on arrays of silver nanoparticles to detect gas-induced refractive index changes in the 10^{-4} range (Fig. 8.5a) [44]. For their gas sensing experiments, the authors fabricated Ag nanoparticle arrays using nanosphere lithography and recorded extinction spectra while alternating the gaseous environment

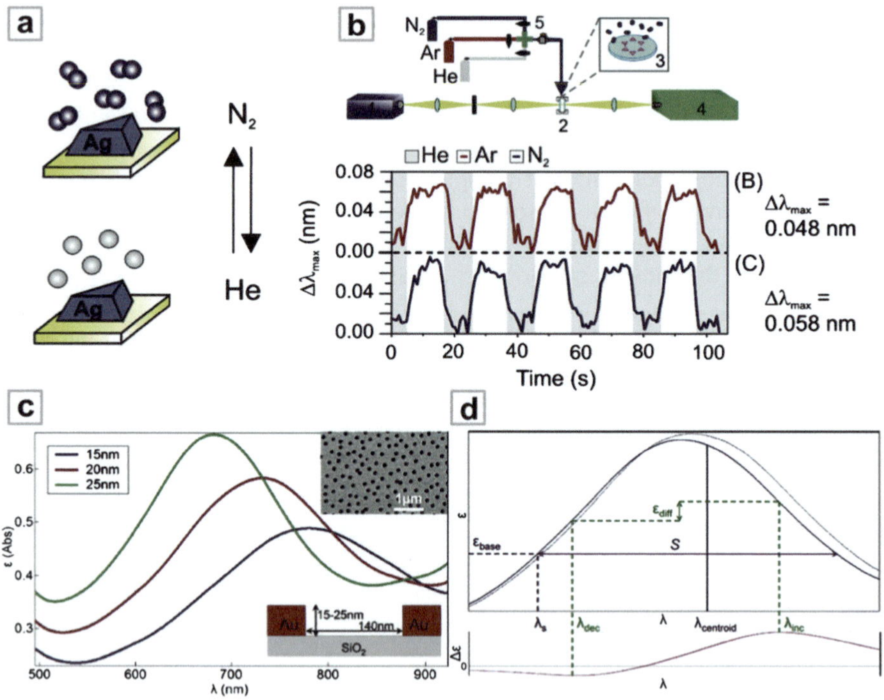

Fig. 8.5 Sensitive reagent detection using high-resolution LSPR spectroscopy. (**a**) Refractive index gas sensing with silver nanoparticles. (**b**) Resonance position of the Ag nanoparticles over time as the gas switches between He/Ar and He/N_2 (**a**, **b** from [44]). (**c**) Extinction spectra of 140 nm-wide nanoholes with constant surface density (~9 μm^{-2}) but different depths. (**d**) Illustration of the parameters used in the centroid detection algorithm using a typical nanohole LSPR peak before and after a small increment in bulk RI. Shown below is the difference between the two spectra. The parameters used to analyze the peak position, centroid wavelength $\lambda_{centroid}$, centroid span S, and baseline intensity ε_{base}, are indicated (**c**, **d** from [45])

repeatedly between helium/argon and helium/nitrogen every 10 s. They observed clear spectral redshifts when switching from He to the target gases, with magnitudes of 0.048 nm and 0.058 nm for Ar and N_2, respectively, which corresponds to a sensitivity of around 200 nm/RIU. In addition, the authors performed gas exchanges every 5 s and again found mass-transport limited rise and fall times, indicating the detection of RI changes in the gaseous environment as opposed to adsorption processes at the Ag surface. The main enabling factors in their high-resolution measurements were the use of silver for plasmonic resonators with low intrinsic damping, the stability of their experimental setup, and the ability to precisely track the plasmon resonance wavelength over multiple measurements.

A reliable and highly accurate technique for determination of the resonance position from (potentially noisy) experimental spectra is the centroid method as introduced by Dahlin et al. for LSPR spectroscopy [45]. As a first application, the authors investigated the sensitivity of nanohole arrays in a thin gold film towards small refractive index changes in the environment (Fig. 8.5c). The centroid method relies on the computation of the first moment (or center of mass) of a spectral intensity distribution $I(\lambda)$, taken over a certain wavelength span S (Fig. 8.5d). In order to reduce measurement noise and provide a continuous analytical representation of the experimental data, the authors first fitted the nanohole absorbance spectra $I(\lambda)$ using a high-degree polynomial $\varepsilon(\lambda)$ of order 20. The starting wavelength λ_s of the span S in the spectra was obtained by minimizing the expression $\varepsilon(\lambda_s + S) - \varepsilon(\lambda_s)$ and the intensity baseline $\varepsilon_{\text{base}}$ was defined as $\varepsilon(\lambda_s)$. The centroid wavelength $\lambda_{\text{centroid}}$ was then calculated using

$$\lambda_{\text{centroid}} = \frac{\int_{\lambda_s}^{\lambda_s+S} \lambda\left(\varepsilon(\lambda) - \varepsilon_{\text{base}}\right) d\lambda}{\int_{\lambda_s}^{\lambda_s+S} \left(\varepsilon(\lambda) - \varepsilon_{\text{base}}\right) d\lambda}$$

Since there is a linear relationship between the resonance wavelength λ_{peak} and $\lambda_{\text{centroid}}$ with a proportionality factor of roughly 1 for most plasmonic systems, the centroid wavelength shift serves as a reliable measure for sensor performance. Overall, the centroid detection approach allows for the sensitive, robust and reproducible extraction of resonance wavelength positions from noisy experimental data and thus enables a plethora of applications in biological and chemical sensing. In this, it is clearly superior to common peak tracking or fitting methods due to the inclusion of the full span S of spectral data and its suitability for, e.g., non-Lorentzian line shapes.

8.5 Engineered Nanoparticles and Smart Dust

Highly engineered or "smart dust" nanoparticles offer many exciting possibilities for the detection of chemical or biological reagents [46]. Furthermore, the combination of multiple materials in complex core-shell nanoparticles and the use of advanced synthesis procedures allow for new functionalities as well as increased flexibility in structure design [47–49]. Implementing this concept for plasmonic gas sensing, Ghodselahi et al. have demonstrated the use of Cu@CuO core-shell nanoparticles for the detection of carbon monoxide [50]. The authors first prepared hexagonal Cu nanoparticle arrays on a glass substrate using codeposition of RF-sputtering and RF-PECVD, and subsequently oxidized them under airflow to obtain the CuO shell structure. This yielded uniform Cu@CuO core-shell nanoparticles with a thickness of 35 nm and a mean lateral dimension of 10 nm. To demonstrate the carbon monoxide sensitivity of their system, the authors recorded absorbance spectra of their sample during exposure to different flow rates of CO. In ambient air, they observed a pronounced plasmon resonance peak at around 600 nm, which underwent a spectral redshift and an increase in the absorbance for increasing CO flow rate (Fig. 8.6a). Spectral redshifts of approx. 3 nm were observed for a low flow rate of 1.6 l/h, with a maximum value of approx. 8 nm and saturation-like behavior for flow rates in the 10 l/h range (Fig. 8.6b). This response could be attributed to CO adsorption on the CuO shell, which leads to a dielectric change in the active material and thus to a spectral shift of the plasmon resonance in the Cu. Since CO molecules are only adsorbed on the particle surface, the CuO shell could potentially be cleaned via a low-temperature annealing step in air. This would reduce sensor performance degradation and enable long-term environmental monitoring applications.

Apart from different metals and oxides, smart dust nanoparticles can also be capped with biological species to facilitate the detection of diverse reagents. Vasileva et al. have demonstrated the use of starch-stabilized silver nanoparticles to detect hydrogen peroxide (H_2O_2) [51]. Hydrogen peroxide plays an important role in the human body, where it is a key factor for oxygen metabolism, enzymatic breakdown of glucose or lactose, and oxidative stress [52]. In their experiments, the authors designed a green synthesis procedure employing only soluble starch, silver nitrate, D-glucose, and sodium hydroxide to obtain an aqueous dispersion of nanoparticles with a mean diameter of 14.4 nm and excellent uniformity. Importantly, they performed ultrasonication and controlled alkalization of the reaction medium to tailor the size and shape of their Ag particles (Fig. 8.6c). A key feature of their synthesis approach is the use of soluble starch as a capping agent, which eliminated the need for toxic chemicals in the reaction process and still allowed for sufficient control over particle morphology. When added to the nanoparticle dispersion, H_2O_2 acted as an active oxidant, leading to the degradation of the starch-stabilized silver particles. The authors could correlate this degradation with the absorbance strength of the silver particles' plasmon resonance and thus were able to reliably differentiate H_2O_2 concentrations ranging from 10^{-1} to 10^{-6} mol/l (Fig. 8.6d).

A different bio-inspired approach for detection of H_2O_2 has been demonstrated by Endo et al. [53]. They used silver nanoparticles in solution to obtain a pronounced plasmonic response, but coated them with polyvinylpyrrolidone (PVP) to improve their stability towards other reagents (Fig. 8.6e). Exposure to H_2O_2 led to an aggregation of the PVP-coated Ag nanoparticles and a subsequent decrease of the

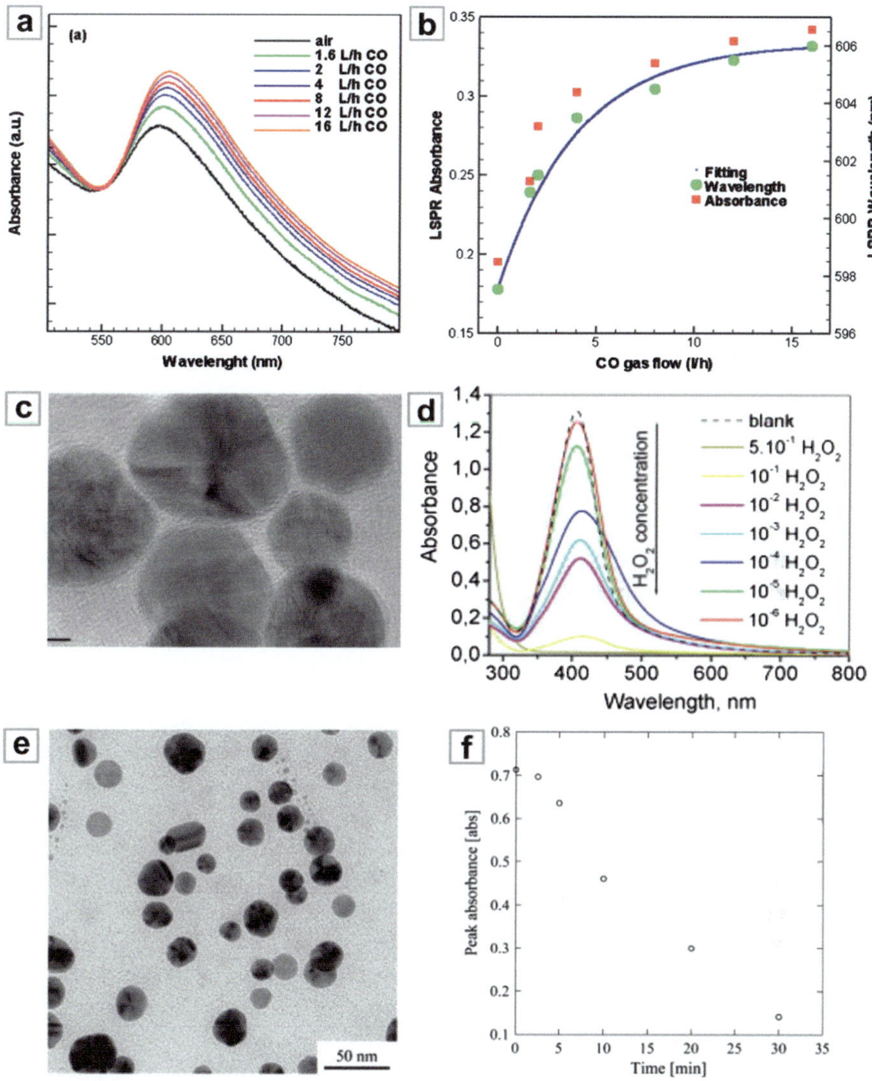

Fig. 8.6 Sensing with complex core-shell nanoparticles. (**a**) Absorbance spectra of Cu@CuO core-shell nanoparticles under air and various CO gas concentrations. (**b**) LSPR absorbance (*left*) and LSPR wavelength (*right*) versus CO concentration (**a**, **b** from [50]). (**c**) TEM image of starch-stabilized silver nanoparticles for hydrogen peroxide detection. (**d**) Absorbance spectra of the Ag nanoparticle solution 15 min after the addition of hydrogen peroxide solution for different H_2O_2 concentrations (**c**, **d** from [51]). (**e**) TEM image of polyvinylpyrrolidone-coated silver nanoparticles in water. (**f**) Absorbance at resonance over time after the addition of 1 mM hydrogen peroxide (**e**, **f** from [53])

absorbance strength of the plasmon resonance at a wavelength of around 405 nm. The authors investigated this behavior in more detail by introducing 1 mM H_2O_2 into a vial of PVP-coated Ag nanoparticles and tracking the absorbance over time (Fig. 8.6f). They again observed a strong decrease of the peak absorbance, with a large enough effect to be visible by naked eye. Thus, such coated nanoparticles could be employed as cheap and biocompatible color markers for medical and pharmaceutical applications.

Due to their unique properties, plasmonic smart dust particles can sensitively probe localized structural features such as size, shape, material, and morphology [54, 55]. Furthermore, the efficiency of many physical, chemical and catalytic processes depends crucially on such nanoscale properties. This is especially true for large-scale catalytic hydrogen generation and storage, which are a key component of green energy production and may facilitate a move away from present, fossil-based technologies.

Palladium constitutes an ideal model system for hydrogen dissociation and subsequent absorption processes in metals [56–59]. During catalytic loading, hydrogen molecules are first chemisorbed on the Pd surface, followed by near barrierless dissociation and subsequent diffusion of atomic hydrogen into the lattice. When increasing the hydrogen concentration, palladium undergoes a phase transition from a mobile, solid solution α-phase to a palladium hydride (PdH) β-phase. This transition causes both an expansion of the Pd lattice as well as a pronounced change in the dielectric function.

To study the influence of the localized particle geometry on this reaction, Tang et al. investigated the hydrogen uptake trajectories for individual Au/Pd core-shell particles with different shapes, faceting, and Pd shell thickness [60]. The authors employed a synthesis method designed for particles with high quality but low uniformity to obtain a mixture of triangular plates, decahedrons, and icosahedrons (Fig. 8.7a). This allowed them to study nanoparticles with different sizes and shapes from the same synthesis process, aiding in reproducibility. The authors first studied hydrogen-induced shifts of the plasmon resonance for an Au/Pd core-shell triangular plate and found a pronounced spectral redshift on the order of 25 nm (Fig. 8.7b). Both this result and similar measurements on icosahedrons were found to be in good agreement with numerical simulations, taking into account the dielectric function change from Pd to PdH. However, when investigating decahedral nanoparticles, the authors measured a pronounced spectral blueshift upon initial H_2 exposure (Fig. 8.7b), which could not be accounted for in numerical simulations using a simple Pd/PdH transition. This contrasting behavior was attributed to two chemical transformations that occur during hydrogen exposure: the interdiffusion of Au and Pd, which leads to an effective decrease of the Au particle size and thus a blueshift, and the formation of palladium silicide due to an reaction of the nanoparticles with silicon atoms of the substrate below. Both effects were found to be especially pronounced in nanoparticles with thicker Pd shells, allowing for the straightforward differentiation of the particle species. Thus, the plasmonic resonance of individual smart dust nanoparticles can serve as a sensitive probe for H_2 storage properties and metal support interactions in nanocatalytic systems, revealing geometry-dependent properties that are not accessible via standard ensemble measurements.

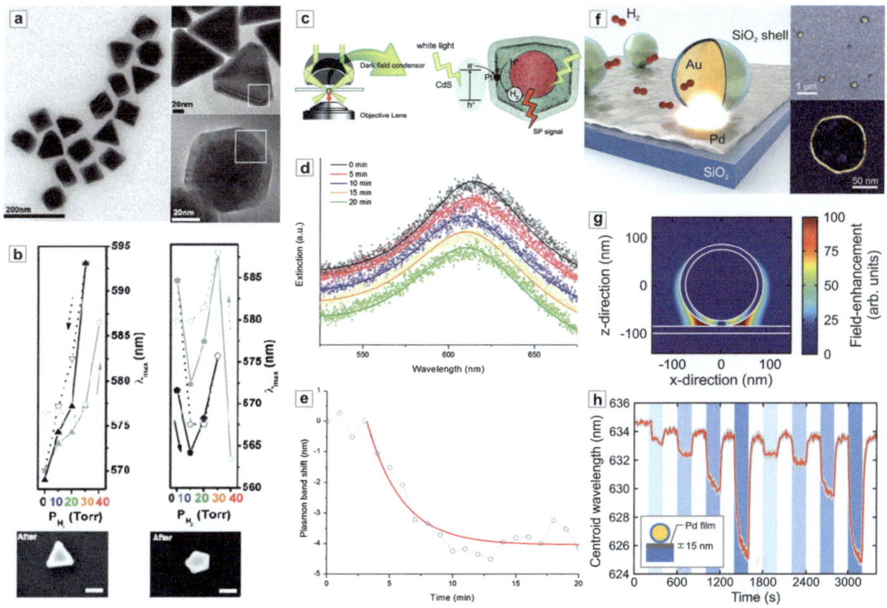

Fig. 8.7 Complex smart dust nanoparticles for hydrogen and chemical sensing. (**a**) Low- and high-resolution TEM images of Au/Pd core-shell nanocrystals (**b**) Plasmon resonance shift upon hydrogen uptake and release for a single Au/Pd core-shell triangular plate (*left*) and decahedron (*right*) (**a**, **b** from [60]). (**c**) Schematic diagram of single-particle measurement and detection on a single Au@Pt/CdS catalyst particle. (**d**) Extinction spectra of a single Au@Pt/CdS cube along the reaction progress. (**e**) Plasmon resonance shift during reaction for the same cube (**c–e** from [61]). (**f**) Schematic of plasmonic probing of local chemical reactions using smart dust nanoparticles (*left*). Exemplary element-resolved TEM and SEM images of the probing platform (*right*). (**g**) Electric near-field distribution at resonance. The intense near-fields are strongly localized at the touching point between the smart dust particle and the Pd film. (**g**) Time-resolved centroid wavelengths in response to hydrogen uptake and release for a single smart dust particle on a 15 nm Pd film (**f–h** from [62])

An ideal application for such smart dust nanoprobes is the catalytic generation of hydrogen, which is a key prerequisite for renewable energy production. Here, Seo et al. investigated the photocatalytic decomposition of lactic acid for hydrogen generation, using platinized cadmium sulfide (Pt/CdS) nanocubes [61]. In order to enable in-site monitoring of the produced H_2 gas, they integrated gold domains into the nanocube reactors and recorded time-resolved dark-field scattering spectra from single structures (Fig. 8.7c). When their nanoreactors were illuminated with white light, photons that match the optical band gap were absorbed in the CdS shells, leading to the generation of photoelectrons. These electrons can transfer to the Pt part and react with protons from the surrounding solution to produce H_2 gas. During the reaction, the authors observed a gradual blueshift of the plasmonic resonance with a maximum value of 14 nm combined with a decrease of the scattering amplitude (Fig. 8.7d). This was found to be consistent with a change of the local environment close to the gold probe from 10 % lactic acid in water ($n = 1.342$)

to hydrogen gas ($n = 1.000$), taking into account that the Au probe has an average distance of 20 nm from the active Pt surface. As expected, the time-resolved spectral resonance position followed a saturation behavior, reaching the maximum blueshift after around 10 min (Fig. 8.7e). By performing further analysis on their spectral time trace data, the authors were able to estimate the hydrogen generation from a single nanoreactor particle at around 1,300 molecules per minute, showing that the use of smart dust probes can yield detailed information on local reaction kinetics from non-invasive far-field measurements.

In order to extend the presented concepts to the detection of chemical or catalytic processes in realistic material systems and environmental conditions, there is a need for a versatile sensing platform that can probe arbitrary surfaces and morphologies. To achieve this, Tittl et al. utilized silica shell-isolated gold nanoparticles to monitor local chemical reactions in real time [62]. The smart dust nanoparticles can focus incident light into subwavelength hot spots of the electromagnetic field and optically report minute environmental changes at their pinning sites on the probed surface during reaction processes (Fig. 8.7f, left). Furthermore, the smart dust particles can easily be dispersed on target surfaces via drop-coating and allow for straightforward all-optical readout via standard dark-field spectroscopy (Fig. 8.7f, right). As a model system, the authors investigated the hydrogen dissociation and subsequent absorption in adjacent palladium thin films. A key feature of their particles is the ultrathin SiO_2 shell with a thickness of 10 nm (Fig. 8.7f, right), which both separates the Au core from direct contact with the probed agents and enables the high localized field-enhancements necessary for the sensitive detection of dielectric changes in the neighboring Pd film (Fig. 8.7g). For their sensing measurements, the authors exposed a "dust on film" sample consisting of smart dust nanoprobes on a 15 nm Pd film to concentrations of 0.5, 1, 2, and 3 vol.% hydrogen in nitrogen carrier gas. They were able to observe different concentration steps as clear spectral blueshifts in the optical scattering response of single smart dust particles, with a maximum value of around 8 nm for 3 % H_2 (Fig. 8.7h). To check the ability of their platform to resolve localized morphology changes, the authors next investigated the "film on dust" system, where the smart dust particles were covered with a highly curved Pd film via tilted angle evaporation and found pronounced redshifts for all concentrations. This shows how the smart dust probing platform can transform small localized morphology changes into a complete reversal of the spectral shift direction in the far-field optical response, allowing it to resolve chemical reactions for nearly arbitrary material placements and configurations. In the future, such techniques could combine two-dimensional imaging and spectroscopy to locally map chemical and catalytic reactions on a subdiffraction-limit scale.

8.6 Large-Area Nanostructured Sensor Chips

Synthesis and deposition-based approaches enable the reliable and high-throughput fabrication of plasmonic nanosensor devices. However, in order to produce large-area sensor chips with a predefined spatial arrangement of plasmonic elements, other fabrication methods such as soft-imprint lithography, nanosphere lithography, or (to some extent) electron-beam lithography often need to be implemented [5, 6].

As a first example of nanosphere lithography, Wang et al. utilized ordered gold nanoparticle arrays to study the chemical adsorption of mercury vapor on the Au surface [63]. They fabricated the particle arrays by evaporating 2 nm chromium (Cr) and 40 nm Au through the interstitial voids of a densely packed monolayer of polystyrene (PS) nanospheres, followed by sonication in ethanol to remove the PS spheres from the substrate. This yielded triangular gold nanoparticles with a hexagonal spatial arrangement, which could be transformed into more circular nanodisks by a subsequent annealing step (Fig. 8.8a). The authors then exposed their sample to 15 ppm of mercury vapor delivered by nitrogen flow and monitored the shift of the plasmon resonance peak over time. They found a clear monotonic spectral blueshift from 565 to 560 nm, which saturated after 2 h and reached 85 % of the maximum value after around 30 min (Fig. 8.8b). This adsorption time was found to be shorter than in the case of an extended Au film, where only 67 % of the maximum response was obtained after 30 min [64]. In addition, the authors were able to regenerate their sensor chip using a heating step at 300 °C for 30 min, resulting in a return of the plasmon resonance position to the original value. The blueshift was attributed to the formation of an adsorbed mercury shell on the hemispherical Au nanoparticles, which could be clearly observed as a volume increase in AFM measurements, leading to a change of the effective refractive index and thus the observed spectral shifts.

Ma et al. used a structurally similar system for the detection of chloroform vapor, a prime example of volatile organic compounds (VOC) that play a role in both industry and medical diagnostics [65, 66]. Their sensor device consisted of an array of triangular silver nanoparticles, covered with a of poly(methyl methacrylate) (PMMA) transduction layer which provided the reagent sensitivity [67]. For fabrication, the authors utilized nanosphere lithography followed by spin-coating of the PMMA layer to produce high-quality ordered nanoparticle arrays over large areas (Fig. 8.8c). PMMA was chosen as a chloroform sensitive material due to its excellent long-term stability, low optical loss and transparency in the visible spectrum. In optical extinction measurements, the system exhibited a clear plasmon resonance at a wavelength of around 670 nm, which decreased in magnitude upon exposure to increasing concentrations of chloroform vapor (Fig. 8.8d). Compared to this pronounced response in extinction magnitude, the authors found only small changes in the resonance wavelength for the different chloroform concentrations. This spectral behavior was found to be consistent with chloroform-induced swelling in the PMMA layer. When switching on the chloroform vapor flow, the system reached the saturated value of maximum extinction change after a fast response time on the order of 10 s and the system returned to the initial state 300 s after switching off the flow. Finally, the authors examined the cross-sensitivity of their sensor device and found only negligible responses for exposures to benzene, toluene, ethanol, propyl alcohol, or hexane. This shows that their system can be utilized as a stable, reliable, and selective chloroform sensor for health and safety or industrial uses.

One key application of large-area nanostructured sensor chips was first demonstrated by Langhammer et al. who utilized palladium nanodisks to perform hydrogen

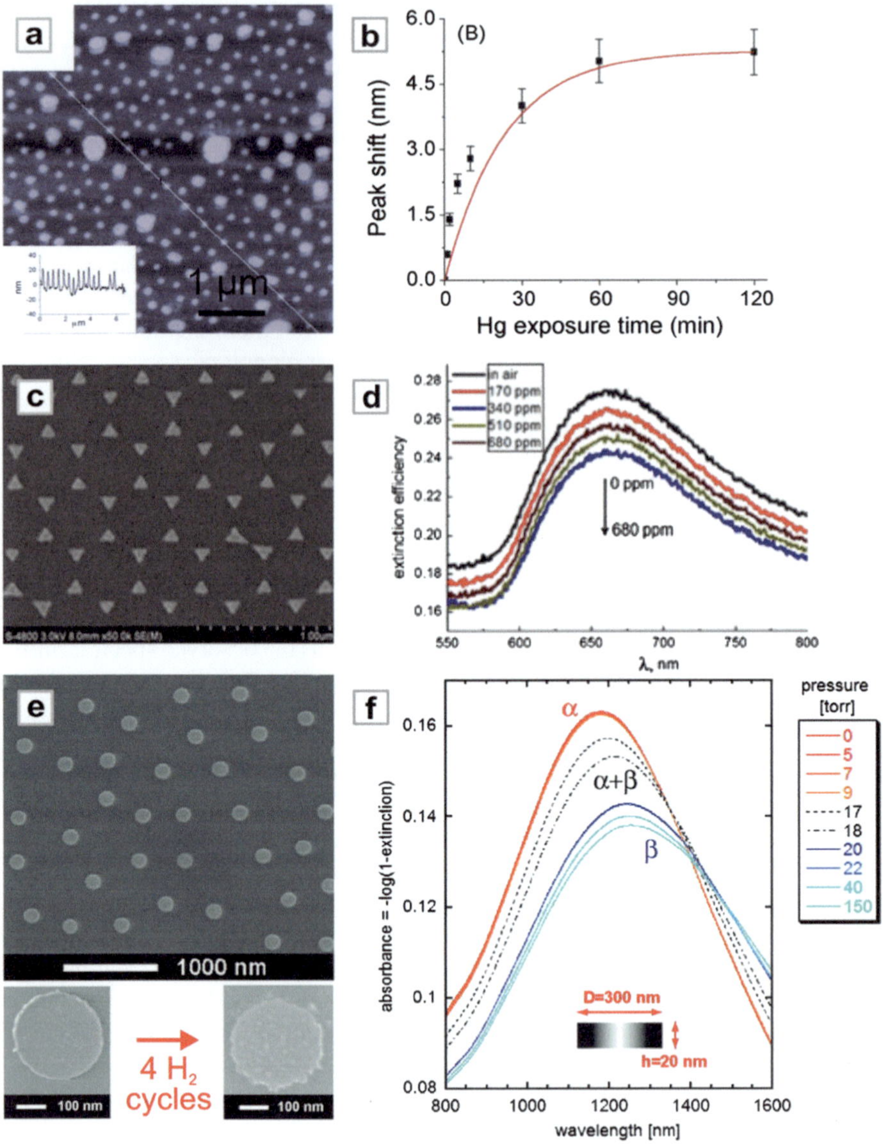

Fig. 8.8 Large-area nanostructured sensor chips. (**a**) AFM image of ordered gold nanoparticle arrays prepared using nanosphere lithography. (**b**) Plasmon resonance wavelength shifts of the gold nanoparticle arrays at different times after exposure to mercury vapor (**a**, **b** from [63]). (**c**) SEM image of plasmonic nanostructures coated with a PMMA transduction film. (**d**) LSPR spectral response of the PMMA-coated structures to chloroform of different concentrations (**c**, **d** from [67]). (**e**) SEM image of a Pd nanodisk array used for hydrogen sensing (*top*) and degradation of the system after several exposure cycles (*bottom*). (**f**) Series of measured extinction spectra for the nanodisk array at successively higher hydrogen pressures (**e**, **f** from [68])

sensing experiments and to study the hydrogen-induced phase transition from palladium (Pd) to palladium hydride (PdH) on the nanoscale [68]. The Pd/PdH transition causes both a strong change of the dielectric function and an expansion of the Pd lattice, allowing it to be observed in resonant plasmonic systems. To achieve this, the authors fabricated a disordered array of supported Pd nanodisks with a diameter of 300 nm using hole-mask colloidal lithography (Fig. 8.8e, top). They then measured extinction spectra for different hydrogen partial pressures ranging from 0 to 150 Torr. The initial plasmon resonance peak at a wavelength of around 1,200 nm underwent a pronounced redshift as well as a broadening and a decrease of the total extinction cross-section for increasing hydrogen pressure (Fig. 8.3f). By correlating the peak shifts with the hydrogen partial pressures, the authors were able to obtain hydrogen absorption and desorption trajectories comparable to the well-know pressure-composition isotherms that characterize the phase transition in the Pd-H system, showing that the nanoscale phase behavior can be effectively retrieved from optical measurements. They also found that it was crucial to use relative spectral shifts instead of absolute peak positions for the derivation of the pressure-composition isotherms, since repeated hydrogen loading and unloading cycles led to irreversible changes in the Pd nanostructures (Fig. 8.8e, bottom). Theses changes are due to the phase-transition-induced 10 % volume increase of the Pd lattice and the associated mechanical stress [56]. For use in highly stable industrial sensor devices, this hydrogen-induced structural deterioration can be overcome by, e.g., alloying Pd with nickel (Ni) and by sandwiching the resulting material in between optimized capping and buffering layers [69].

The presented nanostructured sensor chips exhibit excellent homogeneity over large areas as well as high performance and can thus serve as the reagent-sensitive element in simple, all-optical, and industry-compatible sensor devices. This enables their use in a variety of applications, especially in the field of environmental monitoring, where high numbers of sensor devices are needed.

8.7 Complex Grating-Based Plasmonic Hydrogen Sensors

In order to further improve the performance of plasmonic sensor devices, many recent designs have moved from simple plasmonic resonators towards complex coupled or hybrid nanostructures [16, 70–72]. Coupled plasmonic systems provide a way to tackle two main limiting factors for the sensitivity of such devices: the linewidth (and thus the quality factor) of the plasmon peak and the background signal close to the resonance position.

One approach to reduce the resonance linewidth is to place an array of plasmonic nanowires on top of a photonic waveguide to form a metallic photonic crystal slab [73–76]. To obtain sensing functionality, Christ et al. used Au nanowires on top of a gasochromic WO_3 waveguide, which undergoes a strong modification of its optical properties when exposed to hydrogen gas [77]. The authors also incorporated an

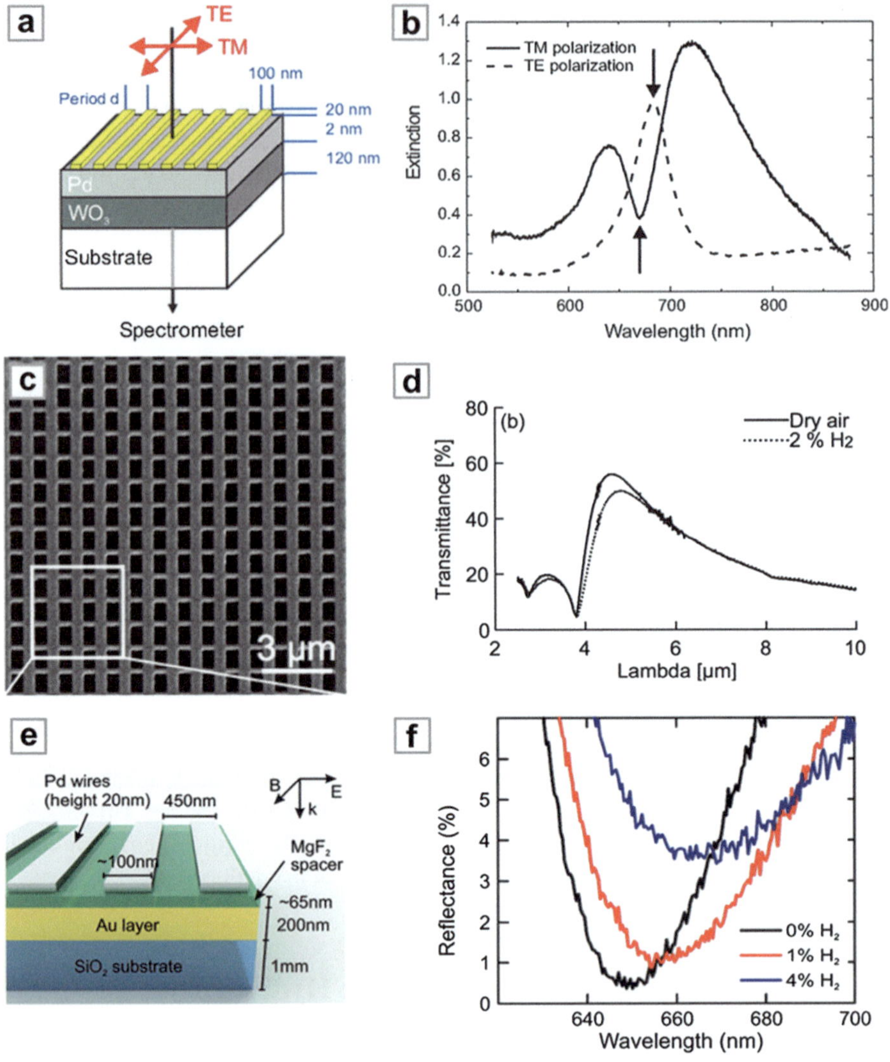

Fig. 8.9 Metallic photonic crystal and perfect absorber hydrogen sensors. (**a**) Schematic of a metallic photonic crystal slab which incorporates WO_3 as the waveguide layer material. (**b**) Extinction spectra of the photonic crystal sensor in TE and TM polarization (**a**, **b** from [77]). (**c**) SEM image of a subwavelength Pd hole array. (**d**) Transmittance change for a hole array with an aspect ratio of 1.6 when exposed to 2 vol.% H_2 in dry air (**c**, **d** from [81]). (**e**) Schematic of the palladium-based perfect absorber hydrogen sensor. (**f**) Reflectance spectra of an optimized perfect absorber structure in N_2 and when exposed to 1 and 4 vol.% H_2 in N_2 (**e**, **f** from [86])

ultrathin (2 nm) catalytic Pd layer between the wires and the waveguide to facilitate the splitting of molecular hydrogen gas into atomic hydrogen for diffusion into the WO_3 waveguide (Fig. 8.9a). When the system is excited perpendicular to the nanowires, the associated plasmon resonance can couple to the quasiguided mode of the WO_3 waveguide, resulting in the formation of a polariton-type coupled system. In extinction measurements, the upper and lower branch of the polariton can be associated with two pronounced maxima in the spectrum. Additionally, the coupling between the broad plasmon peak and the narrow waveguide mode leads to the formation of a sharp extinction dip between the maxima, which is ideally suited for sensing due to its low FWHM (Fig. 8.9b). Upon hydrogen exposure, the optical properties of the WO_3 layer change drastically, which in turn causes pronounced changes in the extinction spectrum of the complete metallic photonic crystal slab. In their sensing experiments, the authors exposed their sample to a constant gas flow with a predefined ratio of H_2 to N_2 flow rates, allowing them to precisely set the hydrogen concentration. Extinction measurements were taken 100 s after each change of the H_2 concentration to allow the system to reach equilibrium. In TM polarization, the authors observed a pronounced response of the extinction dip at a wavelength of 674 nm when going from 0 to 20 vol.% H_2, with an extinction change of 247 % and a resonance blueshift of 13 nm. These results exemplify how the combination of a chemically sensitive waveguide with plasmonic elements can lead to versatile and high-performance sensor geometries. Furthermore, by replacing the Au nanowires with other reactive materials, metallic photonic crystal devices can be expanded to detect two reagents simultaneously, enabling rapid multiplexing.

A different way to obtain narrow spectral features suitable for sensing applications is the coupling of incident light to an array of apertures in an optically thick metal film. In optimized geometries, this leads to a transmittance much greater than predicted from classical aperture theory, a phenomenon known as extraordinary optical transmission (EOT) [78–80]. In this context, Maeda et al. utilized the infrared transmittance through an array of rectangular subwavelength holes in a palladium film to sensitively detect hydrogen [81]. The structures were fabricated using electron-beam lithography, metal sputtering, and wet-chemical lift-off to produce hole arrays with a periodicity of 1.1 μm, a length of 0.8 μm, and varying aspect ratios (length/width) from 1 to 2.6 (Fig. 8.9c, aspect ratio 1.6). In Fourier-transform infrared (FTIR) measurements, the authors found a pronounced transmittance peak that could be tuned from a wavelength of 4 μm up to 5 μm with increasing aspect ratio. Upon exposure to a concentration of 2 vol.% hydrogen, they observed a decrease of the transmittance as well as a spectral redshift of 0.2 μm for a fabricated hole array with an optimal aspect ratio of 1.6. This behavior was confirmed by numerical simulations of the sensing performance, which included both the dielectric change associated with the Pd/PdH transition and the hydrogen-induced expansion of the Pd lattice. The authors were able to reproduce the plasmon resonance positions for all experimental values of the aspect ratio and found spectral redshifts for both the dielectric change and the lattice expansion, adding up to a total redshift consistent with the experimental data. This shows that, compared to simple Pd patches, the use of EOT structures can provide sharp spectral features with large reagent-specific responses and can thus enable highly sensitive sensor devices.

Plasmonic perfect absorber structures, designed to fully absorb incident light of a certain predefined wavelength, can be utilized to reduce the background signals commonly associated with intensity-based sensing approaches [82–85]. This concept was applied to gas sensing by Tittl et al. who utilized a palladium-based perfect absorber structure to detect hydrogen [86]. Their system consisted of an array of Pd nanowires stacked above a dielectric (MgF_2) spacer layer and a gold mirror, which ensures zero transmittance through the structure (Fig. 8.9e). Incident light perpendicular to the wires can excite plasmon oscillations, which in turn cause the oscillation of a mirror plasmon in the thick gold film below. Ideally, these two charge distributions oscillate in antiphase to produce a circular current with an associated magnetic mode. By carefully tuning the width of the Pd wires and the height of the dielectric spacer layer, the authors were able to tailor both the effective electric permittivity ε and the magnetic permeability μ of their structure. This enabled them to match the optical impedance $z = \sqrt{\mu/\varepsilon}$ of the coupled system to the vacuum value, resulting in zero reflectance and thus an absorbance of unity. Following a numerical optimization of the design, the structures were fabricated using electron-beam lithography, thermal evaporation, and wet-chemical lift-off. In optical measurements, the authors observed a pronounced reflectance dip with a minimum value of $R_0 = 0.5$ % at the resonance wavelength of $\lambda_0 = 650$ nm, which corresponds to an absorbance >99 %. Exposure of this sample to 1 and 4 vol.% H_2 in N_2 carrier gas led to a strong increase of the reflectance at resonance combined with a spectral redshift (Fig. 8.9f). For the 4 vol.% concentration, which constitutes the explosion threshold of hydrogen in air, a maximum reflectance value of 4.9 % at λ_0 and a spectral redshift of 19 nm were observed. Since the reflectance at resonance is practically zero, this translated to a very large relative reflectivity change of $\Delta R_{rel} = (R_{H2} - R_0)/R_0 = 8.8$, an increase of one order of magnitude over conventional thin-film geometries [87].

8.8 Antenna-Enhanced Sensor Geometries

Plasmonic nanoantennas can focus incident light into intense hotspots of the electromagnetic field, making them an essential building block for applications ranging from single molecule fluorescence enhancement to nanoscale photodetection [88–90]. The strong nearfields also enable the high sensitivity of such structures towards refractive index changes in, e.g., the nanoscale gap of a plasmonic dimer antenna [91, 92]. Extending this towards catalytic systems, Liu et al. utilized the concept of antenna-enhancement for hydrogen sensing by placing a small palladium disk in the nanofocus of a triangular gold nanoantenna [93]. Compared to previous experiments that studied, e.g., resonant plasmonic nanodisks with diameters in the 200–300 nm range, the authors were able to examine the hydrogen uptake and release trajectories of Pd particles down to 60 nm in size. The Pd/PdH transition in such small particles is very challenging to detect optically, since the high intrinsic damping of Pd causes the Pd nanoparticles to exhibit a very low optical response at visible wavelengths. The hybrid antenna/disk structures were fabricated using two-step electron-beam lithography, and precise alignment between the two exposure processes was achieved through the use of gold alignment marks. This allowed for the reliable positioning of the Pd disk adjacent to the triangular Au antenna with inter-particle gaps down to 10 nm. The authors first exposed this optimized sample to controlled cycles of hydrogen gas while recording the optical scattering response from a single antenna/disk structure. They observed a pronounced plasmonic scattering peak at a wavelength of 638 nm, which underwent a spectral redshift of up to 9 nm for exposure to a hydrogen concentration of 16 Torr, equivalent to a concentration of 2 vol.%. When purging the system with nitrogen, a net redshift of 4 nm remained, demonstrating the presence of hysteresis in a single nanoscale Pd/PdH particle. In a subsequent second exposure cycle, the deterioration of the Pd/PdH particle associated with the hysteresis behavior led to somewhat lower redshifts (Fig. 8.10a). Also, an increase of the antenna-nanoparticle distance caused a severe drop of the sensing performance, again elucidating the important role of strong plasmonic field-enhancement.

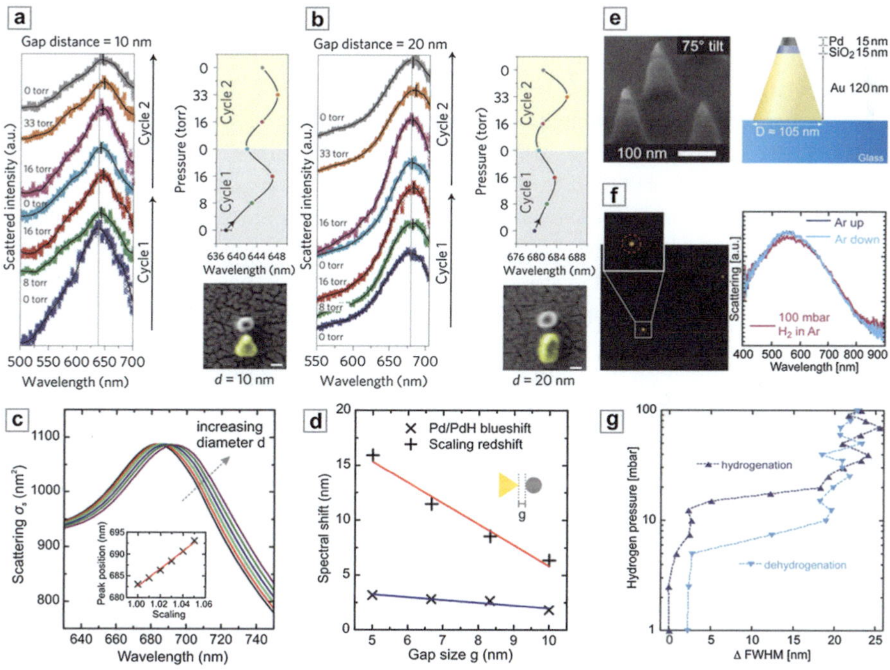

Fig. 8.10 Antenna-enhanced sensor geometries. (**a**) Optical scattering measurements and plasmon resonance shifts for a single palladium-gold triangle antenna during two hydrogen uptake and release cycles. (**b**) Hydrogen-dependent measurements for a single palladium-gold rod antenna (**a**, **b** from [93]). (**c**) Numerically calculated single particle scattering spectra for different diameters of the Pd disk next to the triangular Au antenna (initial gap size g = 7 nm). (**d**) Resonance shifts due to the Pd/PdH transition and the growing of the Pd disk for different initial gap sizes g (**c**, **d** from [94]). (**e**) SEM image and schematic depiction of a 120 nm high, truncated Au nanocone with a 15 nm SiO_2 spacer layer and a Pd particle on the tip. (**f**) Dark-field optical image of a single nanocone (*left*) and its spectral response to 100 vol.% Ar, 10 vol.% H_2 in Ar, and 100 vol.% Ar (*right*). (**g**) Corresponding optical isotherm of the nanocone system during hydrogenation (**e**–**g** from [97])

The crucial influence of antenna shape was demonstrated by replacing the triangular nanoantenna with a simple Au rod antenna. In this geometry, the authors found that the spectral shifts were roughly half smaller compared to those of the triangular system (Fig. 8.10b), showing that the sharp tip of the single bowtie antenna is essential to obtain an intense nanofocus and consequently high sensitivity.

The detailed reason for the experimentally observed spectral shifts was further investigated by Tittl et al. who performed numerical simulations on the discussed nanoantenna-enhanced sensor geometry [94]. They modeled the system using geometrical parameters extracted from the experimental SEM micrographs as well as optical constants for Au and Pd/PdH from literature, and calculated the scattering response of single antenna/disk sensor devices using a commercial finite element (FEM) electromagnetic field solver. This allowed them to investigate the two main features of the Pd/PdH transition, the dielectric change and the lattice expansion, independently from each other. The authors started with the hydrogen-induced dielectric change, since it constitutes the dominant effect for less complex Pd-based sensor geometries [95, 96]. However, when changing the dielectric function of the disk from Pd to PdH, the authors observed a small blueshift of 3 nm instead of the expected pronounced redshift of 9 nm from experiment. To model the effect of lattice expansion, both the diameter and the height of the Pd nanodisk were scaled with a factor s, ranging from 1.00 (no expansion) to 1.04 (maximum expansion from literature [56]). This led to a linear redshift of the plasmon resonance peak, with a maximum value of 13 nm for a scaling of 1.04 (Fig. 8.10c). By combining this scaling with the dielectric function of PdH, the authors were able to perfectly reproduce the experimental results, a hydrogen-induced redshift of the plasmon peak together with significant broadening of the lineshape. Additionally, the two competing spectral effects exhibited a markedly different scaling behavior: the blueshift associated with the dielectric change in the Pd remained mostly constant at 3 nm, whereas the redshift induced by the disk's scaling could be tuned via the gap size (Fig. 8.10d). In a comparison between triangular and rod antenna geometries, the authors also found the reduction of the sensing performance from experiment. However, when considering a double rod antenna geometry by adding a second antenna arm opposite the first, they observed a large redshift of 28 nm, comparable to a double bowtie geometry. This elucidates how the strong field-enhancement in the feed gap of a double antenna becomes dominant over the shape of the single antenna elements, allowing for high sensing performance with easier to fabricate constituent elements.

Still, the precise relative positioning of particles with different material properties and the required multi-layer processing pose significant challenges for high-throughput fabrication. To alleviate this, Langhammer et al. have utilized hole-mask colloidal lithography to fabricate truncated gold nanocones with functionalized tips to detect the hydrogen sorption and desorption kinetics of palladium und magnesium (Mg) nanoparticles [97]. Au nanocone antennas with different truncation height and consequently tip sharpness were fabricated and subsequently covered with a SiO_2 spacer layer and the reactive material in question. For a nanocone height of 120 nm, a 15 nm spacer layer, and a 15 nm thick tip, the authors were able to obtain an antenna-enhanced system with an Pd particle only 30 nm in size (Fig. 8.10e), a factor of 2 smaller than in the experiment by Liu et al. [93]. Still, the authors were able to detect the optical scattering response from single nanocones and could observe clear and reversible shifts of the plasmon scattering peak when exchanging the gaseous environment from 100 vol.% Ar to 10 vol.% H_2 and back to Ar (Fig. 8.10f). To further study the phase-transition behavior in the Pd nanoparticle, complete isotherms hydrogenation and dehydrogenation were recorded at different temperatures. Especially, the authors chose to track the full width at half maximum (FWHM) of the plasmon peak vs. the hydrogen partial pressure instead of the more common spectral shift or peak intensity change, allowing them to observe a clear hysteresis behavior from a single 30 nm Pd particle coupled to a Au nanocone antenna (Fig. 8.10g). Using Van't Hoff analysis, they were able to extract the enthalpy and entropy of hydride formation from the isotherm measurements, indicating a destabilization of the PdH phase compared to the bulk material. Furthermore, the authors extended the nanocone concept to the study of the hydrogenation kinetics of a single Mg nanoparticle. To prevent alloying and isolate the Mg particle from neighboring materials, it was placed between two titanium (Ti) layers, with a thin catalytic Pd layer placed at the top of the structure. This system configuration allowed the authors to trace a full hydrogenation isotherm for a single Mg nanodisk with a diameter of only 35 nm, demonstrating the versatility of their detection scheme. By implementing such advanced, high-throughput fabrication methods, complex plasmonic sensing geometries can be transitioned from high-resolution single-particle spectroscopy towards large-area technological applications, allowing for the sensitive, robust, and cost-effective detection of a variety of reagents and chemical processes.

8.9 Conclusions and Outlook

Plasmonic gas and chemical sensing has evolved from simple functionalized thin films towards increasingly complex and highly sensitive nanodevices for detection of a plethora of chemical reagents. Especially, the development of hybrid plasmonic structures, which combine several materials to incorporate new functionalities and achieve selectivity, has enabled an unprecedented flexibility in the design of sensor nanodevices. Aided by straightforward optical and thus non-invasive readout, plasmonic sensor devices have thus demonstrated sensitivities down to the ppm range combined with extremely small sensing volumes and reproducible operation. Based on these past achievements, plasmonic nanosensors need to overcome two main challenges in the future. In basic science, the ultimate limit of plasmonic chemical detection is still under debate. To resolve this and to extend current approaches towards gas sensing at the single molecule level, a combination of highly stable spectroscopic techniques and novel nanodevices with pronounced reagent-induced optical responses will be highly desirable. In the context of industrial applications, the requirements of high-throughput, cost-effective fabrication and especially long-term stability have to be met, ensuring that the devices can withstand a move from laboratory conditions to realistic and possibly harsh environments. With this, plasmonic nanosensor devices will serve as an essential building block both for nanomaterials science and advanced industrial process monitoring.

Acknowledgements We are grateful to N. Strohfeldt and F. Neubrech for key advice and discussions. A.T. and H.G. were financially supported by the Deutsche Forschungsgemeinschaft (SPP1391, FOR730, GI 269/11-1), the Bundesministerium für Bildung und Forschung (13 N9048 and 13 N10146), the ERC Advanced Grant COMPLEXPLAS, the Baden-Württemberg Stiftung (Spitzenforschung II), and the Ministerium für Wissenschaft, Forschung und Kunst Baden-Württemberg (Az: 7533-7-11.6-8). N.L. was supported by the Sofia Kovalevskaja Award of the Alexander von Humboldt Foundation and Grassroots Proposal M10331 from the Max Planck Institute for Intelligent Systems.

References

1. Valentine J, Zhang S, Zentgraf T, Ulin-Avila E, Genov DA, Bartal G, Zhang X (2008) Three-dimensional optical metamaterial with a negative refractive index. Nature 455:376–379
2. Atwater HA, Polman A (2010) Plasmonics for improved photovoltaic devices. Nat Mater 9:205–213
3. Boisselier E, Astruc D (2009) Gold nanoparticles in nanomedicine: preparations, imaging, diagnostics, therapies and toxicity. Chem Soc Rev 38:1759–1782
4. Kelly KL, Coronado E, Zhao LL, Schatz GC (2003) The optical properties of metal nanoparticles: the influence of size, shape, and dielectric environment. J Phys Chem B 107:668–677
5. Haynes CL, Van Duyne RP (2001) Nanosphere lithography: a versatile nanofabrication tool for studies of size-dependent nanoparticle optics. J Phys Chem B 105:5599–5611
6. Guo LJ (2007) Nanoimprint lithography: methods and material requirements. Adv Mater 19:495–513
7. Larsson EM, Syrenova S, Langhammer C (2012) Nanoplasmonic sensing for nanomaterials science. Nanophotonics 1:249–266
8. Mayer KM, Hafner JH (2011) Localized surface plasmon resonance sensors. Chem Rev 111:3828–3857
9. Anker JN, Hall WP, Lyandres O, Shah NC, Zhao J, Van Duyne RP (2008) Biosensing with plasmonic nanosensors. Nat Mater 7:442–453
10. Willets KA, Van Duyne RP (2007) Localized surface plasmon resonance spectroscopy and sensing. Annu Rev Phys Chem 58:1–33
11. Lal S, Link S, Halas NJ (2007) Nano-optics from sensing to wave guiding. Nat Photonics 1:641–648
12. Maier SA (2007) Plasmonics: fundamentals and applications. Springer, New York
13. Hartland GV (2011) Optical studies of dynamics in noble metal nanostructures. Chem Rev 111:3858–3887
14. Hu M, Novo C, Funston A, Wang H, Staleva H, Zou S, Mulvaney P, Xia Y, Hartland GV (2008) Dark-field microscopy studies of single metal nanoparticles: understanding the factors that influence the linewidth of the localized surface plasmon resonance. J Mater Chem 18:1949
15. Sönnichsen C, Franzl T, Wilk T, von Plessen G, Feldmann J (2002) Drastic reduction of plasmon damping in gold nanorods. Phys Rev Lett 88:077402
16. Halas NJ, Lal S, Chang W-S, Link S, Nordlander P (2011) Plasmons in strongly coupled metallic nanostructures. Chem Rev 111:3913–3961
17. Liu N, Langguth L, Weiss T, Kästel J, Fleischhauer M, Pfau T, Giessen H (2009) Plasmonic analogue of electromagnetically induced transparency at the Drude damping limit. Nat Mater 8:758–762

18. Tassin P, Zhang L, Zhao R, Jain A, Koschny T, Soukoulis C (2012) Electromagnetically induced transparency and absorption in metamaterials: the radiating two-oscillator model and its experimental confirmation. Phys Rev Lett 109:187401
19. Miroshnichenko A, Flach S, Kivshar Y (2010) Fano resonances in nanoscale structures. Rev Mod Phys 82:2257–2298
20. Novotny L, van Hulst N (2011) Antennas for light. Nat Photonics 5:83–90
21. Muhlschlegel P, Eisler H-J, Martin OJF, Hecht B, Pohl DW (2005) Resonant optical antennas. Science 308:1607–1609
22. Zuloaga J, Prodan E, Nordlander P (2010) Quantum plasmonics: optical properties and tunability of metallic nanorods. ACS Nano 4:5269–5276
23. Savage KJ, Hawkeye MM, Esteban R, Borisov AG, Aizpurua J, Baumberg JJ (2013) Revealing the quantum regime in tunnelling plasmonics. Nature 491:574–577
24. Scholl JA, García-Etxarri A, Koh AL, Dionne JA (2013) Observation of quantum tunneling between two plasmonic nanoparticles. Nano Lett 13:564–569
25. Larsson EM, Langhammer C, Zoric I, Kasemo B (2009) Nanoplasmonic probes of catalytic reactions. Science 326:1091–1094
26. Korotcenkov G (2005) Gas response control through structural and chemical modification of metal oxide films: state of the art and approaches. Sens Actuators B 107:209–232
27. Eranna G, Joshi BC, Runthala DP, Gupta RP (2004) Oxide materials for development of integrated gas sensors – a comprehensive review. Crit Rev Solid State Mater Sci 29:111–188
28. Ando M, Kobayashi T, Iijima S, Haruta M (2003) Optical CO sensitivity of Au-CuO composite film by use of the plasmon absorption change. Sens Actuators B 96:589–595
29. Sirinakis G, Siddique R, Manning I, Rogers PH, Carpenter MA (2006) Development and characterization of Au−YSZ surface plasmon resonance based sensing materials: high temperature detection of CO. J Phys Chem B 110:13508–13511
30. Buttner WJ, Post MB, Burgess R, Rivkin C (2011) An overview of hydrogen safety sensors and requirements. Int J Hydrog Energy 36:2462–2470
31. Linic S, Christopher P, Ingram DB (2011) Plasmonic-metal nanostructures for efficient conversion of solar to chemical energy. Nat Mater 10:911–921
32. Buso D, Busato G, Guglielmi M, Martucci A, Bello V, Mattei G, Mazzoldi P, Post ML (2007) Selective optical detection of H_2 and CO with SiO_2 sol – gel films containing NiO and Au nanoparticles. Nanotechnology 18:475505
33. Dharmalingam G, Joy NA, Grisafe B, Carpenter MA (2012) Plasmonics-based detection of H_2 and CO: discrimination between reducing gases facilitated by material control. Beilstein J Nanotechnol 3:712–721
34. Rogers PH, Sirinakis G, Carpenter MA (2008) Direct observations of electrochemical reactions within Au-YSZ thin films via absorption shifts in the Au nanoparticle surface plasmon resonance. J Phys Chem C 112:6749–6757
35. Ohodnicki PR, Wang C, Natesakhawat S, Baltrus JP, Brown TD (2012) In-situ and ex-situ characterization of TiO_2 and Au nanoparticle incorporated TiO_2 thin films for optical gas sensing at extreme temperatures. J Appl Phys 111:064320
36. Della Gaspera E, Guglielmi M, Agnoli S, Granozzi G, Post ML, Bello V, Mattei G, Martucci A (2010) Au nanoparticles in nanocrystalline TiO_2− NiO films for SPR-based, selective H_2S gas sensing. Chem Mater 22:3407–3417
37. Rogers PH, Sirinakis G, Carpenter MA (2008) Plasmonic-based detection of NO_2 in a harsh environment. J Phys Chem C 112:8784–8790
38. Kneipp K, Kneipp H, Itzkan I, Dasari RR, Feld MS (1999) Ultrasensitive chemical analysis by Raman spectroscopy. Chem Rev 99:2957–2976
39. Kneipp K, Wang Y, Kneipp H, Perelman L, Itzkan I, Dasari R, Feld M (1997) Single molecule detection using surface-enhanced Raman scattering (SERS). Phys Rev Lett 78:1667–1670
40. Li JF, Huang YF, Ding Y, Yang ZL, Li SB, Zhou XS, Fan FR, Zhang W, Zhou ZY, De Wu Y, Ren B, Wang ZL, Tian ZQ (2010) Shell-isolated nanoparticle-enhanced Raman spectroscopy. Nature 464:392–395
41. Qian K, Liu H, Yang L, Liu J (2012) Functionalized shell-isolated nanoparticle-enhanced Raman spectroscopy for selective detection of trinitrotoluene. Analyst 137:4644–4646

42. Zhang BQ, Li SB, Xiao Q, Li J, Sun JJ (2013) Rapid synthesis and characterization of ultra-thin shell Au@ SiO_2 nanorods with tunable SPR for shell-isolated nanoparticle-enhanced Raman spectroscopy (SHINERS). J Raman Spectrosc 44:1120–1125
43. Fan X, White IM, Shopova SI, Zhu H, Suter JD, Sun Y (2008) Sensitive optical biosensors for unlabeled targets: a review. Anal Chim Acta 620:8–26
44. Bingham JM, Anker JN, Kreno LE, Van Duyne RP (2010) Gas sensing with high-resolution localized surface plasmon resonance spectroscopy. J Am Chem Soc 132:17358–17359
45. Dahlin AB, Tegenfeldt JO, Höök F (2006) Improving the instrumental resolution of sensors based on localized surface plasmon resonance. Anal Chem 78:4416–4423
46. Jans H, Huo Q (2012) Gold nanoparticle-enabled biological and chemical detection and analysis. Chem Soc Rev 41:2849
47. Yang J, Yang J, Ying JY (2012) Morphology and lateral strain control of Pt nanoparticles via core – shell construction using alloy AgPd core toward oxygen reduction reaction. ACS Nano 6:9373–9382
48. Christian ML, Aguey-Zinsou K-F (2012) Core-shell strategy leading to high reversible hydrogen storage capacity for NaBH4. ACS Nano 6:7739–7751
49. Hsieh Y-C, Zhang Y, Su D, Volkov V, Si R, Wu L, Zhu Y, An W, Liu P, He P, Ye S, Adzic RR, Wang JX (2013) Ordered bilayer ruthenium – platinum core-shell nanoparticles as carbon monoxide-tolerant fuel cell catalysts. Nat Commun 4:2466
50. Ghodselahi T, Zahrabi H, Saani MH, Vesaghi MA (2011) CO gas sensor properties of Cu@CuO core – shell nanoparticles based on localized surface plasmon resonance. J Phys Chem C 115:22126–22130
51. Vasileva P, Donkova B, Karadjova I, Dushkin C (2011) Synthesis of starch-stabilized silver nanoparticles and their application as a surface plasmon resonance-based sensor of hydrogen peroxide. Colloids Surf A Physicochem Eng Asp 382:203–210
52. Halliwell B, Clement MV, Long LH (2000) Hydrogen peroxide in the human body. FEBS Lett 486:10–13
53. Endo T, Shibata A, Yanagida Y, Higo Y, Hatsuzawa T (2010) Localized surface plasmon resonance optical characteristics for hydrogen peroxide using polyvinylpyrrolidone coated silver nanoparticles. Mater Lett 64:2105–2108
54. Hill RT, Mock JJ, Hucknall A, Wolter SD, Jokerst NM, Smith DR, Chilkoti A (2012) Plasmon ruler with angstrom length resolution. ACS Nano 6:9237–9246
55. Ciracì C, Hill RT, Mock JJ, Urzhumov Y, Fernández-Domínguez AI, Maier SA, Pendry JB, Chilkoti A, Smith DR (2012) Probing the ultimate limits of plasmonic enhancement. Science 337:1072–1074
56. Flanagan TB, Oates WA (1991) The palladium-hydrogen system. Annu Rev Mater Sci 21:269–304
57. Jewell L, Davis B (2006) Review of absorption and adsorption in the hydrogen-palladium system. Appl Catal Gen 310:1–15
58. Yamauchi M, Ikeda R, Kitagawa H, Takata M (2008) Nanosize effects on hydrogen storage in palladium. J Phys Chem C 112:3294–3299
59. Khanuja M, Mehta BR, Agar P, Kulriya PK, Avasthi DK (2009) Hydrogen induced lattice expansion and crystallinity degradation in palladium nanoparticles: effect of hydrogen concentration, pressure, and temperature. J Appl Phys 106:093515
60. Tang ML, Liu N, Dionne JA, Alivisatos AP (2011) Observations of shape-dependent hydrogen uptake trajectories from single nanocrystals. J Am Chem Soc 133:13220–13223
61. Seo D, Park G, Song H (2012) Plasmonic monitoring of catalytic hydrogen generation by a single nanoparticle probe. J Am Chem Soc 134:1221–1227
62. Tittl A, Yin X, Giessen H, Tian X-D, Tian ZQ, Kremers C, Chigrin DN, Liu N (2013) Plasmonic smart dust for probing local chemical reactions. Nano Lett 13:1816–1821
63. Wang C, Ma L, Hossain M, Wang H, Zou S, Hickman JJ, Su M (2010) Direct visualization of molecular scale chemical adsorptions on solids using plasmonic nanoparticle arrays. Sens Actuators B 150:667–672

64. Morris T, Szulczewski G (2002) A spectroscopic ellipsometry, surface plasmon resonance, and X-ray photoelectron spectroscopy study of Hg adsorption on gold surfaces. Langmuir 18:2260–2264
65. Ahmadnia-Feyzabad S, Khodadadi AA, Vesali-Naseh M, Mortazavi Y (2012) Highly sensitive and selective sensors to volatile organic compounds using MWCNTs/SnO_2. Sens Actuators B 166–167:150–155
66. Silva LIB, Freitas AC, Rocha-Santos TAP, Pereira ME, Duarte AC (2011) Breath analysis by optical fiber sensor for the determination of exhaled organic compounds with a view to diagnostics. Talanta 83:1586–1594
67. Ma W, Luo J, Ling W, Wang W (2013) Chloroform-sensing properties of plasmonic nanostructures using poly(methyl methacrylate) transduction layer. Micro Nano Lett 8:111–114
68. Langhammer C, Zorić I, Kasemo B, Clemens BM (2007) Hydrogen storage in Pd nanodisks characterized with a novel nanoplasmonic sensing scheme. Nano Lett 7:3122–3127
69. Strohfeldt N, Tittl A, Giessen H (2013) Long-term stability of capped and buffered palladium-nickel thin films and nanostructures for plasmonic hydrogen sensing applications. Opt Mater Express 3:194–204
70. Prodan E (2003) A hybridization model for the plasmon response of complex nanostructures. Science 302:419–422
71. Hentschel M, Dregely D, Vogelgesang R, Giessen H, Liu N (2011) Plasmonic oligomers: the role of individual particles in collective behavior. ACS Nano 5:2042–2050
72. Rahmani M, Lei DY, Giannini V, Lukiyanchuk B, Ranjbar M, Liew TYF, Hong M, Maier SA (2012) Subgroup decomposition of plasmonic resonances in hybrid oligomers: modeling the resonance lineshape. Nano Lett 12:2101–2106
73. Schider G, Krenn JR, Gotschy W, Lamprecht B, Ditlbacher H, Leitner A, Aussenegg FR (2001) Optical properties of Ag and Au nanowire gratings. J Appl Phys 90:3825
74. Tikhodeev SG, Yablonskii AL, Muljarov EA, Gippius NA, Ishihara T (2002) Quasiguided modes and optical properties of photonic crystal slabs. Phys Rev B 66:045102
75. Fan S, Joannopoulos J (2002) Analysis of guided resonances in photonic crystal slabs. Phys Rev B 65:235112
76. Christ A, Tikhodeev S, Gippius N, Kuhl J, Giessen H (2003) Waveguide-plasmon polaritons: strong coupling of photonic and electronic resonances in a metallic photonic crystal slab. Phys Rev Lett 91:183901
77. Nau D, SEidel A, Orzekowsky RB, Lee SH, Deb S, Giessen H (2010) Hydrogen sensor based on metallic photonic crystal slabs. Opt Lett 35:3150–3152
78. Valsecchi C, Brolo AG (2013) Periodic metallic nanostructures as plasmonic chemical sensors. Langmuir 29:5638–5649
79. Brolo AG, Gordon R, Leathem B, Kavanagh KL (2004) Surface plasmon sensor based on the enhanced light transmission through arrays of nanoholes in gold films. Langmuir 20:4813–4815
80. Ebbesen TW, Lezec HJ, Ghaemi HF, Thio T, Wolff PA (1998) Extraordinary optical transmission through sub-wavelength hole arrays. Nature 391:667–669
81. Maeda E, Mikuriya S, Endo K, Yamada I, Suda A, Delaunay J-J (2009) Optical hydrogen detection with periodic subwavelength palladium hole arrays. Appl Phys Lett 95:133504
82. Moreau A, Ciracì C, Mock JJ, Hill RT, Wang Q, Wiley BJ, Chilkoti A, Smith DR (2013) Controlled-reflectance surfaces with film-coupled colloidal nanoantennas. Nature 492:86–89
83. Chen K, Adato R, Altug H (2012) Dual-band perfect absorber for multispectral plasmon-enhanced infrared spectroscopy. ACS Nano 6:7998–8006
84. Liu N, Mesch M, Weiss T, Hentschel M, Giessen H (2010) Infrared perfect absorber and its application as plasmonic sensor. Nano Lett 10:2342–2348
85. Landy N, Sajuyigbe S, Mock J, Smith DR, Padilla WJ (2008) Perfect metamaterial absorber. Phys Rev Lett 100:207402
86. Tittl A, Mai P, Taubert R, Dregely D, Liu N, Giessen H (2011) Palladium-based plasmonic perfect absorber in the visible wavelength range and its application to hydrogen sensing. Nano Lett 11:4366–4369

87. Fedtke P, Wienecke M, Bunescu M-C, Pietrzak M, Deistung K, Borchardt E (2004) Hydrogen sensor based on optical and electrical switching. Sens Actuators B 100:151–157
88. Kinkhabwala A, Yu Z, Fan S, Avlasevich Y, Müllen K, Moerner WE (2009) Large single-molecule fluorescence enhancements produced by a bowtie nanoantenna. Nat Photonics 3:654–657
89. Duan H, Fernández-Domínguez AI, Bosman M, Maier SA, Yang JKW (2012) Nanoplasmonics: classical down to the nanometer scale. Nano Lett 12:1683–1689
90. Knight MW, Sobhani H, Nordlander P, Halas NJ (2011) Photodetection with active optical antennas. Science 332:702–704
91. Zhao Y, Engheta N, Alù A (2011) Effects of shape and loading of optical nanoantennas on their sensitivity and radiation properties. J Opt Soc Am B 28:1266
92. Liu N, Wen F, Zhao Y, Wang Y, Nordlander P, Halas NJ, Alù A (2012) Individual nanoantennas loaded by three-dimensional optical nanocircuits. Nano Lett 13(1):142–147
93. Liu N, Tang ML, Hentschel M, Giessen H, Alivisatos AP (2011) Nanoantenna-enhanced gas sensing in a single tailored nanofocus. Nat Mater 10:631–636
94. Tittl A, Kremers C, Dorfmüller J, Chigrin DN, Giessen H (2012) Spectral shifts in optical nanoantenna-enhanced hydrogen sensors. Opt Mater Express 2:111–118
95. Wadell C, Antosiewicz TJ, Langhammer C (2012) Optical absorption engineering in stacked plasmonic Au-SiO(2)-Pd nanoantennas. Nano Lett 12:4784–4790
96. Dasgupta A, Kumar GVP (2012) Palladium bridged gold nanocylinder dimer: plasmonic properties and hydrogen sensitivity. Appl Opt 51:1688–1693
97. Shegai T, Langhammer C (2011) Hydride formation in single palladium and magnesium nanoparticles studied by nanoplasmonic dark-field scattering spectroscopy. Adv Mater 23:4409–4414

Chapter 9
Planar Hybrid Plasmonic-Photonic Crystals

Sergei G. Romanov

Abstract The invention of hybrid crystals brought about the simultaneous usage of different mechanisms of light transfer into effect in one and the same architecture. We have discussed the preparation, structure and optical properties of planar hybrid metal-dielectric crystals, light in which is carried by photons and plasmons.

Studied hybrids are based on the monolayers of spheres – the planar hexagonal packages of colloidal beads on a substrate. Two basic modifications, the monolayer on a flat metal film and the corrugated metal film on the monolayer have been prepared. Owing to their topology, hybrid crystals respond to the incident light depending on the wavelength, the polarization and the propagation direction with different optical resonances. The respective resonances are the light diffraction in the planar lattice, the diffraction of surface plasmon polaritons in the periodically profiled film, the localized particle and cavity plasmons in metal semishells and Fabry-Perot oscillations. Besides, interpenetration of plasmonic and photonic crystals results in the efficient hybridization of photonic and plasmonic modes. Overlay of different resonances leads to their further modification described by Fano process.

The apparent complexity of the optical properties is paired by their broad variability either by means of tuning the topology and composition of hybrids or through external stimuli. The simple and inexpensive technology in combination with very rich physics ensures the attractiveness of hybrid crystals for fundamental research and practical applications.

S.G. Romanov (✉)
Cluster of Excellence "Engineering of Advanced Materials", Institute of Optics, Information and Photonics, University of Erlangen-Nuremberg, Haberstr., 9a, 91058 Erlangen, Germany

Ioffe Physical Technical Institute RAS, Polythekhnicheskaya ul., 26, 194021 St. Petersburg, Russia
e-mail: Sergei.Romanov@physik.uni-erlangen.de

M. Bardosova, T. Wagner (eds.), *Nanomaterials and Nanoarchitectures*,
NATO Science for Peace and Security Series C: Environmental Security,
DOI 10.1007/978-94-017-9921-8_9

9.1 Introduction

The idea of intentionally structured materials capable of controlling the light emission, propagation and absorption was put forward in 1970–1980s [1]. Following S. John [2], such functionality can be achieved by exploiting the synergy of the scattering of electromagnetic (EM) waves and the following interference of scattered waves. Noteworthy, scattering becomes efficient if the wavelength, λ, approaches the size of a scatterer. In this case, the scattering cross-section is determined by the particle Mie resonances [3], the field distributions in which topologically looks like electron orbitals in atoms. Extending this analogy one can call the scatterers the photonic atoms (PhAs) [4]. The benefit of such discrete approach is the treatment of photonic molecules, photonic glasses and photonic crystals using the analogy with molecules, amorphous and crystalline solids by adopting the tight binding approximation of solid state physics. Then, the properties of photonic architectures can be considered on the basis of hybridization of Mie resonances in arrays of interacting photonic atoms.

In the case of regular periodic arrays of photonic atoms the competing approach of nearly free photons can be applied. In such arrays the scattered EM waves produce the regular interference pattern. On the one hand, this pattern depends on the wavevector and the frequency of waves. On the other hand, it depends on the lattice symmetry that allows to generalize the description of the array in terms of unit cells and to assign it the periodic variation of the dielectric permittivity in one or more dimensions. Thus, we arrive to the concept of photonic crystals (PhCs).

The particular result of collective scattering is the selection of waves possessing the complex lattice-specific field distribution that can propagate in regular ensembles without scattering. They are called the lattice eigenmodes or the Bloch modes. Two eigenmodes with the same wavelength, one of which is located mostly in "heavy" (high index of refraction) and the other – mostly in "light" (low index) dielectrics composing the PhC lattice, possess different energies. Respectively, the frequencies of these modes are different, i.e. they are separated by an energy gap. If no eigenmodes can be found in the frequency gap interval along the line from the centre to the edge of the respective Brillouin zone of the PhC lattice – this interval is called the photonic stop-band. The frequency and the bandwidth of the stop-band vary with the orientation of the wavevector in the Brillouin zone that occurs due to changing of the above mentioned interference pattern in the real space. In order to achieve the omnidirectional bandgap for one particular frequency, all directional stop-bands should overlap at this frequency. Note, that frequently used term, the "complete bandgap", refers to the zero density of photon states in the finite frequency interval. This is a purely theoretical concept because it assumes a PhC lattice of infinite stretching in all three dimensions. Finally, the pool of eigenmodes in the momentum space is called the photonic bandgap structure (PBG) by analogy to the electron energy band structure of atomic crystals.

Summarizing, owing to scalability of EM waves, the properties of which are independent on the wavelength, the PBG structure can be adopted to any desired spectral range by fitting the period of the lattice to the target wavelength, $a \sim \lambda$, whereas the scattering efficiency or the photonic strength can be controlled by a purposive choice of the geometry and the material of unit cells. As expected, the most attractive artificial PhCs operate in the visible, thus mimicking the structural coloration that is widely demonstrated by living creatures. The most illustrative example of photonic crystal is the lattice of spherically shaped scatterers, which can be equally well described using both approaches [5]. The drawback of nearly free photon model is its limitation in description of the lattice irregularities to the perturbative approach and incapability of describing photonic glasses and molecules.

Coming back to reality, we should note that many factors like the finite size, the finite dielectric contrast between photonic atoms and the surrounding medium and the inevitable lattice disorder lead to infilling of photonic gaps with photon modes, whereas the lower than spherical symmetry of achievable lattices makes this gap strongly anisotropic. Thus, the practically studied PhCs are far from possessing the ideal bandgap. Consequently, the most attractive PhC properties, like the suppression of spontaneous emission, the exceptionally high fidelity of resonators, the non-Markovian dynamics of photon-dressed atom states, the thresholdless lasing and so on require tremendous efforts of material scientists to be realized experimentally. That is why none of them, perhaps with the exception of high-Q resonators, were demonstrated so far. What is left? Undoubtedly, these are the manipulations with the light flow on a smaller length scale compared to that of conventional optical elements and the control on the strength of light-matter interaction.

Since early days of PhC research, numerous attempts were made to expand the functionality of PhCs. These attempts include, in particular, designing the complex architectures, like the photonic quasicrystals for enlarging the photonic gaps [6] and the photonic hetero-crystals for exploiting the interfaces between photonic crystals [7]. Substantially, these approaches are based on increased complexity of PhCs that leads to the sophisticated interference pattern and the local disruption of the photon transport.

Alternatively, in the race for novel functionalities the researches look for physical effects beyond the interference paradigm. One way is to prepare PhCs from materials having electron resonances of the high oscillator strength, which dielectric constant dramatically changes at bandgap frequencies. If the dielectric response of the PhC material shows a pole in the vicinity to the Bragg resonance, such PhC is called the resonant PhC (see review [8]). In this case light is carried by so-called polaritons, which are the quasiparticles resulting from strong coupling of EM waves with electric or magnetic excitations (Fig. 9.1a). Assuming the same lattice topology, the most evident distinction of resonant PhCs is the splitting of the diffraction resonance dispersion [9]. Examples of resonant PhCs are the Bragg superlattices of semiconductor quantum wells [10] or lattices of holes that are infiltrated with semiconductors exhibiting strong exciton resonances [11]. Thus, the interaction of light with material resonances is the method to alter the way of light flow processing in PhCs. The common drawback of resonant PhCs is the light losses

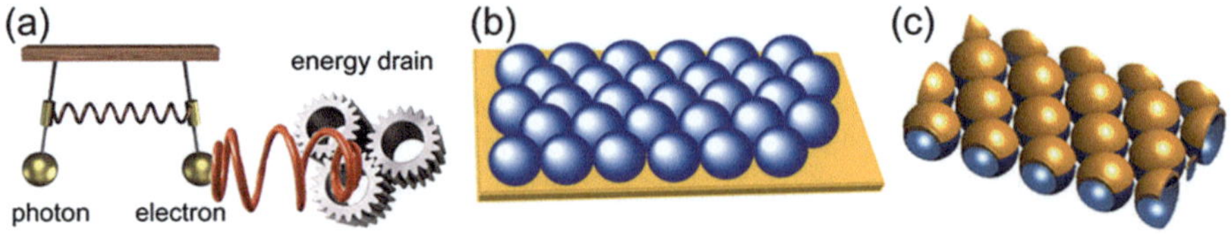

Fig. 9.1 (**a**) The idea of polaritonic excitation. In such coupled oscillators, the energy is dissipated due to electron motion. Schematics of the (**b**) monolayer of spheres on a flat metal film and (**c**) corrugated metal film on a monolayer of spheres

due to absorption in the PhC framework. However, the losses can be also exploited for some applications, e.g., for materials with superabsorption or superradiance properties [12].

Another class of polaritonic PhCs is represented by periodic metal-dielectric (MD) architectures, where electronic resonances are the localized or the propagating plasmons. The classical example of such structures is the metallic diffraction gratings demonstrating the Wood anomalies occurring due to light coupling to surface plasmon polaritons (SPPs) [13]. To our opinion, plasmonic-photonic crystals symbolize very efficient energy transducers allowing to tune the balance between scattering and absorption.

Investigations of metal-dielectric PhCs deal with many interesting effects such as the extraordinary optical transmission (EOT) through subwavelength-size holes in opaque metal films [14], guided plasmon-polaritons in dielectric waveguides decorated with metal nanostructures [15], plasmonic crystals (PlCs) possessing the SPP energy band structure [16] and the plasmon-assisted absorption in regularly structured metal films [17]. The powerful trend in optical materials is towards the metamaterials the optical properties of which can be designed well beyond the limits achievable with natural materials. Metamaterials are nanostructured but optically homogeneous materials, because they are assembled from nanostructures that are much smaller compared to the operation wavelength, $a \leq 0.1\lambda$. The ultimate goal of this technology is to achieve the control on the sign of the dielectric permittivity and the magnetic permeability, because functionalities like super-resolution imaging, optical illusions [18] and transformation optics [19] require materials with positive and negative indices of refraction [20].

Further development of our ideas is to combine the plasmonic and photonic crystals into one heterostructure. In our notations, the photonic heterocrystal is a sequence of seriously connected PhCs possessing different PBG diagrams [7]. The hybrid hetero-crystal, in turn, is the assembly of electromagnetic crystals possessing not only the different PBGs but, principally, the different light carriers [21, 22]. Essentially, a hybrid plasmonic-photonic structure replicates itself in hybrid physical properties: it acquires three pools of eigenmodes – photon modes of a dielectric crystal, plasmon modes of a metal crystal and hybrid modes of a metal-dielectric interface. These interface modes are the most interesting consequence of heterostructuring.

Opals are naturally occurring semiprecious stones, whose attractive play of colours is based on light diffraction in the ordered ensemble of silica spheres [23]. Their artificial counterparts nowadays are made with very high accuracy. They have been used as templates for designing new materials since early 1980s [24] and, particularly, as photonic crystals since 1995 [25]. Naturally, opals can be treated as the crystals of photonic atoms.

To date the self- and force-assembling of colloidal spheres is arguably the most frequently used technique for the preparation of 2D slab and 3D PhCs [26]. Moreover, colloidal crystals are conveniently used for impregnating with dielectric and metal "guest" materials [27]. On the other hand, it is very important to develop the technology leading to a formation of opal-like ensembles of metal spheres. This design is driven by the theoretical prediction of achieving a high refractive index contrast between the metal spheres (or metal-coated dielectric particles called nanoshells) and air as a necessary condition for creating the broad robust omnidirectional PBG [28]. Unfortunately, the optical properties of such structures prepared so far are impaired by the strong light absorption [29]. Other attempts to modify PBG properties of opals by integrated metal components, namely, to assemble dielectric spheres with small metal cores [30, 31] or to infiltrate metal in opal voids [32], were only partially successful. Opals with disconnected metal particles demonstrated the optical properties, which were comparable with those of all-dielectric opals, whereas the opals with interconnected metal particles demonstrated the functionality of a cut-off filter at the best.

More successful experiments have been made with metal containing 2D slab PhCs based on either monolayers of metal nanoshells [33] or metal replicas of monolayers of dielectric spheres [34]. Overall, the optical response of such architectures is dominated by localized particle and/or cavity plasmons.

The key strategy, which preserves the properties of both the dielectric and the metal components in the optical response of metal-dielectric PhCs, is to design the hybrid architectures containing interconnected metal pieces but to maintain a low volume fraction of metal. In recent publications we argued in favour of hybrids assembled from thin metal films that are located in the near-field vicinity to planar two- or bulk three-dimensional colloidal PhCs. Metal can be shaped as a flat film deposited on the substrate that supports the PhC (Fig. 9.1b) or as a periodically profiled film spattered on the PhC surface (Fig. 9.1c). The absorption losses in the case of flat 20–50 nm thick Au and Ag films remain within 2–5 % for normally propagating light of $\sim$1 micron wavelength, whereas the most of incident light is reflected. Towards shorter wavelengths the metal absorption increases rapidly and peaks by approaching the frequency of interband electron transitions.

The phase matching conditions necessary for excitation of plasmons at the surface of the flat metal film are fulfilled with a help of a dielectric grating coupler. This can be a monolayer of spheres, which dictates the frequency and propagation direction SPPs in the case of architecture in Fig. 9.1b. Alternatively, the metal film adopts the periodicity of a dielectric substrate in the architecture shown in Fig. 9.1c. In this case phase matching conditions remain the same, but the deep profile of metal corrugation enhances the SPP scattering. Since the transverse dimensions of such planar architectures are about the operating wavelength, the dielectric and plasmon modes are prone to hybridization. It is worth to note that hybridization takes place at the metal-dielectric interface, because the plasmon field decays rapidly away from the metal surface. Luckily, the efficient mixing of dielectric and plasmon modes reduces the absorption losses of in-plane guided light.

Engineering of the field distribution in hybrid crystals can be achieved either by changing the thickness and profile of the metal film or by changing the geometry of the monolayer of dielectric beads. The specific properties of such hybrids are: (i) the reasonably low absorption and high reflectance, (ii) the reach spectrum of quasi-guided modes; (iii) the strong Fabry-Perot resonances; (iv) the support of localized resonances of dielectric beads and metal cavities in profiled films.

9.2 Grating-Coupled Hybrid Crystals

The straightforward method to realize the hybrid structure is to place the non-resonant flat metal film in the near-field zone of the PhC. In this case, the plasmons will be excited at the PhC-metal interface via so-called grating coupling. The advantage of such approach is the guaranteed matching of photonic and plasmonic eigenmodes, moreover, all of them are of hybrid character to larger or smaller extent. Let us discuss the eigenmodes of such planar hybrid PhCs.

The optical properties of monolayers (ML) of spheres are well documented and their correspondence to those of 2D slab PhC has been established [35]. The preparation of hybrid 2D slab PhCs (h-PhCs) by self-assembling of colloidal beads on a metal-coated substrate (Fig. 9.2 left) is also known since early 1990s [36]. A number of sharp dips in reflectance spectra of ML samples deposited on the

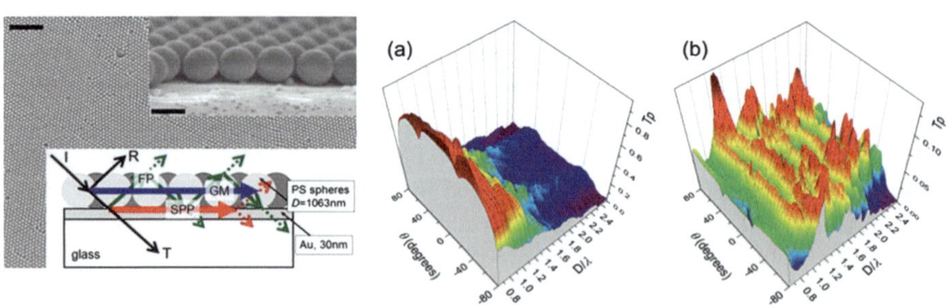

Fig. 9.2 *Left panel – top* view scanning electron micrograph of the ML-Au sample (scale bar is 10 μm). Diameter of spheres $D \approx 1{,}070$ nm, metal film thickness $t_{Au} = 30$ nm. *Inset top*: *side* view of this monolayer (scale bar is 1 μm). *Inset bottom*: schematics of ML-Au hybrid. Numbers show the dimensions of structural elements. The incidence light intensity (I) is distributed between the reflected (R), transmitted (T) and guided light. The guided light is transported by the index-guided modes, the SPP modes and the Fabry-Perot modes (FP). *Dashed arrows* indicate escaped light. *Right panels* – angle-resolved transmission spectra of (**a**) ML and (**b**) ML-Au samples in p-polarized light, which were collected in the zero diffraction order. Frequency is shown in dimensionless units D/λ

thick opaque metal films were assigned to guided modes (GMs) [37]. The quality factors of these resonances are in the range from 80 to 150 that is much higher compared to those of monolayers on glass substrates. Numerical modelling of the field distribution allows to assign the observed reflectance minima either to index-guided eigenmodes, which propagate along the planar colloidal crystal and localised in the dielectric, or to propagating plasmons, which are localised at the ML-metal interface. If the former can be considered as the eigenmodes of 2D slab PhC, the explanation of the latter is based on the concept of grating coupler.

If the metal film is semi-transparent, the light confinement and, hence, the quality factor of resonances is reduced, but the investigator gains access to very informative transmission spectra. To highlight the role of a metal substrate, it is instructive to compare the spectra of monolayer of spheres on a glass substrate and on the Au-coated glass substrate. The differences in spectra of angle-resolved transmission are gigantic: the transmission surface of monolayer on dielectric is moderately patterned by angle-dependent (dispersive) resonances (Fig. 9.2 right, a) in contrast to spectra of the monolayer on the gold-coated substrate, which are dominated by sharp non-dispersive resonances (Fig. 9.2 right, b).

The transmission of a monolayer is high and the reflectance is low, but both are decreased towards shorter wavelengths. The resonances are seen as minima in transmission and reflectance spectra (curves 1, 2 in Fig. 9.3a). Since these minima are almost in registry with each other, they indicate losses of transmitted and reflected light. In the past we argued that the transmission drop at $\lambda < D$ (λ – wavelength, $D = 1070$ nm – diameter of sphere) occurs due to surface diffraction, which produces diffraction orders channelling light along different directions [38]. Here we will consider only a special case of diffraction, where the diffraction order propagates along the plane of the grating in a manner similar to the Wood anomaly.

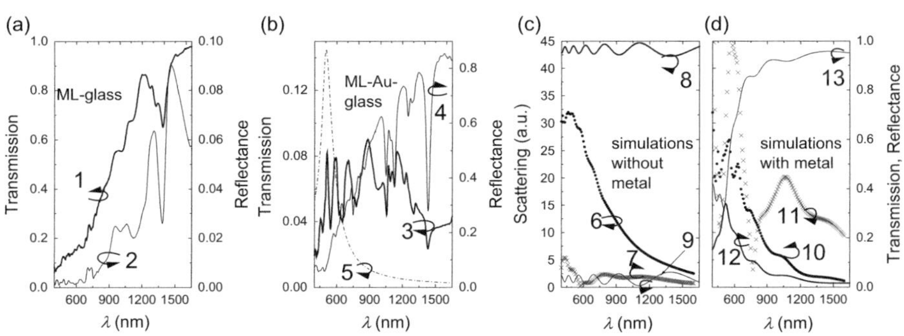

Fig. 9.3 Transmission (*curves* 1, 3) and reflectance (2, 4) spectra of (**a**) ML-glass sample and (**b**) ML-Au-glass hybrid at $\theta = 6^{\circ}$ in p-polarized light. *Curve* 5 represents the transmission of 30 nm Au film. Calculated transmission (6, 10) and reflectance (7, 11) spectra of (**c**) 50 L-glass and (**d**) 50 L-Au-glass samples at $\theta = 0^{\circ}$. Calculated spectra of forward (8, 12) and backward (9, 13) scattered light of (**c**) sphere on a glass substrate and (**d**) sphere on Au-coated glass substrates at $\theta = 0^{\circ}$

On the other hand, the reflectance of ML-Au hybrids is high and transmission is low that nicely correlates the reflectivity of a metal mirror, moreover, the resonances in transmission are better resolved compared to that in ML-glass sample. Early studies of reflectance spectra revealed resonances at $D/\lambda \sim 0.77;\ 0.83;\ 0.89$ for the normal light incidence. They were assigned to SPP, dielectric and hybrid guided modes, respectively [37]. In Fig. 9.3b the corresponding transmission minima can be observed at 1,434, 1,296 and 1,125 nm, but there are additional features in this sequence, with a major one at ~1,240 nm (curve 3 and 4). Important, the minimum-minimum correlation persists at the long-wavelength part of reflectance and transmission spectra, but it is replaced with minimum-maximum correlation at shorter wavelengths. Surprisingly, the transmission spectrum of ML-Au neglects the rapidly changing transmission of the bare Au film (curve 5, Fig. 9.3b), provisionally, due to the random scattering that rapidly increases to shorter wavelength. Summarizing these observations, one can suspect that the light transport mechanism in ML-Au-glass samples changes along the wavelength decrease. In what follows we will describe this behavior as a superposition of three simple mechanisms [39].

1. Guided Modes Multilayer model is useful to describe the index-guided (GM) and evanescent modes. In order to mimic the dielectric permittivity profile of ML of spheres we performed the homogenization of a monolayer. The Maxwell's equations for a system with periodic permittivity, $\varepsilon(\mathbf{r}, z) = \sum_{\mathbf{G}} \varepsilon_{\mathbf{G},z} e^{i\mathbf{G}\mathbf{r}}$, where $\mathbf{r}$ is the in-plane spatial vector, $\mathbf{G}$ is the reciprocal vector and $\varepsilon_{\mathbf{G},z}$ is the Fourier component of permittivity, can be written using the envelope function in the form $\mathbf{E_k}(\mathbf{r}, z) = \tilde{\mathbf{E}}(\mathbf{r}, z)\, e^{i(\mathbf{kr}-\omega t)}$ as

$$(\mathbf{k}+\mathbf{G}) \times (\mathbf{k}+\mathbf{G}) \times \tilde{\mathbf{E}}_{\mathbf{G},z} = -\left(\frac{\omega}{c}\right)^2 \sum_{\mathbf{G}'} \varepsilon_{\mathbf{G}-\mathbf{G}',z} \tilde{\mathbf{E}}_{\mathbf{G}'}, \tag{9.1}$$

where the electric field is also a periodic function $\tilde{\mathbf{E}}(\mathbf{r}, z) = \sum_{\mathbf{G}} \tilde{\mathbf{E}}_{\mathbf{G},z} e^{i\mathbf{G}\mathbf{r}}$. If $|\varepsilon_{\mathbf{G}=0}| > |\varepsilon_{\mathbf{G}\neq 0}|$, the solution can be found in the following form

$$(\mathbf{k}+\mathbf{G}) \times (\mathbf{k}+\mathbf{G}) \times \tilde{\mathbf{E}}_{\mathbf{G},z} = -\left(\frac{\omega}{c}\right)^2 \varepsilon_{\mathbf{G}=0,z} \tilde{\mathbf{E}}_{\mathbf{G},z}, \tag{9.2}$$

which corresponds to a solution for a homogeneous medium with permittivity $\varepsilon_{\mathbf{G}=0}$ and wave vector $\mathbf{k}+\mathbf{G}$. We used the 50 slices (50 L) of the sphere ε-profile to approximate the zero Fourier component (Fig. 9.4a). The light confinement without a metal layer is weak and energy leaks from the layer of low refractive index into the substrate of the higher index, hence, resonances are broad and shallow. But introduction of a metal film firmly isolates guided modes.

Seven eigenmodes of 50 L-stack on Au film-coated glass substrate can be found in the spectral range of interest in the s-polarized light (Fig. 9.4b). Due to dispersion, these guided modes eventually fall at lower energies in the light cone,

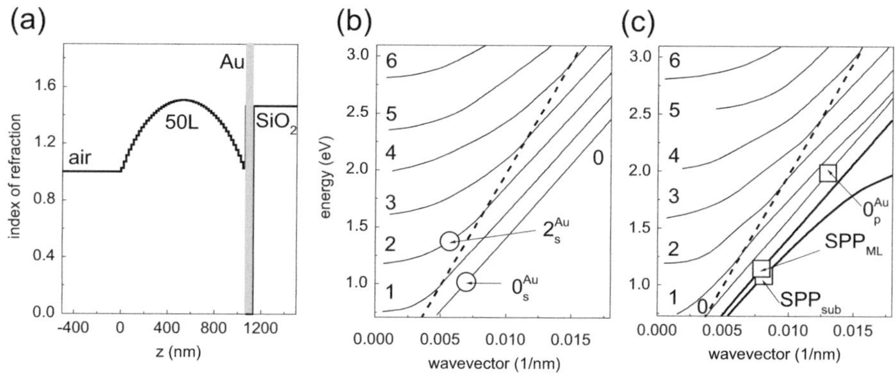

Fig. 9.4 (**a**) Refractive index profile of the model 50 L-Au-glass structure. The complex refractive index of gold is displayed as a *grey bar*. Eigenmodes of 50 L-Au-glass structure are shown for (**b**) s- and (**c**) p-polarized light and numbered consecutively. *Dotted lines* show the light line in air. *Empty circles* and *squares* mark the points, at which the Poynting vector profiles have been calculated. Modes SPP_{ML} and SPP_{sub} are located at 50 L and glass interfaces of the Au film, respectively

where they become familiar Fabry-Perot (FP) resonances and where they can be directly coupled to the incident light. Apart from these modes, in p-polarized light two additional evanescent SPP modes can be found at interfaces of the metal film with the ML of spheres and with the substrate (Fig. 9.4c). Obviously, evanescent modes do not interact with incident light without the aid of coupling devices. One should note that the mode dispersions in 50 L-Au-glass structure differ from that of the 50 L-glass structure owing to the strong frequency dependence of the dielectric permittivity of gold in the studied spectral range [see 40].

Let us estimate the effect of a gold film on the light confinement. In Fig. 9.4 several characteristic points, for which we compared the in-plane energy flux are indicated. Let a plane wave to come from infinity and to possess an electrical field of unity strength at the structure boundary $z = 0$. And let the projection of its wavevector on the slab waveguide plane $k_x^2 = (\omega/c)^2 - k_z^2$ to be equal to the wavevector of an eigenmode of 50 L waveguide. If $\mathbf{k_x}$ is in the light cone, then $\mathbf{k_z}$ is real and the mode is a radiative one, otherwise $\mathbf{k_z}$ is complex and the eigenmode is evanescent.

The flux transported by the s-polarized index-guided 0_s mode in the 50 L structure is concentrated in the middle of the waveguiding layer. Adding a metal mirror leads to $\sim 10^4$ times increase of the energy coupled to 0_s^{Au} mode (Fig. 9.5a). For the radiative 2_s mode one can find several nodes accommodated across the 50 L-structure that is the property of a radiative mode. After adding the metal film, the energy coupled to the delocalized 2_s^{Au} mode is increased by a factor of 3 (Fig. 9.5b).

The p-polarized 0_p mode remains the index-guided one, the flux of which is concentrated in the slab waveguide (Fig. 9.5c). Adding a metal film only marginally enhances the light flow guided by the 0_p^{Au} mode, but 1/10 fraction of mode power is bounded to the ML-Au interface, i.e., this mode is hybridized with the SPP mode.

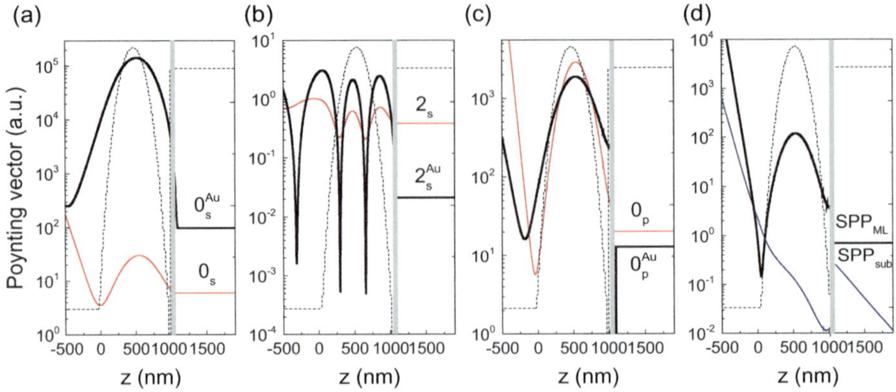

Fig. 9.5 Distributions of electromagnetic flux across the 50 L-glass and 50 L-Au-glass structures calculated for characteristic points at dispersion *curves* shown in Fig. 9.4. *Dashed lines* show the refractive index profile as in Fig. 9.4a. The *grey bar* indicates the Au film. Au film thickness is 30 nm

Both SPP_{ML} and SPP_{sub} plasmon modes are evanescent ones (Fig. 9.5d), they are bounded to opposite interfaces of the Au film and their Poynting vector peaks at metal-dielectric interfaces [41]. SPP_{ML} mode acquires the hybrid character because it possesses the main maximum in the middle of a planar dielectric waveguide and the fractional maximum resides at the waveguide-metal interface. The SPP_{sub} mode is the only true plasmon mode, which flux is bounded to the metal-substrate interface. The fact that SPP_{sub} carries much smaller energy compared to modes guided in a dielectric relates to the exploited excitation configuration, in which the monolayer acts as a gap in the Kretschmann coupling geometry.

Using the transfer matrix formalism, we calculated transmission and reflectance spectra of model planar waveguides. The 50 L-glass structure is highly transparent one. Its spectra are modulated by Fabry-Perot oscillations so that the reflectance maximum corresponds to the transmission minimum (curves 6, 7 in Fig. 9.3c). In contrast, the 50 L-Au-glass structure is weakly transparent, but highly reflective at $\lambda > 800$ nm (curves 10, 11 in Fig. 9.3d). At shorter wavelengths, the transmission increases following the transmission of the Au film (curve 5 in Fig. 9.3b), and the reflectance decreases towards the onset of the interband transitions in the gold that occurs at $\lambda \leq 500$ nm. However, the magnitude of Fabry-Perot oscillations remains small in both 50 L-glass and 50 L-Au-glass structures for incident directions close to the waveguide normal that leaves a little chance for them to be detected.

The Fabry-Perot modes are preserved in the broad range of angles of light incidence (Fig. 9.6). Moreover, their relative magnitude in 50 L-Au-glass structure becomes much higher for oblique light incidence compared to that of 50 L-glass structure. This observation agrees with the enhanced confinement of light coupled to Fabry-Perot modes along the increase of the metal film reflectivity. The interband transitions in a gold film affect the spectra at wavelength of ~500 nm (Fig. 9.6b).

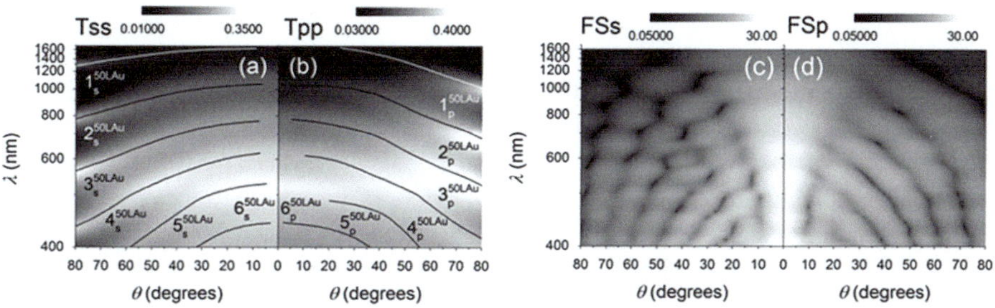

Fig. 9.6 Calculated transmission maps of 50 L-Au-glass multilayers in (**a**) s- and (**b**) p-polarized light. Calculated maps of forward scattered light intensity in (**c**) s- and (**d**) p-polarized light for the 1,063 nm PS sphere on the glass substrate coated with 30 nm Au film

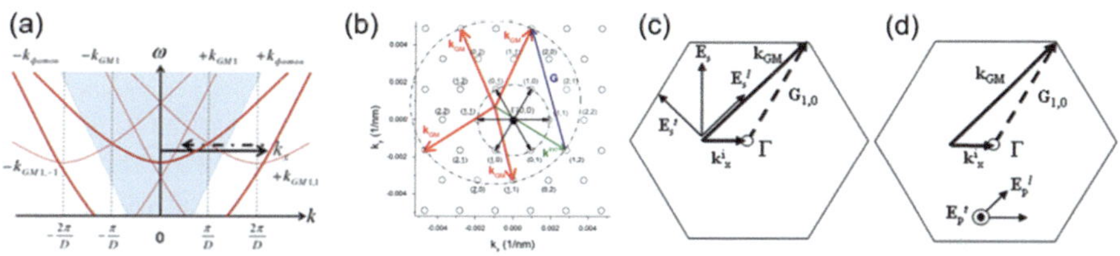

Fig. 9.7 (**a**) Folding of WG modes back in the light cone in periodically profiled waveguide. (**b**) Reciprocal lattice of a 2D hexagonal array. Numbers indicate the lattice sites. *Dashed lines* are the projections of two Ewald spheres arbitrarily chosen for the normal and the oblique incident light. *Arrows* are the wavevectors of incident wave $\mathbf{k}^{inc}$, the vector of the reciprocal lattice **G** and its diffraction orders $\mathbf{k}_{GM}$. (**c**, **d**) Sketches of in-plane diffraction orders with transverse and longitudinal orientation of the electric field for s- and p-polarized guided modes (indices *t* and *l*) respectively. $\mathbf{E}_{s,p}^{t,l}$ is the vector of the electric field, indices s and p indicate the polarization. Γ is the center of the Brillouin zone. $\mathbf{k}_{GM}^{i,k}$ is the in-plane projection of the wavevector of diffraction order

Additional consequence of Au mirrored substrate is the elimination of the Brewster angle-related suppression of reflectance in p-polarized light.

2. Diffractive Coupling In the presence of a periodic waveguide modulation each GM of the 50 L-structure becomes expanded into a series of diffraction orders, wavevectors of which satisfy the momentum conservation with vectors of the reciprocal lattice of the corrugation (Fig. 9.7a, b) [42]. These orders may fall in the light cone, where they are called quasi-guided modes and can be directly coupled to the incident light. This mechanism is called the grating coupling to evanescent modes outside the light cone. Quasi-guided modes are of higher quality factor than Fabry-Perot oscillations because they originate from evanescent modes.

Since the grating coupling in ML-Au-glass differs from the coupling method used to calculate Poynting vector of non-radiative modes of the 50 L-Au-glass structure, the actual flux distribution associated with quasi-guided modes may differ considerably.

The symmetry of the hexagonal package of spheres in monolayer is *p6mm*. In this group symmetry one can find the principal 6th-order rotation axis, vertical mirror reflectance plane and two non-collinear translations of equal length with $\pi/3$ (or $\pi/6$) angle between them [43]. *6mm* (C_{6v}) subgroup assumes six irreducible states in the center of the Brillouin zone: four non-degenerate 1D ($\Gamma_1, \Gamma_2, \Gamma_3, \Gamma_4$) and two doubly degenerate 2D (Γ_5, Γ_6) states. Hence, in the dipole approximation the optical transitions at $\mathbf{k} = 0$ should obey the selection rules. For the p-polarized light ($\mathbf{e}_p \parallel C_6$) only transitions between configurations of the same symmetry are allowed $\Gamma_n \leftrightarrow \Gamma_n \quad (n = 1, 2, 3, 4, 5, 6)$, i.e., modes are excited with translations that are coplanar to the plane of incidence only (Fig. 9.7c). In this case, the SPPs can be excited with different efficiencies depending on the orientation of the $\mathbf{E}_\mathrm{p}$ vector with respect to the lattice vector. For the s-polarized light ($\mathbf{e}_s \perp C_6$) only transitions between configurations of different symmetries are allowed, like $\begin{matrix} \Gamma_1 \leftrightarrow \Gamma_5 \leftrightarrow \Gamma_2 \\ \updownarrow \\ \Gamma_3 \leftrightarrow \Gamma_6 \leftrightarrow \Gamma_4 \end{matrix}$, i.e., modes are only excited with translations that are non-coplanar with the plane of incidence (Fig. 9.7d). In the case of the strictly transverse wave, no SPP excitation is expected. As a result, one should expect the polarization anisotropy of diffractively coupled eigenmodes in transmission/reflectance spectra and, hence, the polarization anisotropy of ML-Au-glass spectra.

3. Mie Resonances of Spheres We numerically calculated the spectra of light scattered by a dielectric sphere using the transfer matrix formalism [44]. The electric field $\mathbf{E}_s$ scattered by a sphere on a substrate can be written as $\mathbf{E}_s = (1 - \mathbf{T}_s\mathbf{R})^{-1}\mathbf{T}_s\left(\mathbf{E}_p + \mathbf{E}_p^r\right)$, where $\mathbf{E}_p$ is the incident plane wave written in the basis of electromagnetic multipoles, $\mathbf{E}_p^r$ is a plane wave reflected by the substrate, $\mathbf{T}_s$ is a diagonal T-matrix of the sphere and $\mathbf{R}$ is a matrix, which relates the scattered-reflected field $\mathbf{E}_s^r$ with the scattered field $\mathbf{E}_s$ through $\mathbf{E}_s^r = \mathbf{R}\mathbf{E}_s$ [45]. We assume that no absorption occurs in a sphere. The electric field of the back-scattered wave can be written in the far field as $\mathbf{E} = (1 + \mathbf{R})\,\mathbf{E}_s$, whereas the electric field of forward-scattered wave and the one transmitted into the substrate can be expressed as $\mathbf{E} = \mathbf{T}\mathbf{E}_s$. Here $\mathbf{T}$ denotes a matrix representation of a transmission operator suggested in [46]. A sufficiently large number of multipoles was used to ensure the convergence of the simulation. The dielectric constant of gold was adopted from [47]. The scattering geometry was fitted to the layout of angle-resolved transmission and reflectance experiment (Fig. 9.8a).

In the case of sphere-on-glass configuration, the forward scattered (FS) light intensity is ten times higher compared to the back-scattered one. These spectra contain multipole Mie resonances, which become noticeable only at $\lambda < 800$ nm (curves 8, 9 in Fig. 9.3c). The average level of scattered light increases with decreasing wavelength. If the substrate is coated by a gold film, the intensity of the forward-scattered light becomes lower whereas the intensity of back-scattered light increases by a factor of $10 \div 40$ (curves 12, 13 in Fig. 9.3d). In our case the Mie resonances of sphere-Au-glass structure appear in counter-phase to these resonances in the sphere-glass case and their magnitude is strongly enhanced.

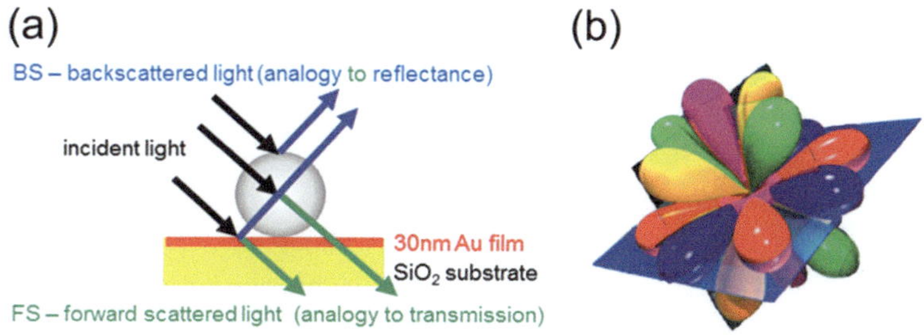

Fig. 9.8 (**a**) Incident and scattered fields taken into consideration. (**b**) Schematic explanation of the origin of false angle dispersions of Mie resonances

The calculated maps of forward scattered light for the "sphere-on-metal" show the complex angle dependence (Fig. 9.6c, d). Naturally, the frequencies of Mie resonances do not depend on the observation angle, but their intensities are modulated according to the radial distribution of multipole scattering diagrams as illustrated schematically in Fig. 9.8b. Moreover, the intensity of scattered light demonstrates the polarization anisotropy, which affects the resolution of scattering lobes. It is worth noting, that the angle distribution of transmitted/reflected light in a ML-Au-glass hybrid should not necessarily follow that of a single sphere because of in-plane hybridization of resonance states in a close-packed ensemble.

Below 1,000 nm the resonances in experimental spectra of ML-Au-glass hybrid (Fig. 9.3b) are quite close to Mie resonances of a sphere on a metal (Fig. 9.3d), hence we can conclude that the spectral profile of the ML-Au-glass hybrid (Fig. 9.2, right, b) is determined by Mie resonances of individual spheres in a monolayer.

The experimentally observed reflectance and transmission maps of the ML-Au-glass sample show the superposition of all discussed above components. In agreement with the transmission of the 50 L structure, the FP oscillations can be traced in spectra of the ML-Au-glass sample (Fig. 9.9). These modes can be recognized by their angle dispersion (Fig. 9.9d). The spectral positions, dispersions and magnitudes of Fabry-Perot modes are in remarkable agreement with those of 50 L-Au-glass model structure (Fig. 9.6a, b).

The estimate of the onset of diffractively coupled bands in ML-glass sample is $\lambda_{GM} = n_{effML}D \approx 1,335$ nm for the normal light incidence. Correspondingly, at shorter wavelengths there exist a high number of quasi-GMs. Quasi-GMs dominate the reflectance/transmission maps at long wavelengths and their traces are much narrower compared to FP modes (Fig. 9.9). The dispersions of these modes satisfy the momentum conservation law, as demonstrated by overlapping the calculated quasi-GM dispersion curves in Fig. 9.9d. These diffraction bands are conveniently associated with band-like ballistic energy transport in the PhC. However, along the wavelength decrease, the contribution of guided modes vanishes and becomes replaced by scattering resonances of spheres.

As was argued above, the multipole Mie resonances overlay the diffraction bands in the middle of studied spectral range and dictate the intensity profiles at shorter

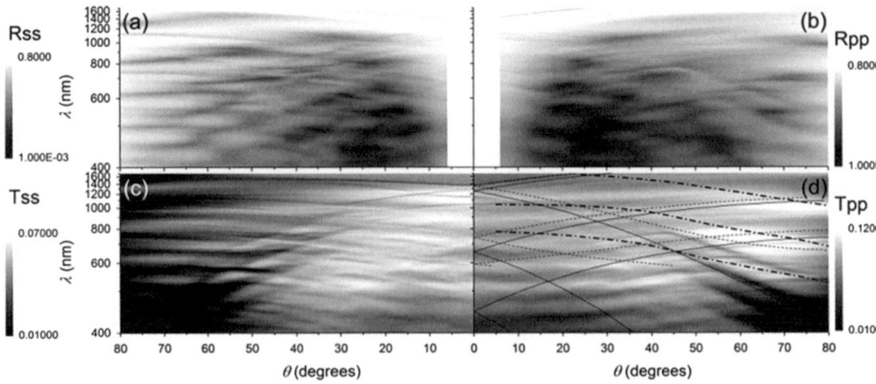

Fig. 9.9 Experimental (**a**, **b**) reflectance and (**c**, **d**) transmission spectra of ML-Au-glass hybrid. *Left* and *right* columns are for s- and p-polarized light, respectively. Intensities are plotted in the logarithmic scale. *Solid* and *dashed lines* in panel (**d**) show the dispersions of diffraction bands of SPP_{ML} and SPP_{sub} modes, respectively. Thick *dash-dotted lines* show the Fabry-Perot modes 0_p, 1_p and a calculated transmission map of 50L-Au-glass multilayer

wavelengths. Remarkably, the appearance of Mie resonances in reflectance of ML-Au-glass sample (Fig. 9.9a, b) closely resembles the calculated backscattering spectra of the single sphere sitting on Au-coated glass substrate (Fig. 9.6c, d). In particular, one can distinguish not only the Mie resonances of the single sphere but also the radial modulation of their scattering efficiency.

In transmission maps of ML-Au-glass the multipole Mie resonance are also clearly seen, but their angle modulation is suppressed (Fig. 9.9c, d). In fact, in p-polarized light at $\theta < 35^o$ the Mie resonance-related features appear almost angle-independent. Apparently, this is a collective effect in a lattice of touching spheres, where the cross-sections of multipole Mie resonance states efficiently overlap each other, but the interaction is not strong enough to hybridize them into a continuous band structure.

It is instructive to mention the interesting but not yet studied interplay of ballistic and hopping light transport in lattices of spheres. The close analogy of the hopping transport is the energy transfer between coupled resonators. Important, the hopping transport combines the long dwell time of light on a sphere (Mie resonance) with the slow drift of this excitation across the monolayer. These features are exactly the key requirements for an efficient light-matter interaction. Judging from the similar magnitude of the diffractive and Mie resonances one can derive that both mechanisms provide the same strength of such interaction. However, it is yet to be proved that Mie resonances can provide a certain in-plane flux, but not merely scatter the light out of a waveguide plane. Obviously, these two transport mechanisms are not independent from each other, since the particle scattering can contribute to the energy exchange between GMs inside and outside the light cone.

Along the wavelength decrease, the band-like transport of EM energy becomes more vulnerable with respect to the Rayleigh scattering at lattice defects.

At the same time, Mie resonances become stronger localized and less susceptive to hybridization. Eventually, the hopping transport will be also suppressed, but this may happen in a different spectral range (shorter wavelengths) compared to the suppression of ballistic transport.

9.3 Crystals of Plasmonic-Photonic Atoms

The contribution of plasmons to optical properties of ML-on-metal hybrids is rather weak, because the metal is not a plasmonic crystal itself. One way to realize the idea of hybrid hetero-PhC architecture as the series connection of photonic and plasmonic crystals is obtain a structure consisting of the periodic dielectric profile and the patterned metal film. Each part of such architecture can be considered as a lattice of PhAs with their own eigenmodes. If the near-field zones of these crystals overlap in space and their resonances overlap in frequency, one can expect the hybridization of their eigenmodes.

To prepare the corresponding architectures we exploited the concept of templated ordered arrays of nanostructures in porous dielectric matrices, which can be traced back to 1970s [48]. A particular implementation of this strategy is the evaporation of a metal on a ML of spheres to form the metal-capped spheres (Fig. 9.10) [49]. Templating provides the same period of spatial corrugation of Ag plasmonic crystal as that of ML PhC on the one hand, and the interpenetration of plasmonic and photonic crystals, on the other hand. So the conditions for eigenmode hybridization are fulfilled. Since a thickness of $t \approx 30$ nm is sufficient for the electrical connectivity of individual Ag caps, plasmonic excitations can propagate along this hybrid crystal. In fact, it might be incorrect to discuss the optical properties of Ag-ML in terms of the linear superposition of separate entities. In this regard, a more adequate approach is

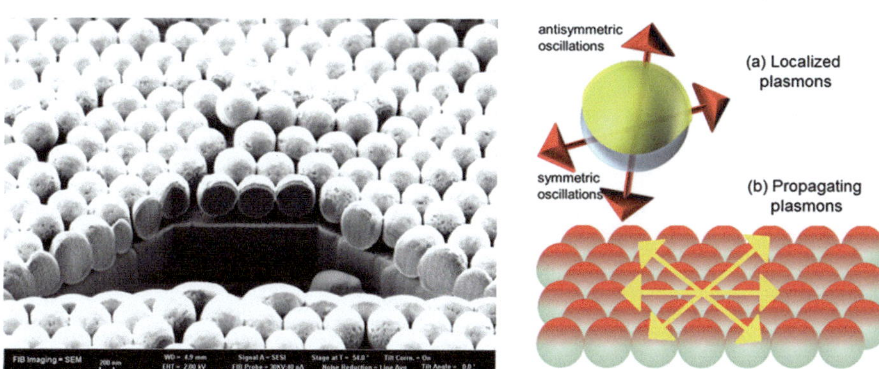

Fig. 9.10 *Left panel*: SEM image of Ag-coated (thickness ≈ 30 nm) ML of 625 nm SiO_2 spheres. *Right panel*: schematics of (**a**) the metal semishell with its localized plasmons and (**b**) ML of semishells with its propagating plasmons

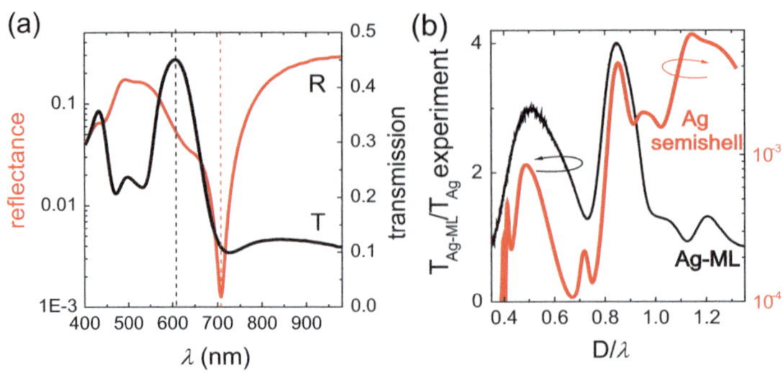

Fig. 9.11 (**a**) Transmission and reflectance spectra of Ag-ML at the normal light incidence. $D_{sphere} = 520$ nm, $t_{Ag} = 30$ nm. *Dashed lines* indicate detuning of EOT and SPP resonances. (**b**) Calculated extinction spectrum of the single semishell in comparison to the transmission spectrum of the Ag-ML (the sample of panel (**a**)) assembled from such semishells

to consider metal semishells (Fig. 9.10 right, a) as the photonic-plasmonic atoms. By analogy to metal nanoshells, the Mie resonances are the hybrid localized plasmon resonances [50], which originate from the cavity resonance provided by a sphere and the particle resonance provided by a metal semishell. In the tightly assembled ML of semishells, these resonances are hybridized in the collective propagating modes (Fig. 9.10 right, b). Thus, one can expect the hybridization of both localised and collective resonances.

Figure 9.11a shows the transmission and reflectance spectra of Ag-ML at the normal light incidence with dips in the reflectance spectra and peaks in transmission ones. The mechanisms leading to these features were intensively studied since the discovery of the extraordinary transmission (EOT) in the metal films perforated with holes of subwavelength size [51]. Later, the same effect was discovered in continuous metal films possessing a periodic profile [52]. In the case of periodically corrugated structures, the extraordinary transmission is conveniently related to the tunnelling of SPPs through the metal film. The diffraction of SPP in the lattice results in the formation of SPP eigenmodes (Figs. 9.7, and 9.10 right, b). Since the higher probability of tunnelling can be naturally associated with peaks in the spectrum of the density of plasmon states, the transmission subsequently peaks at the crossings of plasmon mode dispersions.

Extraordinary transmission was thoroughly investigated in the monolayers of metal-on-sphere architectures [53]. But, as we have noticed, no consensus has been achieved so far regarding the physical mechanisms of light transfer and different reports on angle and polarization dependence of transmitted light disagree considerably with each other. Apparently, it is the complexity of field propagation in such architectures, where the metal films is both corrugated and perforated (Fig. 9.10 left) that hinders the explanation. In such topology it is impossible to separate the light transit via corrugation and perforation of a metal film that appeals for the systematic study. Also, numerical simulations demonstrated that

Bloch modes in a monolayer promote the extraordinary transmission by the spatial concentration of the field at special places along the metal film. Thus, the actual light propagation depends on synergy of many factors, in particular, on the metal film topology, the dielectric function profile and the mode hybridization [54].

The reflectance minimum in Fig. 9.11a is a clear evidence of the SPP, because its position satisfies the momentum conservation law. The extraordinary transmission peak is red-shifted with respect to the SPP that is usually explained by the Fano resonance [55]. The Fano model considers the interference of two light flows: the one supported by a continuum of non-resonant states (here, the scattering of incident radiation) and another supported by a resonant state (here, the diffraction resonance) related to the periodic structure. Due to flipping of the phase of the EM wave occurring at crossing of the resonance frequency, the constructive interference of the two fields switches to the destructive one. As a result, the spectrum of the resonance becomes highly asymmetric, i.e., it combines the frequency intervals of the transmission enhancement with that of the transmission suppression. More specifically, the transmission minimum, which coincides with the reflection minimum, corresponds to the Wood's anomaly that is the diffraction order propagating along the corrugated metal film [56]. Thus, the Fano approach requires the assumption concerning the position of the sharp resonance and explains only the interaction strength of different transport mechanisms.

Alternatively, there is an argument that supports the tight-binding approach in the case of the Ag-ML hybrid. In Fig. 9.11b we show the experimental transmission spectrum of the Ag-ML in comparison with the extinction spectrum of the single semishell of the same topology, which was calculated by the finite difference method. Upon conversion of the extinction spectrum in the transmission one, it is readily seen that the scattering minimum corresponds to the maximum of the extraordinary transmission of Ag-ML, whereas the scattering maximum corresponds to the SPP-related transmission minimum of Ag-ML. This observation can be interpreted in terms of the tight binding model, which postulates that the energy bands of the hybrid plasmonic-photonic crystal are formed due to hybridization of the Mie resonances of the respective photonics atoms.

In the case of ML-Ag architecture, Mie resonances of dielectric spheres became observable owing to metal substrate that reduces the light leakage from the monolayer. In contrast, Mie resonances of semishells in Ag-ML architecture are inherently bounded to metal caps. Generally, monolayers assembled from semishells look more prospective for engineering the hybrid's optical response due to the broader range of balancing between individual and collective resonances.

Figure 9.12a compares numerically calculated and experimental transmission spectra of Ag-ML. The general shape and the position of spectral features are very similar, but in the experimental spectrum the transmission magnitude is strongly reduced and the spectral features are smeared away. This difference is in line with published results [56] and can be assigned to the defects of the lattice ordering. The deep transmission minimum at ~900 nm coincides in both spectra, whereas the experimental extraordinary transmission band centred at 747 nm overlaps two theoretical bands centred at 680 and 765 nm.

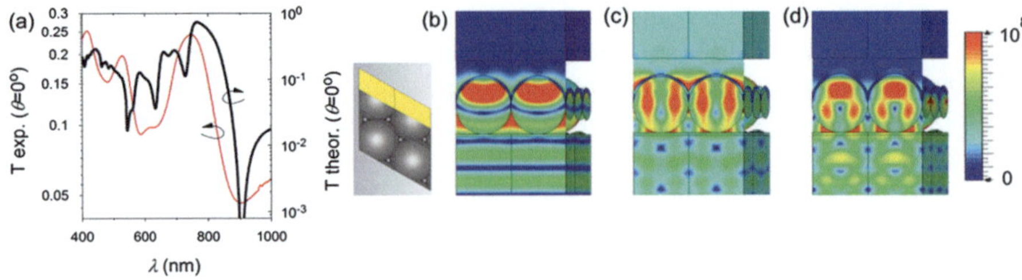

Fig. 9.12 (**a**) Numerically calculated (*thick line*) versus experimental transmission spectra of Ag-ML sample on a glass substrate ($D = 625$ nm, $t_{Ag} = 30$ nm) at the normal light incidence. Calculated absolute value of the electric field distribution in the Ag-ML hybrid at the frequencies of (**b**) the transmission minimum at 901 nm and (**c**) the EOT at 765 nm and (**d**) the minimum at 727 nm. Inset shows the cutting plane in the lattice unit cell. Linear scale of intensity is given in V/m. The illuminating plane wave approaches Ag-ML through the substrate (from the *bottom* of the picture) (Courtesy of N. Arnold)

In order to illustrate the mechanism of light propagation in the opposite extrema, the field distributions in the extraordinary transmission maximum and adjacent transmission minima were numerically modelled. It is readily seen that 901 nm transmission minimum corresponds to the field lobe right behind the metal semishell and looks like the cavity resonance of the metal semishell (Fig. 9.12b). At this frequency the light is back reflected and a standing wave is formed along the incidence wavevector that reveals itself as the second lobe and the node in the centre of the sphere. Hence, this transmission minimum correlates nicely the maximum of scattering efficiency of the individual semishell in Fig. 9.11b.

In contrast, at the extraordinary transmission peak we see the uniform field strings across each nanoshell, which indicate the transversely propagating mode, as well as the field maxima in between semishells, which couple semishells in the array (Fig. 9.12c). Such field configuration supports the transversal light flow. Thus, extraordinary transmission peak in Ag-ML is a product of lateral semishell coupling [57] that is, in fact, the diffraction effect.

The next minimum at 725 nm corresponds to almost the same field distribution like in the extraordinary transmission maximum, but the longitudinal mode is shifted by ½-period and aligned with semishells (Fig. 9.12d). This difference in the field distribution correlates with the idea of the Fano shape of the extraordinary transmission resonance. Therefore, this minimum can be assigned to the Fano resonance, which is not resolved in the experiment due to defects of the lattice package.

The diffraction nature of the extraordinary transmission also agrees with the fact that the guided mode-related transmission minimum occurring in the bare monolayer (similar to that in Fig. 9.3a) shares the same spectral range with the SPPs, i.e., both Bloch waves relate to the same lattice periodicity. In the spectrum of Ag-ML the transmission peak substitutes the transmission minimum of the bare monolayer owing to the tunnelling effect.

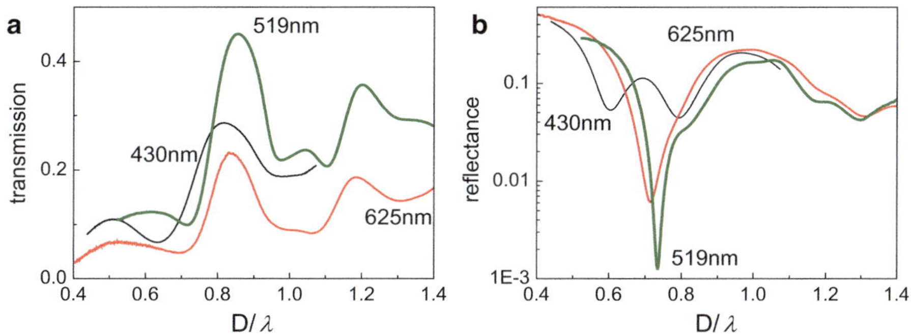

Fig. 9.13 Transmission (**a**) and reflectance (**b**) spectra of Ag-ML samples possessing different periodicities. Sphere diameters are indicated at curves

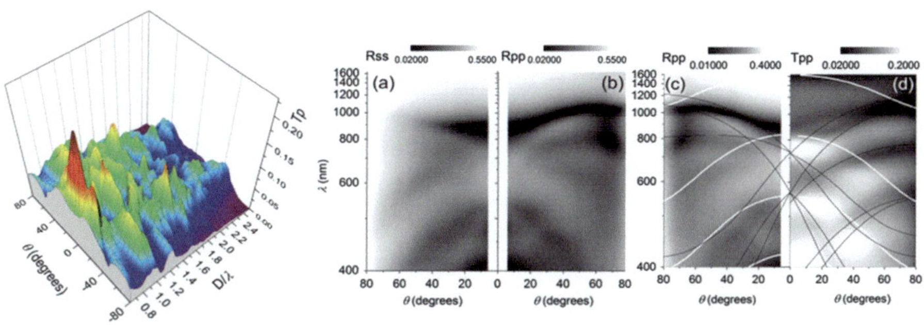

Fig. 9.14 *Left panel*: angle-resolved transmission spectra of Ag-ML sample in p-polarized light dominated by the EOT peak. $D = 1{,}070$ nm, $t_{Ag} = 30$ nm. *Right panel*: (**a, b**) Reflectance maps of Ag-ML in s- and p-polarized light. (**c, d**) Comparison of reflectance and transmission maps of Ag-ML in p-polarized light. $D = 625$ nm, $t_{Ag} = 30$ nm. In panels (**c, d**) *white lines* show the dispersions Fabry-Perot modes and *black lines* – dispersions of selected SPP modes

It is important to mention that Ag-ML samples based on spheres of different diameters all show qualitatively the same behavior (Fig. 9.13) in spite of the lack of wavelength scalability for the dielectric function of silver. However, extraordinary transmission peak disappears for samples of smaller diameters, if its respective position falls in the range of the metal absorption caused by its intraband transitions. Naturally, the better the ordering of a monolayer package – the sharper these resonances.

The transmission of the Ag-ML sample (Fig. 9.14 left) differs remarkably from that of the ML-Ag hybrid (Fig. 9.3b): (a) the overall transmission of the Ag-ML is higher for the same nominal Ag film thickness; (b) The domination by EOT peak (Fig. 9.14); (c) Plasmon bands are seen in both polarizations; (d) The elaborated profile of the transmission and reflectance spectra results from mixing of dispersive and non-dispersive modes.

In Ag-ML the metal film, which reaches $t \approx 30$ nm thickness on top of spheres, possesses up to 50 % corrugation depth. Since the very existence of the long-range SPPs requires less than 10 % of such depth [58], they cannot survive and we deal with the short-range ones. Hence, the short-range ordering of semishells in a monolayer is sufficient to preserve diffraction features in the optical spectra, but at the cost of spectral broadening of the respective resonances. This makes the resonance structure robust with respect to monolayer disordering. The immediate consequence of the deep corrugation profile is the short SPP lifetime, which becomes shorter compared to the characteristic SPP absorption time. Overall, one can expect that the losses due to the light scattering in planar semishell lattices overcome the absorption losses.

It is instructive to explore the transmission and reflectance spectra as a function of the light polarization. To reveal the polarization dependence, the analysis of angle-resolved spectra is required. First, the Fabry-Perot modes are spectrally broad and show themselves as respective minima/maxima in transmission and reflectance for both polarizations. Second, the plasmonic modes demonstrate complex behaviour with respect to the angle of light incidence (Fig. 9.14). At small angles the s- and p-polarized reflectance spectra look alike. At around 900 nm the plasmon-related minimum can be thought as a superposition of dispersive and non-dispersive features. The non-dispersive one is the localized plasmon resonance of a semishell, as we discussed above [59]. The dispersive one satisfies, roughly, the dispersion of SPP modes. Third, at grazing incidence the plasmonic features vanish in the spectra of s-polarized light but persist in p-polarization (Fig. 9.14a, b). In transmission spectra one can see the extraordinary transmission peak immediately above the SPP minimum at 747 nm (Fig. 9.14d). This peak is continued by transmission dip along the same dispersion curve for higher incidence angles. But the latter assignment is not unique, because it disregards the Fano effect. Moreover, the dispersions depend on the azimuth orientation of the monolayer lattice, whereas experimental spectra average the minima over many such orientations, so they may not correlate with the calculated curve made for one particular lattice orientation. Thus, this issue remains open.

Angle-resolved reflectance spectra of Ag-ML in Fig. 9.15a, b compare the responses from its uncoated monolayer side and the Ag-coated side under p-polarized light. Two sets of observed modes are distinctly different. According to the difference between refractive indices of surrounding media, the long wavelength set of modes relates to SPPs at the metal-ML interface, whereas the other set refers to modes that face the air. A satisfactory match between the reflectance dips and dispersions of SPP modes is observed in both reflectance maps.

When the Ag-ML hybrid is observed from the Ag film side, the reflectance map is dominated by the SPP minimum for Ag-ML interface. Vice versa, the modes at the metal-air interface are seen in the reflectance map obtained from the opposite side of the Ag-ML hybrid at low angles of incidence. This controversy can be resolved, if the interplay of scattering and diffraction resonances is taken into account. If the ML-side is under illumination, the light is efficiently scattered and absorbed at the wavelength of the localized plasmon resonance of the semishell $\lambda \simeq 710$ nm.

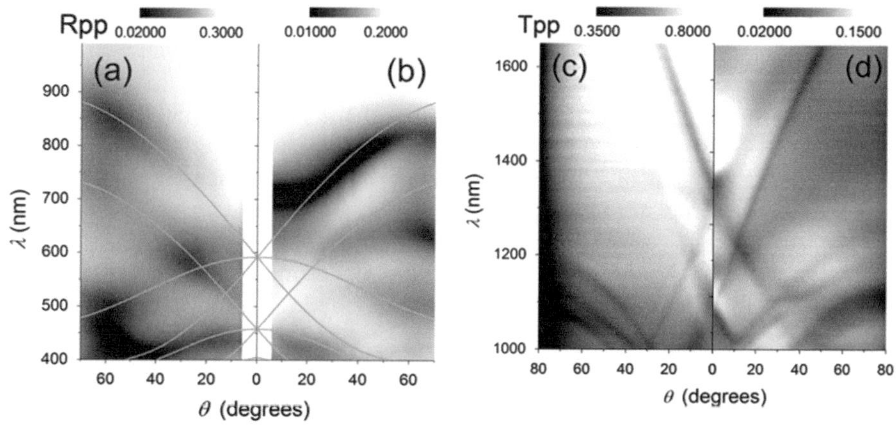

Fig. 9.15 Reflectance maps of Ag-ML in p-polarized light from (**a**) ML and (**b**) Ag sides. $D = 520$ nm, $t_{Ag} = 30$ nm. *Lines* show the dispersions of SPP modes. Transmission maps of (**c**) ML and (**d**) Ag-ML in p-polarized light. $D = 1{,}063$ nm, $t_{Ag} = 30$ nm

In particular, it is backscattered to the monolayer side (Fig. 9.12b) and, hence, produces the reflectance maximum in the map in Fig. 9.15a. The same resonance excited from the air-metal interface favours the forward scattering due to the difference of indices between air and monolayer that leads to the reflectance minimum. Alternatively, the sharp reflectance minimum in plasmonics is explained by achieving phase matching conditions. In the case of the grating coupler – to the diffraction resonance, which directs the respective diffraction order along the plane of hybrid at $\lambda \approx n_{eff}D$. The above mentioned Fano resonance helps to explain the splitting of the scattering and diffraction resonances. Here we meet again the example of the tight binding approximation that shows the collective resonance (diffraction order) originating from the scattering resonance of a single photonic atom – the nanoshell.

It is also instructive to elaborate the scenario of absorption. The scattering resonance at $\lambda \approx 870$ nm shows the minimum in transmission spectrum due to absorption of light that is bounced in a metal cap. The same consideration can be applied to the reflectance spectrum obtained from the metal-air interface (compare panels c and d in Fig. 9.14). The mode hybridization means, in particular, that EM field of plasmons extends deeper into dielectric spheres compared to the extension of plasmon field in air at the metal-air interface. Hence, the latter are more susceptive to the absorption and their lifetime is shortened by both the scattering and absorption. So the respective peak of the density of plasmon states at the metal-monolayer interface is high enough to provide the extraordinary transmission peak at $\lambda \approx 725$ nm, but is insufficient to establish the same for SPP modes at the metal-air interface at $\lambda \approx 595$ nm.

The final note in this section deals with the difference between PhC Bloch modes of a monolayer photonic crystal and SPP modes of plasmonic crystal. If one compares the two transmission maps of bare ML and Ag-ML (Fig. 9.15c, d), the similarity is striking. With the exception of the overall transmission level and the extraordinary transmission peak in Ag-ML spectra, it is hard to find a difference in the quasi-guided modes. As we demonstrated, the resonances that control the light transport in Ag-ML hybrid are definitely of the plasmonic nature. Thus, the dispersion of SPPs is fully determined by the diffraction in the lattice of spheres likewise the dispersion of quasi-guided modes in their corresponding monolayer photonic crystal template. Thus, the metal-on-ML architecture represents a case of the strong plasmonic-photonic mode hybridization.

To conclude, we want to highlight again the higher number of the independent mechanisms, the interplay of which determines the hybrid's optical response.

9.4 Summary and Outlook

Colloidal crystals are frequently used as materials for nanophotonics and, in particular, for preparing photonic crystals. In contrast to 3D opal crystals, the planar crystals or monolayers of spheres lend themselves for unambiguous tracing of the changes that follow manipulations with the symmetry, ordering, refractive index contrast and so on. We used this openness to separate localized and collective resonances in the optical response and to illustrate our findings using the model of the hybridization of individual optical resonances of photonic atoms. Overall, planar colloidal photonic crystals represent a versatile instrument for studies of various fundamental aspects of nanophotonics. An added value of the colloidal crystals is a simple sizes. From the application point of view, industrial up-scaling of colloidal crystal-based devices was proved possible and can be less expensive compared to other mass-production techniques like nanoimprinting.

A new momentum was given to colloidal photonic crystals by the invention of hetero-opals, the light propagation in which is controlled by the interfaces between crystals with different photonics energy band structures. Now is the time of hybrid crystals, which can be considered as a separate generic class, the specific functionality of which is based on involving different optical excitations in the light transfer and their coupling at the interfaces. This novelty came from the encounter of the metal and dielectric electromagnetic crystals, but this approach is not limited to the consecutive light transport by plasmons and photons. The important role can be given to hybrid excitations. By purposive structure engineering, the strong light confinement offered by plasmons can be combined with the lossless propagation of photons. Such materials bear prospects for efficient all-optic light processing, highly desirable wavelength-size photon management architectures for light harvesting and emitting devices, advanced sensors as well as for long-awaiting nanosize electronic-photonic interconnects.

We demonstrated that 2D slab photonics crystals can be converted into hybrid metal-dielectric plasmonic-photonic crystals merely by assembling a monolayer of spheres onto a semitransparent flat metal film. The major reason for the observed strong modification of the monolayer optical response is a tighter light confinement in the plane of crystal, the generation of diffractively coupled propagating plasmons and the enhancement of local resonances of colloidal spheres. We interpreted the optical response of the ML-Au-glass hybrid in terms of the superposition of Fabry-Perot resonances, quasi-guided modes and multipole Mie resonances. Based on the interplay of these excitations we may suggest that the light transport in a plane of a hybrid 2D slab photonic crystal is supported by two mechanisms: (i) the ballistic propagation of eigenmodes in the spectral range $D/\lambda < 1$ and (ii) the hopping of Mie excitations from sphere to sphere in the range $D/\lambda > 1$. It is expected that in the intermediate frequency range both mechanisms will cooperate in the in-plane light transfer.

It is reasonable to expect that the light-matter interaction will be enhanced, if light is coupled to long-living Mie resonances as compared to light coupled to ballistically propagating modes. By validating this hypothesis, one can engineer the devices operating with coupled localized-ballistic modes to outperform current light sources and gas sensors based on diffractive modes of photonic crystals.

Another planar architecture was prepared by coating the surface of a monolayer by a thin metal film. Such metal film adopts the profile of the spheres, i.e. it is deeply corrugated and also perforated with a lattice of through holes. In this case we deal with a dielectric photonic crystal, which is interpenetrated by a plasmonic crystal. As the result, the eigenmodes of such superimposed crystals are almost completely hybridized, i.e. the fields of these modes are spread across the monolayer depth. Contrary to the previous case of ML-Ag hybrids, in the Ag-ML architecture the separation of propagating modes on dielectric-like and plasmon-like modes should be modified in terms of symmetric and antisymmetric combinations of eigenmodes.

The semishell topology of metal-coated spheres is an important qualifying factor of this architecture. On the one hand, it is inherently related to the properties of metal nanoshells, the spectrum of which is defined by the hybridization of particle and cavity modes. On the other hand, the half-sphere openings make semishell resonances more vulnerable to external stimuli. In particular, this topology makes feasible the hybridization of quasi-guided nodes of photonic and plasmonic crystals.

Depending on the ordering, the resonances can be sharper or smoother but the deep profile allows only short-range SPPs to survive, i.e., the observed propagating plasmons relay on the short-range lattice ordering. Nevertheless, the hybridization of plasmonic and photonic modes reduces both scattering and absorption compared to the case of flux transfer by highly evanescent modes along the metal-dielectric interface.

Finally, it is worth mentioning the usefulness of the photonic atom concept. Firstly, it appears very instructive for interpretation of the optical properties of architectures studied in the $D/\lambda > 1$ spectral range. By replacing a dielectric photonic atom with a hybrid plasmonic-photonic atom we can use the same tight-binding approach to describe the behavior of dielectric and metal-dielectric colloidal crystals. Secondly, local plasmon resonances naturally explain the optical resonances at the spectral range below the onset of in-plane diffraction. Finally, this approach explains the idea of hopping light transport as the coupling of identical resonators. Its useful consequence is the idea of combining the efficient light-matter interaction with high optical flux throughput in the spectral range, where the ballistic and hopping transport mechanisms overlap.

References

1. Bykov VP (1972) Spontaneous emission in a periodic structure. Sov Phys JETP 35:269; Zengerle R (1987) Light propagation in singly and doubly periodic planar waveguides. J Mod Optics 34:1589; Yablonovitch E (1987) Inhibited spontaneous emission in solid-state physics and electronics. Phys Rev Lett 58:2059; John S (1987) Strong localization of photons in certain disordered dielectric superlattices. Phys Rev Lett 58:2486
2. John S (1995) Localization of light. In: Soukoulius CM (ed) Photonic band gap materials. Kluwer, Dordrecht, pp 563–735
3. Mie G (1908) Beitrarge zur Optik trueber Medien, speziell kolloidaler Metalloesungen. Annalen der Physik, 330, IV Folge, 25(3):377–445; Debye P (1909) Der Lichtdruck auf Kugeln von beliebigem Material. Annalen der Physik, 335, IV Folge, 30(11):57–136
4. Arnold S (2001) Microspheres, photonic atoms and the physics of nothing. Am Sci 89:414; Arnold S, Folan LM (1989) Energy transfer and the photon lifetime within an aerosol particle. Opt Lett 14:387
5. Lidorikis E, Sigalas MM, Economou EN, Soukoulis CM (1998) Tight-binding parametrization for photonic band gap materials. Phys Rev Lett 82:1405
6. Kohmoto M, Sutherland B, Iguchi K (1987) Localization of optics: quasiperiodic media. Phys Rev Lett 58:2436–2438; Valy Vardeny Z, Nahata A, Agrawal A (2013) Optics of photonic quasicrystals. Nat Photonics 7:177–187
7. Romanov SG, Yates HM, Pemble ME, De La Rue RM (2000) Opal-based photonic crystal with double photonic bandgap structure. J Phys: Cond Matter 12:8221; Istrate E, Sargent EH (2006) Photonic crystal heterostructures and interfaces. Rev Mod Phys 78:455
8. Ivchenko EL, Poddubnyi AN (2013) Resonant diffraction of electromagnetic waves from solids (a review). Phys Solid State 55:905
9. Toader O, John S (2004) Photonic band gap enhancement in frequency-dependent dielectrics. Phys Rev E 70:46605
10. Ivchenko EL, Nezvizhevskii AN, Jorda S (1994) Bragg reflection of light from quantum-well structures. Phys Solid State 36:1156; Kochereshko VP, Pozina GR, Ivchenko EL, Yakovlev DR, Waag A, Ossau W, Landwehr G, Hellmann R, Göbel EO (1994) Giant exciton resonance reflectance in Bragg MQW structures. Superlatt Microstruct 15:471; Merle d'Aubigne Y, Wasiela A, Mariette H, Dietl T (1996) Polariton effects in multiple-quantum-well structures of $CdTe/Cd_{1-x}Zn_xTe$. Phys Rev B 54:14003
11. Fujita T, Sato Y, Kuitani T, Ishihara T (1998) Tunable polariton absorption of distributed feedback microcavities at room temperature. Phys Rev B 57:12428; Yablonskii AL, Muljarov EA, Gippius NA, Tikhodeev SG, Fujita T, Ishihara T (2001) Polariton effect in distributed feedback microcavities. J Phys Soc Jpn 70:1137; Shimada R, Yablonskii AL, Tikhodeev SG, Ishihara T (2002) Transmission properties of two-dimensional photonic crystal slab with an excitonic resonance. IEEE J Quantum Electron 38:872
12. Matsko AB, Savchenkov AA, Liang W, Ilchenko VS, Seidel D, Maleki L (2009) Collective emission and absorption in a linear resonator chain. Opt Express 17:15210
13. Wood RW (1902) XLII. On a remarkable case of uneven distribution of light in a diffraction grating spectrum. Philos Mag 4:396; Fano U (1941) The theory of anomalous diffraction gratings and of quasi-stationary waves on metallic surfaces (Sommerfeld's waves). J Opt Soc Am 31:213

14. Ebbesen TW, Lezec HJ, Ghaemi HF, Thio T, Wolff PA (1998) Extraordinary optical transmission through sub-wavelength hole arrays. Nature 391:667
15. Christ A, Tikhodeev SG, Gippius NA, Kuhl J, Giessen H (2003) Waveguide-plasmon polaritons: strong coupling of photonic and electronic resonances in a metallic photonic crystal slab. Phys Rev Lett 91:183901
16. Barnes WL, Kitson SC, Preist TW, Sambles JR (1997) Photonic surfaces for surface-plasmon polaritons. J Opt Soc Am A 14:1654; Kretschmann M, Maradudin AA (2002) Band structures of two-dimensional surface-plasmon polaritonic crystals. Phys Rev B 66:245408
17. Teperik TV, Popov VV, García de Abajo FJ (2005) Void plasmons and total absorption of light in nanoporous metallic films. Phys Rev B 71:085408; Teperik TV, García de Abajo FJ, Borisov AG, Abdelsalam M, Bartlett PN, Sugawara Y, Baumberg JJ (2008) Omnidirectional absorption in nanostructured metal surfaces. Nat Photonics 2:299
18. Chen H, Chan CT, Sheng P (2010) Transformation optics and metamaterials. Nat Mater 9:387
19. Leonhardt U (2006) Optical conformal mapping. Science 312:1777
20. Veselago VG (1968) The electrodynamics of substances with simultaneously negative values of e and μ. Sov Phys Usp 10:509
21. Romanov SG, Regensburger A, Korovin AV, Peschel U (2011) Hybrid colloidal plasmonic photonic crystals. Adv Mater 23:2515
22. Romanov SG (2012) Hybrid metal-dielectric plasmonic-photonic crystals on colloidal platforms. In: Limonov M, De La Rue R (eds) Optical properties of photonic structures: interplay of order and disorder, ch. 8. Taylor & Francis, Oxford, UK
23. Sanders JV (1964) Colour of precious opal. Nature 204(4964):1151
24. Bogomolov V, Kumzerov Y, Romanov S (1992) Fabrication of three-dimensional superlattices of nanostructures. In: Davies JH, Long AR (eds) Physics of nanostructures. IOP Publishing, Bristol, pp 317–321
25. Astratov VN, Bogomolov VN, Kaplyanskii AA, Prokofiev AV, Samoilovich LA, Samoilovich SM, Vlasov Yu. A (1995) Optical spectroscopy of opal matrices with CdS embedded in its pores: quantum confinement and photonic band gap effects. Il Nuovo Cimento 17D:1349; Romanov SG, Fokin AV, Tretiakov VV, Butko VY, Alperovich VI, Johnson NP, Sotomayor Torres CM (1996) Optical properties of ordered 3-dimensional arrays of structurally confined semiconductors. J Cryst Growth 159:857
26. Galisteo-López JF, Ibisate M, Sapienza R, Froufe-Pérez LS, Blanco Á, López C (2011) Self-assembled photonic structures. Adv Mater 23:30; Bardosova M, Pemble ME, Povey IM, Tredgold RH (2010) The Langmuir-Blodgett approach to making colloidal photonic crystals from silica spheres. Adv Mater 22:3104
27. Romanov SG, Sotomayor Torres CM (2000) Three-dimensional lattices of nanostructures – template approach. In: Nalwa HS (ed) Handbook of nanostructured materials and technology, vol 4, ch 4. Academic Press, Waltham, pp 231–323
28. Wang Z, Chan CT, Zhang W, Ming N, Sheng P (2001) Three-dimensional self-assembly of metal nanoparticles: possible photonic crystal with a complete gap below the plasma frequency. Phys Rev B 64:113108
29. Romanov SG, Susha AS, Sotomayor Torres CM, Liang Z, Caruso F (2005) Surface plasmon resonance in gold nanoparticle infiltrated dielectric opals. J Appl Phys 97:086103; Jiang Y, Whitehouse C, Li J, Tam WY, Chan CT, Sheng P (2003) Optical properties of metallo-dielectric microspheres in opal structures. J Phys: Condens Matter 15:5871
30. Moroz A (2002) Metallo-dielectric diamond and zinc-blende photonic crystals. Phys Rev B 66:115109
31. Lu Y, Yin Y, Li Z-Y, Xia Y (2002) Synthesis and self-assembly of Au@SiO_2 core−shell colloids. Nano Lett 2:785; Rodrıguez-Gonzalez B, Salgueirino-Maceira V, Garcıa-Santamarıa F, Liz-Marzan LM (2002) Fully accessible gold nanoparticles within ordered macroporous solids. Nano Lett 2:471
32. Eradat N, Huang JD, Vardeny ZV, Zakhidov AA, Khayrullin I, Udod I, Baughman RH (2001) Optical studies of metal-infiltrated opal photonic crystals. Synth Metals 116:501; Wang D, Li J, Chan CT, SalgueiriÇo-Maceira VA, Liz-Marzan LM, Romanov S, Caruso F (2005) Optical

properties of nanoparticle-based metallodielectric inverse opals. Small 1:122; Perez N, Huls A, Puente D, Gonzalez-Vinas W, Castano E, Olaizola SM (2007) Fabrication and characterization of silver inverse opals. Sensors and Actuators B 126:86; Kozlov ME, Murthy NS, Udod I, Khayrullin II, Baughman RH, Zakhidov AA (2007) Preparation, structural, and calorimetric characterization of bicomponent metallic photonic crystals. Appl Phys A 86:421; Lytle JC, Stein A (2006) Recent progress in synthesis and applications of inverse opals and related macroporous materials prepared by colloidal templating. In: Cao G, Brinker CJ (eds) Annual reviews of nano research, vol 1, ch 1. World Scientific Publishing, New Jersey

33. Le F, Brandl DW, Urzhumov YA, Wang H, Kundu J, Halas NJ, Aizpurua J, Nordlander P (2008) Metallic nanoparticle arrays: a common substrate for both surface-enhanced Raman scattering and surface-enhanced infrared absorption. ACS Nano 2:707
34. Pennington RC, D'Alessandro G, Baumberg JJ, Kaczmarek M (2009) Spectral properties and modes of surface microcavities. Phys Rev A 79:043822
35. Inoue M (1987) Enhancement of local field by a two-dimensional array of dielectric spheres placed on a substrate. Phys Rev B 36:2852; Kurokawa Y, Miyazaki H, Jimba Y (2004) Optical band structure and near-field intensity of a periodically arrayed monolayer of dielectric spheres on dielectric substrate of finite thickness. Phys Rev B 69:155117
36. Dushkin CD, Nagayama K, Miwa T, Kralchevsky PA (1993) Colored multilayers from transparent submicrometer spheres. Langmuir 9:3695
37. Shi L, Liu X, Yina H, Zi J (2010) Optical response of a flat metallic surface coated with a monolayer array of latex spheres. Phys Lett A 374:1059; Yu X, Shi L, Han D, Zi J, Braun PV (2010) High quality factor metallodielectric hybrid plasmonic–photonic crystals. Adv Funct Mater 20:1910; Lopez-Garcıa M, Galisteo-Lopez JF, Blanco A, Sanchez-Marcos J, Lopez C, Garcıa-Martın A (2010) Enhancement and directionality of spontaneous emission in hybrid self-assembled photonic–plasmonic crystals. Small 6:1757
38. Romanov SG, Bardosova M, Povey I, Pemble M, Sotomayor Torres CM (2008) Understanding of transmission in the range of high-order photonic bands in thin opal film. Appl Phys Lett 92:191106; Romanov SG, Bardosova M, Pemble M, Sotomayor Torres CM (2006) (2+1)-dimensional photonic crystals from Langmuir-Blodgett colloidal multilayers. Appl Phys Lett 89:43105
39. Romanov SG, Vogel N, Bley K, Landfester K, Weiss CK, Orlov S, Korovin AV, Chuiko GP, Regensburger A, Romanova AS, Kriesch A, Peschel U (2012) Probing guided modes in a monolayer colloidal crystal on a flat metral film. Phys Rev B 86:195145
40. Jonson PB, Christy RW (1972) Optical constants of the noble metals. Phys Rev B 6:4370
41. Nikoma J, Loudon R, Tilley DR (1974) Elementary properties of surface polaritons. J Phys: Solid State Phys 7:3547
42. Mueller KG, Veith M, Mittler-Neher S, Knoll W (1997) Plasmon surface polariton coupling with dielectric gratings and the thermal decomposition of these dielectric gratings. J Appl Phys 82:4172; Li X, Han D, Wu F, Xu C, Liu X, Zi J (2008) Flat metallic surfaces coated with a dielectric grating: excitations of surface plasmon-polaritons and guided modes. J Phys: Condens Matter 20:485001
43. Vainshtein BK, Fridkin VM, Indenbom VL (2000) Modern crystallography series, vol 2. Springer, Berlin, p 520. ISBN 3-540-67474-8
44. Bobbert PA, Vlieger J (1986) Light scattering by a sphere on a substrate. Phys A 137:209
45. Fucile E, Denti P, Borghese F, Saija R, Sindoni OI (1997) Optical properties of aggregated spheres in the vicinity of a plane surface. J Opt Soc Am A 14:1505
46. Arnoldus HF (2005) Reflection and refraction of multipole radiation by an interface. J Opt Soc Am A 22:190
47. Weber MJ (2003) Handbook of optical materials. CRC Press LLCV, Boca Raton, p 536. ISBN 0-8493-3512-4
48. Bogomolov VN (1978) Liquids in ultrathin channels (filament and cluster crystals). Sov Phys Usp 21:77
49. Takei H (1999) Surface-adsorbed polystyrene spheres as a template for nanosized metal particle formation: optical properties of nanosized Au particle. J Vac Sci Technol B 17:1906;

Himmelhaus M, Takei H (2000) Cap-shaped gold nanoparticles for an optical biosensor. Sens Actuators B Chem 63:24
50. Aden AL, Kerker M (1951) Scattering of electromagnetic waves from two concentric spheres. J Appl Phys 22:1242; Prodan E, Lee A, Nordlander P (2002) Electronic structure and polarizability of metallic nanoshells. Chem Phys Lett 360:325
51. Martin-Moreno L, Garcia-Vidal FJ, Lezec HJ, Pellerin KM, Thio T, Pendry JB, Ebbesen TW (2001) Theory of extraordinary optical transmission through subwavelength hole arrays. Phys Rev Lett 86:1114
52. Bonod N, Enoch S, Li L, Popov E, Neviere M (2003) Resonant optical transmission through thin metallic films with and without holes. Opt Express 11:482
53. Farcau C, Astilean S (2007) Probing the unusual optical transmission of silver films deposited on two-dimensional regular arrays of polystyrene microspheres. J Opt A: Pure Appl Opt 9:S345–S349; Landström L, Brodoceanu D, Piglmayer K, Bäuerle D (2006) Extraordinary optical transmission. Appl Phys A 84:373–377; Zhan P, Wang Z, Dong H, Sun J, Wu J, Wang H-T, Zhu S, Ming N, Zi J (2006) The anomalous infrared transmission of gold films on two-dimensional colloidal crystals. Adv Mater 18:1612–1616; Lezec HJ, Thio T (2004) Diffracted evanescent wave model for enhanced and suppressed optical transmission through subwavelength hole arrays. Optics Express 12:3629
54. Landstrom L, Brodoceanu D, Bauerle D, Garcia-Vidal FJ, Rodrigo SG, Martin-Moreno L (2009) Extraordinary transmission through metal-coated monolayers of microspheres. Opt Express 17:761
55. Fano U (1961) Effects of configuration interaction on intensities and phase shifts. Phys Rev 124:1866–1875; Genet C, van Exter MP, Woerdman JP (2003) Fano-type interpretation of red shifts and red tails in hole array transmission spectra. Opt Comm 225:331; Miroshnichenko AE, Flach S, Kivshar YS (2010) Fano resonances in nanoscale structures. Rev Mod Phys 82:2257
56. Ghaemi HF, Thio T, Grupp DE, Ebbesen TW, Lezec HJ (1998) Surface plasmons enhance optical transmission through subwavelength holes. Phys Rev B 58:6779
57. Maaroof AI, Cortie MB, Harris N, Wieczorek L (2008) Mie and Bragg plasmons in subwavelength silver semi-shells. Small 4:2292
58. Ursu I, Mihailescu IN, Prokhorov AM, Konov VI, Tokarev VN (1985) On the role of the periodical structures induced by powerful laser irradiation of metallic surfaces in the energy coupling process. Phys B+C 132:395
59. Charnay C, Lee A, Man S-Q, Moran CE, Radloff C, Bradley RK, Halas NJ (2003) Reduced symmetry metallodielectric nanoparticles: synthesis and plasmonic properties. J Phys Chem B 107:7327

Chapter 10
Fundamentals and Applications of Organised Molecular Films

Osvaldo N. Oliveira Jr., Felippe J. Pavinatto, and Débora T. Balogh

Abstract In this Chapter we describe the experimental procedures and main features of solid organised films produced with three techniques: Langmuir-Blodgett (LB), electrostatic layer-by-layer (LbL) and self-assembled monolayers (SAMs). Emphasis is placed on possible applications in which molecular control of the film architectures is exploited. In particular, the use of organised films in sensing units for electronic tongues (e-tongues) and biosensors is highlighted not only in terms of the nanotech-based methods but also in connection with computational methods. The latter are employed in sensing and biosensing data analysis, and becoming increasingly important for generating fully-fledged clinical diagnosis systems. With regard to basic science involved in organised films, we discuss the use of Langmuir monolayers in cell membrane models, which is important for drug design and drug delivery systems.

10.1 Introduction

Interfacial effects have been studied for many decades, initially as an important subject of physical chemistry of surfaces, but over the last few years they have become even more prominent owing to the increasing use of nanomaterials with large surface/volume aspect ratio. Particularly important in this context are the nanostructured films, which may serve as coatings for supports of varied shapes and composition. In this chapter we shall discuss briefly some of the fundamental properties of nanostructured films obtained with three methods, namely Langmuir-Blodgett (LB), layer-by-layer (LbL) and self-assembly.

The variety of possible uses of these nanostructured films is immense, beyond the scope of the chapter. We shall therefore concentrate on a major application in sensing and biosensing, in which the interesting features of the films are exploited. We also include a discussion on the use of Langmuir monolayers as cell membrane

O.N. Oliveira Jr. (✉) • F.J. Pavinatto • D.T. Balogh
São Carlos Institute of Physics, University of São Paulo, CP 369, 13560-970 São Carlos, SP, Brazil
e-mail: chu@ifsc.usp.br

M. Bardosova, T. Wagner (eds.), *Nanomaterials and Nanoarchitectures*, NATO Science for Peace and Security Series C: Environmental Security,
DOI 10.1007/978-94-017-9921-8_10

models. The chapter is organised as follows. Section 10.2 brings a brief description of the organised nanostructured films, whereas Sect. 10.3 is dedicated to Langmuir monolayers, particularly to the experimental methods to characterise them. The applications in sensing and biosensing are discussed in Sect. 10.4.

10.2 Different Types of Organised Films

The majority of the methods to produce organised films were conceived for organic materials, but now inorganic compounds and hybrids are also used. Since most nanostructured films are not self-sustainable, they can be classified according to the type of adsorption mechanism. For example, the so-called self-assembled films may be produced via chemical absorption [1] or via physical interactions such as electrostatic attraction or H-bonding [2]. The labels for such types of film have changed over time: the term self-assembled monolayers (SAMs) is used for chemically adsorbed monolayers, normally on metallic substrates [3]. Multilayers assembled with electrostatic or other physical interactions are termed layer-by-layer (LbL) films [4]. This list is completed with the Langmuir-Blodgett (LB) films which are adsorbed on solid substrates upon transfer of Langmuir films from the air/water interface [5].

In this Chapter we shall provide a brief description of these types of films, starting with the experimental procedures to produce them and then discussing their many possible uses and applications.

10.2.1 Langmuir and Langmuir-Blodgett (LB) Films

These are the oldest type of organised films, whose procedures were developed in the first few decades of the twentieth century by Langmuir [6, 7]. He established a sound scientific basis for monolayers at the air/water interface, which were already known much before Langmuir's time. The so-called Langmuir monolayers are obtained by spreading a solution of amphiphilic molecules dissolved in an organic, volatile solvent on the surface of an aqueous solution. After solvent evaporation, the film-forming molecules are compressed by movable barriers until they are tightly packed and can be transferred onto solid substrates to form the Langmuir-Blodgett (LB) films [5, 8–12]. The adsorption mechanisms include hydrophobic and dispersions forces [10], and therefore transfer requires surface pressure gradients at the meniscus, which is performed in a special apparatus, referred to as Langmuir trough. With the surface pressure being held constant, a solid substrate is dipped into or raised out of the water, and multilayers can be obtained by repeating this dipping/lifting procedure as many times as desired. An important feature in this process is the possible control at the molecular level, not only by varying the molecules used to produce the Langmuir monolayer but also by depositing different materials on the same substrate, thus forming multicomponent LB films. In the latter case, mixed monolayers with controlled proportion of materials or multilayers of alternated composition can be transferred, and synergy may be reached between properties of distinct materials, particularly by controlling the film architecture [13].

The LB technology was conceived for amphiphilic, organic materials such as fatty acids, fatty esters and phospholipids [8], but now many other types of molecules are used. Indeed, Langmuir and LB films can be obtained from relatively large molecules [14], polymers [15, 16] and non-amphiphilic materials [17]. Organic-inorganic hybrids can also be used for fabricating LB films [18]. In many of these cases, some of the film-forming materials are dissolved into the solution and then made to adsorb onto the monolayer of a surfactant, in order to be co-deposited on a solid substrate [19]. It should be stressed that for large molecules and non-amphiphilic materials, the film at the air/water interface may not be a true monolayer, being thus named Langmuir film, instead of Langmuir monolayer.

Many are the methods to detect and characterise the presence of a Langmuir film at the interface, the most common of which is the measurement of the surface pressure, defined as the change in surface tension of the liquid owing to the film. We shall return to these methods when a possible use of Langmuir monolayers is discussed in Sect. 10.3. The LB films, in their turn, are characterized by the same conventional methods to study thin films, though in some instances adaptations are made to take account of the ultrathin nature of the films. Significantly, LB films are important ingredients in testing and developing surface-specific techniques that are essential for understanding functionalisation and coating of surfaces.

For decades there was the hope that LB films would eventually be employed in commercial applications, especially because of the ability to control film properties at the molecular level and possible organisation of the films in well-controlled film architectures. However, this expectation was not materialised owing to the various limitations of LB films. Such limitations stem from the poor thermal and mechanical stability of the films, in addition to the requirement of time-consuming, relatively sophisticated experimental procedures to fabricate the LB multilayers. Nevertheless, the Langmuir monolayers and the LB films have been proven extremely useful for proof-of-principle experiments, where the applicability of a given material can be probed with molecular-level accuracy. Particularly important in this regard has been the research into materials of biological interest since the LB technology is suitable for immobilisation of biomolecules with their activity preserved [20]. This is why LB films are still used in many types of biosensors published in the literature [21]. But perhaps the most relevant application of the LB technology today is in using Langmuir monolayers to mimic a cell membrane [22]. Section 10.3.3 will be dedicated to this subject.

10.2.2 Self-Assembled Monolayers (SAMs)

The self-assembly of organic molecules was pioneered by Sagiv and coworkers in the 1980s [23], with multilayers being deposited via chemical adsorption of molecules with functional groups [12, 24–27]. An important example of this type of film is that of functionalised alkanethiols adsorbed on gold [28]. The obvious advantage is in the stability afforded by the covalent bonds between the film-forming materials and the substrate, and among the molecules of adjacent layers. However, establishing covalent bonds between layers is restrictive in terms of the materials and experimental conditions that may be employed in film fabrication. This is perhaps the reason why in recent times chemisorption has been mainly used in building the so-called self-assembled monolayers (SAMs) [29], for there are less restrictions for a one-layer film. SAMs are advantageous in comparison to LB and LbL films in that the position and density of the molecules adsorbed can be controlled both vertically and laterally [30]. This is particularly important for biosensing because ligands may be attached to allow for immobilisation of various types of biomolecules, as is the case of DNA detection [31].

10.2.3 Layer-by-Layer (LbL) Films

An alternative way to produce multilayers with molecular control and in a layer-by-layer fashion is to make use of electrostatic interactions among film-forming molecules in adjacent layers. This concept was introduced by Iller [32] in 1966 for coating nanoparticles, but apparently did not receive much attention. A very similar procedure was proposed by Decher et al. for fabricating multilayers from oppositely charged polyelectrolytes in the early 1990s [4, 33, 34], which then was immediately recognised as a very versatile method. In the first few years the films so obtained were referred to as self-assembled films, but the most used term in recent times has been layer-by-layer (LbL) films. Though originally conceived for polyelectrolytes, the method turned out to be suitable for any charged species with a minimum size, and is now used for a wide variety of materials, including metal and semiconductor particles [35, 36], dendrimers [37], azopolymers [38], polymeric nanocrystals [39], light-emitting polymers [40, 41], organic dyes [42, 43] proteins and other biomolecules [44–47]. A more complete list of materials can be found in the review in Ref. [48].

The LbL film technique has been increasingly used owing to three of its features, namely experimental simplicity, versatility in the type of materials and supports to be covered, and suitability for biomolecules since film fabrication is performed under mild conditions. As for the experimental simplicity, LbL films may be obtained using only three beakers containing the aqueous solutions of the oppositely charged materials and a washing solution. The sequential adsorption of oppositely charged materials is carried out by immersing alternately the solid substrate into the two beakers containing the film-forming molecules. Adsorption is, in this case, driven by electrostatic attraction, and the whole adsorption process is very simple. In a more recent procedure, the layers may be sprayed with simple apparatus [50], with the added advantage of a reduced time for adsorption of each layer. The versatility in terms of materials that can be used has already been mentioned, and the list of possible materials is long and growing. The LbL films may be deposited on solid supports of any shape and chemical constitution; though most of the literature is dedicated to films adsorbed on plane, solid substrates, such as glass, metals and semiconductors, these films may be adsorbed on nanoparticles and nanotubes. They can even be used to produce capsules [51] where the LbL films would initially cover a core that is removed at the end of the fabrication process. Though the majority of LbL films are explored adsorbed on a solid support, free-standing films have also been produced [52]. The procedures to obtain the latter are invariably based on dissolution or melting, after film growth, of the solid support or of a first sacrificial layer in order to render the free-standing film [53, 54], with the hollow capsules mentioned above being a good example.

The third feature related to the suitability for biomolecules is shared with the other methods to obtain organised films from organic materials because they are all based on the presence of water. In fact, it seems that the main reason why activity is preserved for immobilised biomolecules in solid films is the entrained water that stays in the film, even after drying. Furthermore, the aqueous solutions for deposition of LbL films may have the temperature and pH controlled in order to suit the optimised conditions for the biomolecule. As a consequence, the number of possible applications of LbL films in the biological and biomedical fields is increasing fast [48, 55]. Some of the examples are the encapsulation of cells with LbL films to eliminate immunosuppressive therapy during transplants [56], the fabrication of self-healing films for tissue engineering [57], use of LbL films for vascular therapy [58] and as drug delivery vehicles [59, 60]. Also important in this context is the possible control of molecular architecture since various materials may be deposited on the same substrate, thus seeking synergy in the properties among the different components.

In this Chapter we shall single out the application in sensing and biosensing, an area in which the authors have contributed for more than a decade.

10.3 Why Langmuir Monolayers Are Still Important

We have already mentioned that commercial applications of LB films are not pursued with the same emphasis as in the past, for it became clear that other methods – including the LbL technology – are more suitable for such applications. This does not mean that the interest in Langmuir monolayers has subsided. On the contrary, with recent surface-specific experimental methods to study interfaces and the prominence of the biological and biomedical fields, Langmuir monolayers have received a lot of attention. As always, they are studied to identify materials suitable for LB films, but perhaps their most relevant use has been in mimicking interfaces, especially biointerfaces. In this Chapter, we shall highlight the use of Langmuir monolayers as simplified cell membrane models. Before we discuss this use, we shall introduce the various methods employed to characterise Langmuir monolayers.

10.3.1 Characterisation of Langmuir Monolayers

The most used method to detect a Langmuir monolayer at the air/water interface consists in measuring the surface pressure, which is the change in surface tension due to the monolayer presence. Formally, the surface pressure, π, is defined as the decrease in surface tension, $\pi = \gamma_o - \gamma$, where γ_o and γ are the surface tensions in the absence and presence of the monolayer, respectively. Since π depends on the packing density of the monolayer-forming molecules, it varies upon compression leading to the so-called surface pressure-area (π-A) isotherms. A typical π-A isotherm is illustrated in Fig. 10.1 for the phospholipid dipalmitoylphosphatidylcholine (DPPC), an important component of plasma membranes [61]. It features distinct phases, similarly to a 3D system, referred to as gaseous, liquid-expanded (LE), liquid-condensed (LC) and solid phases, and includes a phase transition. The latter corresponds to the coexistence of LE and LC phases, where the hydrophobic chains gradually change from a disordered to an ordered state in the LC phase. In the gaseous phase, the molecules do not interact strongly with each other, leading to a negligible surface pressure. If the monolayer is compressed beyond its condensed, solid phase, collapse occurs with irreversible non-monolayer clusters being formed. From the surface pressure isotherms, one can also estimate the surface compressional modulus, Cs^{-1}, defined as $Cs^{-1} = -A.d\pi/dA$. This elastic property serves not only to define the physical state of the monolayer, but it is also relevant for phenomena in cell membranes, as it will be exemplified later on.

The mechanical properties of Langmuir monolayers may also be investigated with rheometry experiments, besides the surface pressure measurements just mentioned. Interfacial rheology can be measured by employing rotational or dilatational methods, which can be performed, for instance, using a floating magnetic needle or

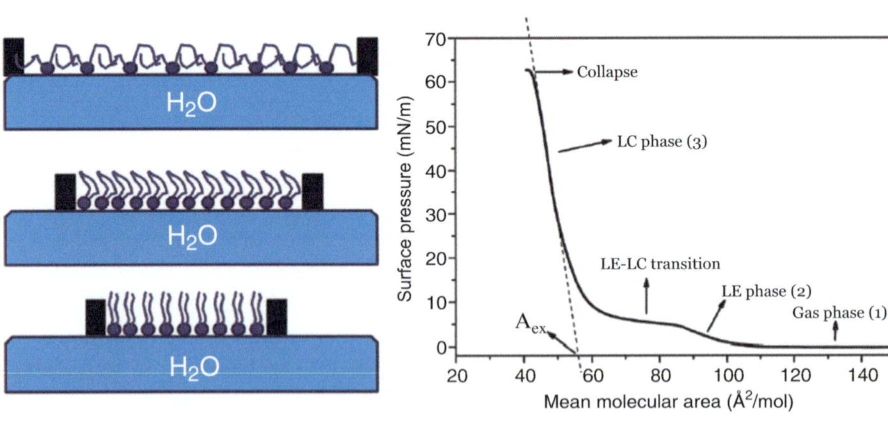

Fig. 10.1 The surface pressure-area isotherm on the right is typical of DPPC, which undergoes a LE-LC phase transition. The phases are indicated in the figure, as well as the extrapolated area to zero pressure, which corresponds roughly to the cross section area occupied by each molecule (Reproduced with permission from Ref. [49])

by simply oscillating the barriers in a Langmuir trough, respectively [62]. Another powerful technique to investigate these properties is the pendant drop method [63], with which static and dynamic characteristics can be measured for a Langmuir or Gibbs monolayer on the surface of a drop. The image of the drop is digitalized to obtain the geometrical parameters to be fed into the Young-Laplace equation:

$$\Delta P = (\rho_d - \rho_l)\, gh = (\gamma/R_1) + (\gamma/R_2) \quad (10.1)$$

where ΔP is the difference in pressure at the interface, ρ_d and ρ_l are the densities for the most and least dense phases, respectively, g is the acceleration of gravity, h is the height of the liquid column at the drop. R_1 and R_2 are the main curvature radii for the drop, with R_1 being the value for a circumference and R_2 for an ellipsoid. The dynamic properties are inferred from the response of the drop subjected to a sinusoidal mechanical stimulus at a specified frequency. It is worth mentioning that the parameters and main rheological properties for interfacial monolayers have to be treated with theoretical formalism slightly different from used for 3D systems, since geometrical features are different and dissolution of film material can play a significant role in the 2D setup. This has been competently done by a few groups [62, 64].

Since Langmuir monolayers are formed from amphiphilic molecules, some of which have strong polar groups and that could be electrically charged, techniques to investigate the interfacial electrical properties have been widely used. The most popular is the measurement of the surface potential [65], whereas the measurement of surface conductance has been much less used. The surface potential is defined as the difference in potential between a monolayer-covered surface and a clean subphase, i.e. $\Delta V = V_m - V_o$. Note that V_o, the surface potential of a clean water surface, is non-zero owing to the organisation of the molecules at the air interface, being taken as the reference. Surface potential is normally measured with the Kelvin probe technique, which is currently used in Kelvin Probe Microscopy [66]. A typical surface potential isotherm for DPPC is shown in Fig. 10.2, whose main features are the zero potential at large areas per molecule and the abrupt increase at a critical area. A vanishing surface potential at large areas indicates the lack of significant aggregates that could lead to a measurable potential. Indeed, for Langmuir films of non-amphiphilic compounds, where aggregation may be inevitable, the surface potential is non zero even at large areas per molecule [67]. The critical area corresponds to the coming together of monolayer domains, which causes a measurable potential. It is usually larger than the onset area at which the surface pressure rises, and therefore with the surface potential measurement one may probe the monolayer structuring at an earlier stage than with the surface pressure measurement.

For simple amphiphilic molecules, the monolayer surface potential can be interpreted with phenomenological models based on the Helmholtz equation [68], in which the monolayer is taken as a parallel plate capacitor. The surface potential is

$$\Delta V = \mu_n / A\varepsilon_r\varepsilon_0 \quad (10.2)$$

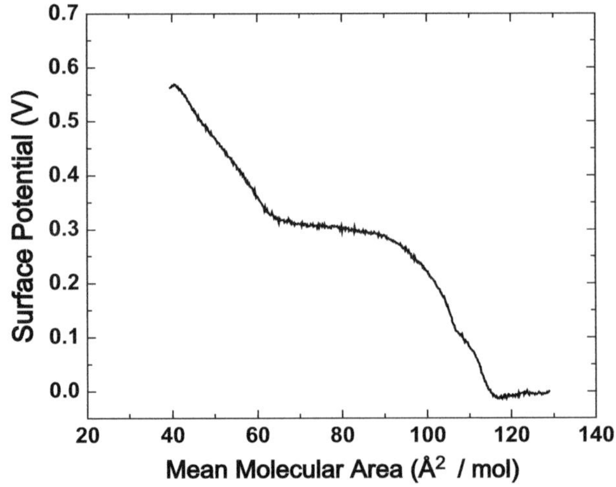

Fig. 10.2 Surface potential isotherm for DPPC monolayer on ultrapure water

where μ_n is the normal component of the dipole moment per molecule, ε_r and ε_0 are the relative dielectric constant and the permittivity of free space, respectively, and A is the area per molecule. Improvements to the Helmholtz equation have been made to account for dipole moment contributions from molecular groups at distinct parts of the monolayer-forming molecules and from reorientation of water molecules. A successful model was proposed by Demchak and Fort [69], who introduced effective relative permittivities with the surface potential, for a non-ionised monolayer, being given as

$$\Delta V = (\mu_1/\varepsilon_1 + \mu_2/\varepsilon_2 + \mu_3/\varepsilon_3)\,/A\varepsilon_0 \tag{10.3}$$

where μ_1/ε_1 is due to reorientation of water molecules, μ_2 and μ_3 are the normal components of the dipole moments for the groups at the hydrophilic headgroups and hydrophobic tails, respectively, while ε_2 and ε_3 are the corresponding dielectric constants at these regions. Figure 10.3 shows a schematic diagram for the Demchak-Fort model [69].

Oliveira et al. [71] used the Demchak-Fort model to obtain the parameters for condensed Langmuir monolayers of long-chain aliphatic compounds, which are $\mu_1/\varepsilon_1 = -65$ mD, $\varepsilon_2 = 6.4$, and $\varepsilon_3 = 2.8$. Note that the estimated dielectric constant at the headgroup region (6.4) is considerably lower than that of bulk water (78.5). This lowering is consistent with the ordering of water at interfaces, and has important implications due to the relevance of interfacial water. In theoretical simulations of the properties of water the dielectric constant normally employed is that of the bulk value, and perhaps the time has come to introduce changes in order to better capture the behaviour at the interface. From the study of Oliveira et al. [71], one should expect the dielectric constant of interfacial water to vary from 5 to 10 (6.4 was just the average to account for the surface potential data).

When the monolayer is at least partially ionised, an electrical double-layer is formed which also contributes to the surface potential, so that

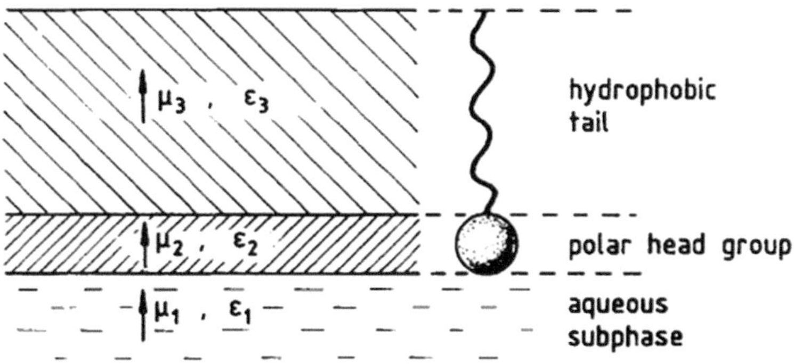

Fig. 10.3 Schematic diagram for the Demchak-Fort model (Reproduced with permission from Ref. [70])

$$\Delta V = \frac{1}{A\varepsilon_0}\left(\frac{\mu_1}{\varepsilon_1} + \frac{\mu_2}{\varepsilon_2} + \frac{\mu_3}{\varepsilon_3}\right) + \psi_0 \tag{10.4}$$

where ψ_0 is the double-layer potential. For Langmuir monolayers of fatty acids, the simple Gouy-Chapman model [72] has been proven suitable to explain the Ψ_0, given by

$$\Psi_0 = (2\mathrm{kT/e}) \sin \mathrm{h}^{-1} \left[\sigma / (2\mathrm{n}\varepsilon\mathrm{kT}/\pi)\right] \tag{10.5}$$

where k is the Boltzmann constant, T is temperature, e is the charge of a proton, n is the number of ions per cm^3 in the bulk phase, σ is the charge density (in $C\ cm^{-2}$) and ε is the dielectric constant of bulk water. In the Gouy-Chapman model, Ψ_0 is derived by solving the Poisson-Boltzmann equation assuming that the surface is uniformly charged and that the counterions are point charges that do not adsorb on the surface. In spite of its simplicity, the model was successful in explaining the surface potential data for a stearic acid monolayer with various ionic concentrations [73], provided that the degree of ionisation of the molecules was considered dependent on such concentration. This dependence takes place because the pH at the interface differs from that of the bulk when the surface is charged, since

$$\mathrm{pH^i} = \mathrm{pH^b} + \mathrm{e}\Psi_0/(2.3\mathrm{T}) \tag{10.6}$$

where pH^i is the interfacial pH and pH^b is the bulk value. The degree of ionization α is then calculated from the Henderson-Hasselbalch equation:

$$\mathrm{pK} = \mathrm{pH^i} - \log\left[\alpha/(1-\alpha)\right] \tag{10.7}$$

Though this difference between interfacial and bulk pHs has been mentioned for more than 100 years, it is sometimes overlooked. One should always bear in mind that the degree of ionisation of molecules at the air/water interface will differ from that expected for solvated molecules.

In addition to surface pressure and surface potential measurements, which are two of the most simple, widespread methods for characterising Langmuir monolayers, several others are routinely applied. For probing monolayer morphology, Brewster angle microscopy (BAM), fluorescence microscopy, and optical spectroscopy [74–76] have been used. In BAM (Brewster Angle Microscopy), use is made of the optical property according to which *p*-polarized light (parallel to the incidence plane) is not reflected from an interface when impinged at the Brewster angle [77, 78]. The latter is peculiar for each interface and is given by the ratio between the refraction indexes of the two media (n_1 and n_2), according to equation (10.8) (Brewster angle $= \theta$). Imaging is done usually by using a *p*-polarized laser beam and a CCD camera, with the background (zero) signal collected from a bare interface. When the film is present, even being as thin as a monolayer, interface and refractive indexes are changed, and the system is no longer set to the proper new Brewster angle. As a consequence, light is reflected and images are formed by contrast in the CCD camera, which can arise either from regions with and without film, or regions of the film with different fluidity and thickness.

$$\tan\theta = n_1/n_2 \tag{10.8}$$

Determining the molecular arrangement in Langmuir monolayers requires sophisticated methods owing to the very small number of molecules. In recent times, this is done with either X-ray diffraction or scattering with synchrotron light or with neutron scattering [65].

In Grazing Incidence X-ray Diffraction (GIXRD), a monochromatic X-ray beam is made to impinge on the water surface at a grazing angle close to that corresponding to total reflection, and the diffracted radiation is detected by a position-sensitive detector for variable vertical scattering angles [9]. With this type of measurement it is possible to verify whether the alkyl chains in monolayers of some amphiphilic molecules are ordered, and even determine the crystal structure and lattice parameters [79]. One may also discover the tilting angle adopted by the chains [80] and the occupied area per molecule can be compared to the values measured with surface pressure isotherms. The X-ray and neutron scattering techniques are based on measuring the intensity of a reflected beam made to impinge at the air/water interface at a glancing angle [81], with which changes in the thickness of Langmuir monolayers as small as 1–2 Å can be detected. For example, structural changes induced by photoisomerisation of a Langmuir film made with azobenzene side chains could be probed with X-ray reflectivity measurements [82]. With deuterated amphiphiles neutron reflectivity studies may obviate the problems from poor contrast between the hydrocarbon chains and the aqueous subphase in X-ray reflectivity measurements. This allows for determining the structure of Langmuir films, including the adsorption of ions [83].

10.3.2 Vibrational Spectroscopy Techniques

As far as the use of Langmuir monolayers like cell membrane models is concerned, two vibrational spectroscopy techniques have gained increased attention because they allow for determining the functional groups of monolayer-forming molecules that interact with guest molecules at the interface. These techniques are the polarization-modulated reflection absorption spectroscopy (PM-IRRAS) [84] and sum-frequency generation (SFG) spectroscopy [85].

PM-IRRAS is a variation of the IRRAS (infrared reflection-absorption spectroscopy) method, based on the interference of incident and reflected components of the electric field, attained when the light beam is impinged at incidence angles of ca. 80° [86]. Under such condition, only *p*-polarized light is reflected from the surface of a metal, but both polarizations (*p*- and *s*-) are reflected from the water surface. By alternating the incidence beam polarization between *s* and *p* at a frequency of tens of kHz [87], one obtains the differential reflectivity (S), given by

$$S = \frac{R_p - R_s}{R_p + R_s} \tag{10.9}$$

where R_p and R_s are the reflectivities for *p* and *s* polarizations, respectively.

PM-IRRAS for Langmuir monolayers is surface specific because the film reflectivity may be subtracted from the spectrum of the aqueous subphase reference, and therefore contributions can be filtered out from that of isotropically oriented molecules, as is the case of CO_2 and H_2O. The water molecules can be those from vapour or from the subphase. Furthermore, the difference in reflectivity ($R_p - R_s$) corresponds to the signal from the interface, while the sum ($R_p + R_s$) is the reference spectrum. Due to this minus sign, the bands can either point upward (positive bands) or downward (negative bands). For incidence angles normally used (ca. 80°), positive bands correspond to transition moments preferentially on the surface plane, while negative bands indicate preferential orientation perpendicular to the surface. Some of the uses of PM-IRRAS for cell membrane models will be exemplified in the next section.

SFG is a non-linear optical method with intrinsic selectivity for interfacial contributions [88, 89], being also therefore surface specific. Experimentally, in SFG two high-intensity laser beams at frequencies ω_{vis} and ω_{IR} are made to overlap at the interface, thus generating an output beam whose frequency is $\omega_{\mathrm{SFG}} = \omega_{\mathrm{vis}} + \omega_{\mathrm{IR}}$. The visible beam has a fixed frequency, but the frequency of the infrared beam is varied in order to obtain a vibrational spectrum. Since this is a second-order phenomenon, the SFG signal intensity is proportional to the square of the effective second-order nonlinear susceptibility of the interface, $\chi^2_{eff}\,(\omega_{SFG} = \omega_{wis} + \omega_{IR})$. Hence, the SFG signal vanishes in media with inversion symmetry, and therefore only molecules at the interface, with a non-random organisation, contribute to the signal. Formally, the SFG intensity can be written as

$$I_{SFG} \propto \left|\chi_{eff}^{(2)}\right|^2 = \left|\chi_{NR}^{(2)} + \chi_{R}^{(2)}\right|^2 \left|\chi_{NR}^{(2)} + N_s \sum\nolimits_q \frac{A_Q}{\omega IR - \omega_Q + \Gamma_q}\right|^2 \tag{10.10}$$

Where $\chi_{NR}^{(2)}$, $\chi_{R}^{(2)}$, N_s, A_q, ω_q, Γ_q are the nonresonant and resonant contributions to $\chi_{eff}^{(2)}$, the number of molecules per unit area, and the oscillator strength, resonant frequency and linewidth of the q-th vibrational mode, respectively [90]. Information on the organisation of functional groups of the molecules is obtained from the SFG spectra, similarly to any other vibrational spectroscopy method. Of particular relevance to biological studies, the structuring of water at interfaces can be probed [91], as will be commented upon later on.

10.3.3 Cell Membrane Models

Cell membranes are essential for virtually all biological processes because they serve as a barrier for the cells, thus controlling the trafficking of nutrients, ions and drugs or toxic molecules in and out of the cells. According to the widely used Singer-Nicholson model [92] a cell membrane is made of a bilayer of phospholipids and other lipids such as cholesterol and cardiolipin, onto which proteins and other biomolecules are attached, in some cases even forming channels through the membrane. The lipid constitution of the membrane depends on the type of cell, and can vary in a range from 30 % to 80 %, with phospholipids always being responsible for a major portion of these lipids [61]. Because experiments to obtain molecular-level interaction in real cell membranes are very difficult to execute, cell membrane models have been extensively employed. These models include vesicles, lipid bilayers adsorbed on solid substrates and Langmuir monolayers [93]. The latter are advantageous owing to the possible control of composition and packing of the molecules, which is important to probe the elastic properties. On the other hand, since a Langmuir monolayer mimics only half of the membrane, it is not suitable to investigate processes such as diffusion of ions and molecules through the membrane.

Langmuir monolayers have been used to simulate phenomena in cell membranes for several decades, and this is likely to be intensified in the near future due to the development of surface-specific techniques to interrogate effects at the air/water interface (see e.g. [90]) in addition to the use of molecular dynamics simulations [94]. More specifically, it has become possible to identify functional groups through which biologically-relevant molecules, such as proteins, polysaccharides and pharmaceutical drugs, interact with the cell membrane. In some cases it is even possible to determine whether biomolecules maintain their native structure [95].

The main features that can be investigated with Langmuir monolayers are related to the interaction between biologically-relevant molecules (referred to here as guest molecules) with the model membrane. What one normally does is to attempt to answer the following questions:

1. Are the guest molecules coupled to the membrane? Do they penetrate into the hydrophobic region?
2. How is the membrane affected by the guest molecules? Are the monolayers expanded or condensed? Are the monolayer elasticity properties changed?
3. Do the effects depend on the concentration of the guest molecules?
4. Do the effects depend on the charged state and composition of the model membrane?
5. How are the guest molecules affected by the interaction with the model membrane?
6. Do biological guest molecules preserve their structure and bioactivity while interacting with the model membrane?
7. For guest molecules initially dissolved in the subphase, how is the kinetics of adsorption? Is it diffusion-controlled?

The guest molecules may have surface activity, and therefore they co-adsorb on the Langmuir monolayer. In this case, such guest molecules may be co-spread with the monolayer-forming molecules since they are probably soluble in organic solvents. The effects on the Langmuir monolayer are thus analysed as in mixed films, where one important factor is miscibility of the film components. The area per molecule is calculated taking into account the molecular weight of all components, and may be plotted against the composition in the mixture. A linear additivity line can then be established to connect the areas of the individual components, and indicates whether there is ideal mixing of non-interacting molecules or their complete immiscibility. If the plot is not linear, one infers that molecular-level interactions take place [96], with positive deviations meaning repulsive interactions between molecules of different compounds. Negative deviations obviously point to attractive interactions [97]. In these co-spread monolayers, the guest molecules normally protrude into the air phase, then occupying an effective area at the interface. They can also affect the packing of the lipid layer representing the membrane, particularly the elasticity properties.

The majority of the biologically-relevant guest molecules are at least partially soluble in aqueous solutions, with little or no surface activity. Then, the first question to be answered with monolayer experiments is whether they couple to the membrane, and how the latter is affected by their presence. A first indicative of incorporation can be obtained by simply monitoring surface pressure for a solution of the guest molecules, thus obtaining the so-called adsorption kinetic curves. This can be done for a bare interface and when a lipid monolayer is already spread on the subphase surface. In the latter case, such film can have different initial pressures, and guest molecules can be either pre-dissolved in the subphase or injected after monolayer formation. For these systems, the surface pressure isotherms are plotted in terms of area per lipid molecule, so that in most cases the monolayers are expanded owing to the presence of the guest molecules. The increase in area per lipid molecule depends strongly on whether the guest molecules penetrate into the hydrophobic region. Other reasons for monolayer expansion come from repulsive interactions induced by the guest molecules, or interactions with the monolayer-forming molecules that disrupt their packing. There are cases, however, in which the monolayer is actually condensed with a shift toward lower areas in the surface pressure isotherms. One such example is the introduction of charged guest molecules that shield the repulsive interactions between charged headgroups of the monolayer-forming molecules.

An interesting example of concentration-dependent effects of guest molecules was presented by Caetano et al. [98] for the drug dipyridamole (DIP). When interacting with dipalmitoylphosphatidyl choline (DPPC) monolayers, DIP induced a decrease in the surface pressure at which the LE-LC phase transition occurred when in low concentrations, but this was reverted for higher DIP concentrations, as shown in Fig. 10.4.

It is seen that the maximum effect appeared for a very small DIP concentration, which can only be explained by cooperativity effects whereby DPPC molecules far away from DIP molecules are also affected [98]. This inference was reinforced by the surface potential data for the DIP-containing DPPC monolayers. As indicated in Fig. 10.4b, the surface potential increased at low DIP concentrations, but then decreased at the higher concentrations. The maximum effect took place exactly at the same DIP concentration where the maximum change in pressure occurred in Fig. 10.4, which is relevant because the surface pressure and the surface potential are very distinct properties of the monolayer. Cooperativity was later observed in the interaction of drug and other types of molecules with phospholipid monolayers [99, 100].

With molecular dynamics simulations, it was possible to attribute the cooperativity effects to changes in packing of phospholipid molecules away from the drug molecule [101].

The dependence on the charge of the model membrane is studied in detail by using zwitterionic phospholipids such as DPPC and phosphatidylethalonamines (PEs), in comparison with the negatively charged phosphatidic acids (PAs), phosphatidylglycerols (PGs) and phosphatidylserines (PSs). The charge state is taken for these molecules at the air/water interface with ultrapure water whose pH is ca. 5.5–6.0. As one can predict, positively charged guest molecules usually interact strongly with the negatively charged phospholipids (see details of experiments with chitosans below), but the monolayer behaviour also depends on other types of interaction, including H-bonding and hydrophobic interactions. Furthermore, charged guest molecules may have particularly strong interaction toward one of the charges in the zwitterion for the neutral phospholipids.

Strong charge effects were observed in a study of interactions between capped gold nanoparticles and phospholipid monolayers [102]. Figure 10.5 shows the changes induced by negatively charged gold nanoparticles, which were capped with citrate ions, on DPPG and DPPC monolayers. The effects were larger on DPPC than on DPPG, because the latter was negatively charged. Moreover, the capping appeared to be even more important than the nanoparticles themselves, for a larger effect was observed on DPPG with citrate ions, with no nanoparticles.

This is relevant for investigation of nanotoxicity, since it implies that the effects of nanoparticles on membranes vary considerably depending on the capping.

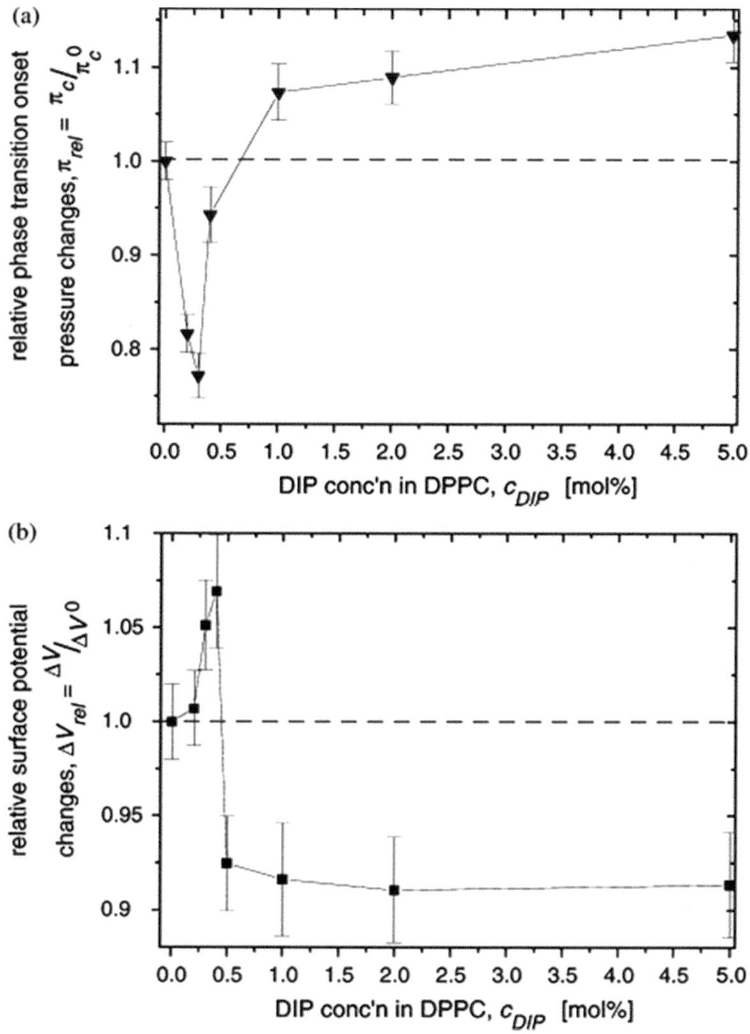

Fig. 10.4 (**a**) Changes in the relative surface pressure at the critical area, A_c; (**b**) Relative surface potential changes at the same areas, A_c. Values for the pure DPPC monolayers were used to normalize the surface pressure and potential values taken at the LE/LC phase transition onset, A_c, in six independent runs (Reproduced with permission from Ref. [98])

Indeed, positively charged gold nanoparticles had larger effect on DPPG than on DPPC, as indicated in Fig. 10.6. Though the binding of the positive nanoparticles to DPPG should be expected, it again indicates that the effects from nanoparticles are largely dependent on their capping.

Studies based on surface pressure and surface potential isotherms mentioned so far may provide crucial information on the interaction between guest molecules and monolayer-forming materials. However, even simple inferences such as those

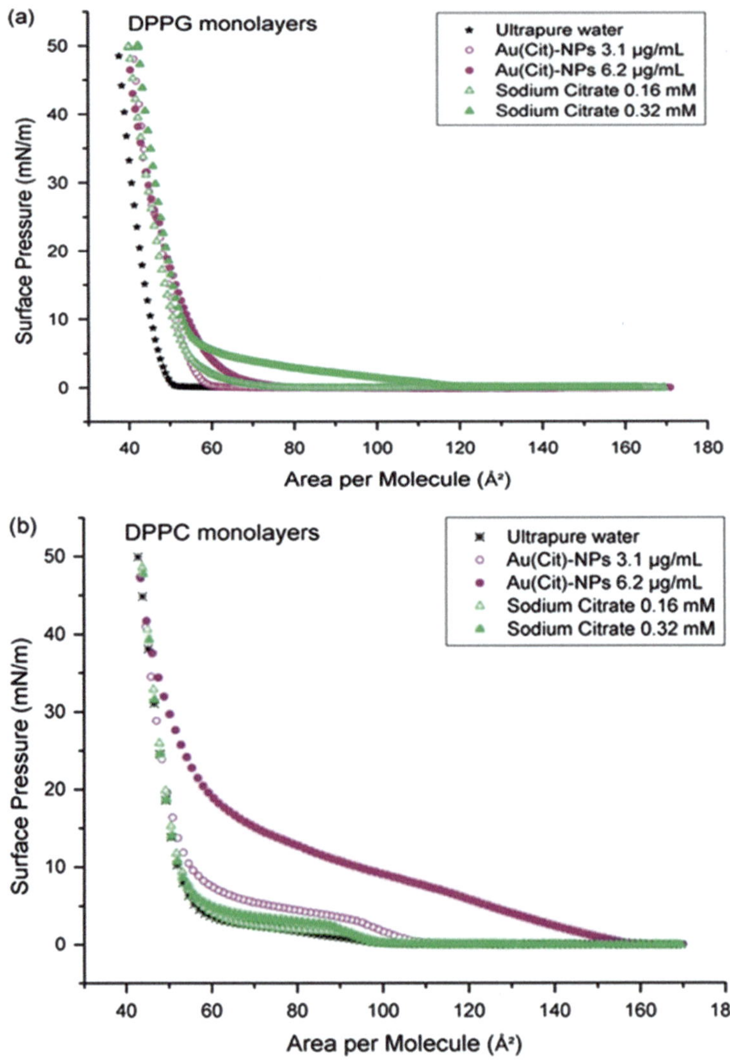

Fig. 10.5 Surface pressure – area isotherms of phospholipid monolayers on subphases containing Au(Cit)-NPs and sodium citrate aqueous solutions. (**a**) DPPG and (**b**) DPPC (Reproduced with permission from [102])

related to possible penetration into the hydrophobic region are made indirectly because there are always more than one possible reason for the changes in the surface pressure isotherms. This limitation has served as further motivation for the use of surface-specific spectroscopic techniques, such as PM-IRRAS and SFG introduced in Sect. 10.3.2. With these two techniques one may determine the binding site for the guest molecules and whether the hydrophobic tails were affected, among other things.

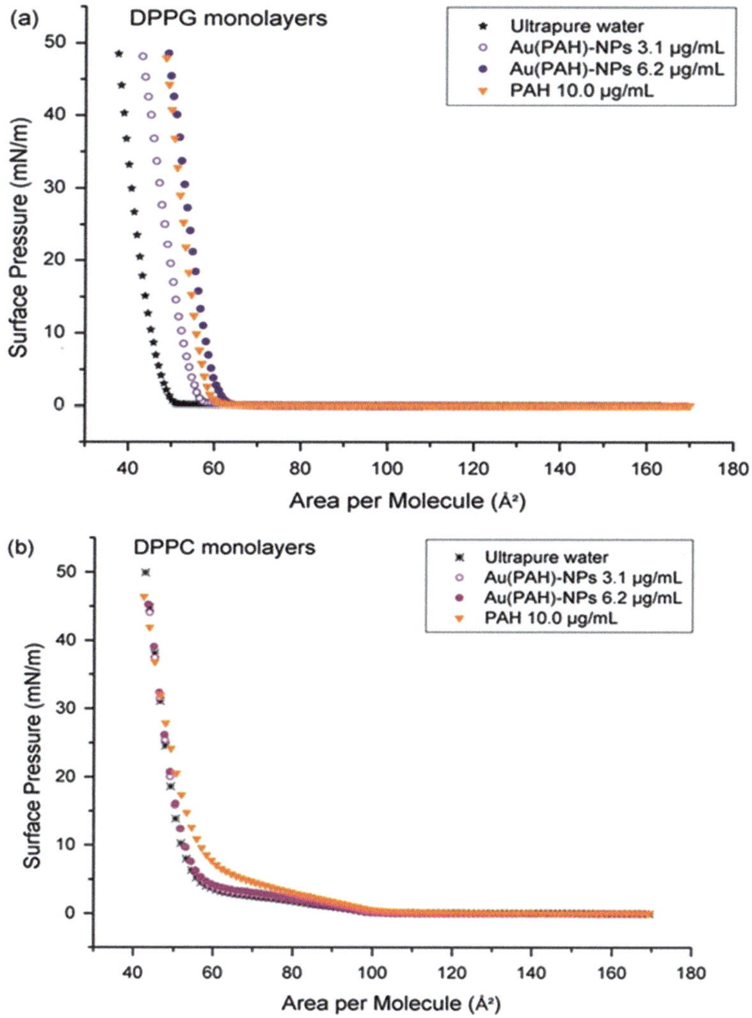

Fig. 10.6 Surface pressure – area isotherms of phospholipid monolayers on subphases containing Au(PAH)-NPs and PAH aqueous solutions. (**a**) DPPG and (**b**) DPPC (Reproduced with permission from Ref. [102])

Perhaps one of the most illustrative examples of the use of PM-IRRAS has been the confirmation that chitosan may remove β-lactoglobulin [103] from negatively charged phospholipid monolayers. The protein as well as chitosan adsorb onto either zwitterionic or negatively charged monolayers. When they are both introduced into the subphase, the protein initially adsorbs, thus causing an increase in the monolayer surface pressure but it is then progressively removed by chitosan, provided that the phospholipid is negatively charged. These statements could be unequivocally proven by inspecting the functional groups at the interface via

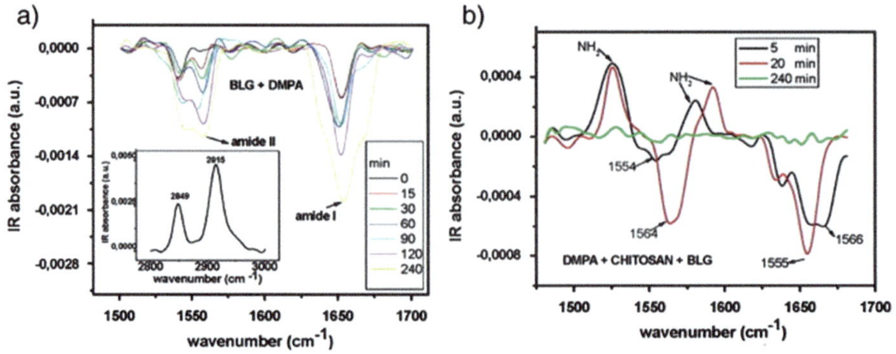

Fig. 10.7 PM-IRRAS spectra, taken at initial surface pressure of 20 mN m^{-1}, of β-lactoglobulin adsorbing onto DMPA monolayers without (**a**) and with 0.2 mg mL^{-1} chitosan (**b**) in the subphase (Reproduced with permission from Ref. [103])

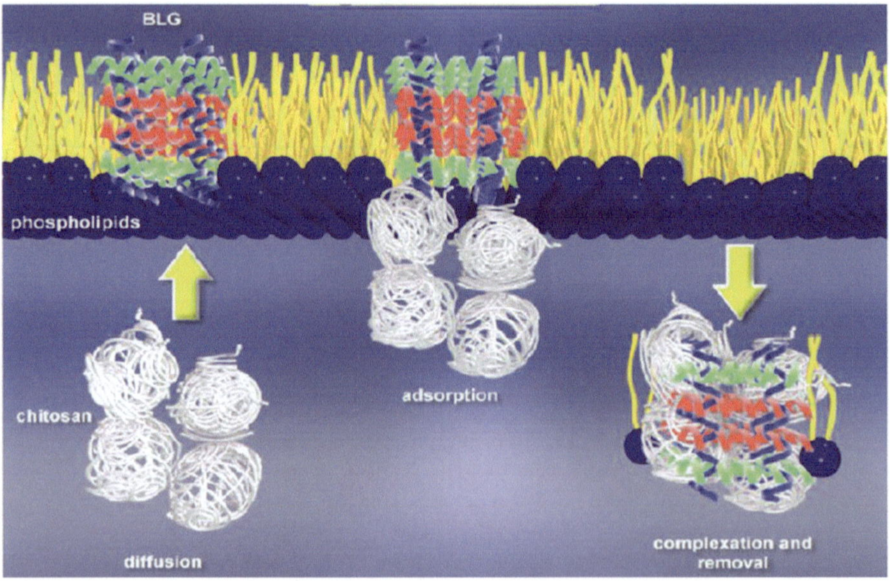

Fig. 10.8 Scheme showing the diffusion, adsorption and complexation stages for removing the BLG from phospholipid monolayers (Reproduced with permission from [103])

PM-IRRAS. Figure 10.7 shows that the amide bands assigned to β-lactoglobulin are readily apparent in the spectrum taken in the beginning of the experiment. However, for longer periods of waiting time these bands vanish, until no trace of either β-lactoglobulin or chitosan is noted in the PM-IRRAS spectrum.

The removal process of β-lactoglobulin by chitosan is depicted in the cartoon of Fig. 10.8, where they are believed to form a complex. Such a formation of complex had actually been suggested in experiments for removing the protein from milk samples [104].

Another important application of PM-IRRAS is to verify if proteins preserve their native structure upon interacting with cell membrane models, which is usually done by monitoring the amide bands. The latter are shifted when the protein secondary structure is affected. The protein septin 2 was found to have its structure preserved when interacting with the lipid PtdIns(4,5)P2 at the air/water interface, but this did not happen when the monolayer was made of DPPC [105]. This specificity is important because upon losing its structure septin 2 forms aggregates, as in the amyloid-like fibers associated with Alzheimer's disease. The denaturing of septin 2 was attributed to its hydrophobic portion being inserted into the hydrophobic tails of DPPC, while for PtdIns(4,5)P2 septin 2 was kept hydrated at the water surface owing to the larger headgroup of this lipid.

SFG is in many aspects complementary to PM-IRRAS and other vibrational spectroscopic methods. Though SFG is experimentally more involved (and costly) than the other methods, it is advantageous in many ways. The most relevant advantage is the surface selectivity and high sensitivity owing to its very stringent selection rules, since the modes need to be both IR- and Raman-active to be observed in SFG. In many cases, SFG also provides a quantitative analysis of molecular orientation [90]. With regard to biomimetic systems, SFG allows one to investigate lipid chain conformation by analysing the intensity ratio of CH_2 and CH_3 stretches, in addition to being unique for the study of the structuring of water at an interface. It also serves to probe amide bands, analogously to PM-IRRAS, and then determine changes in protein structure. These bands are normally shifted to higher wavenumbers when changes from β-sheets to α-helices occur. For instance, changes in secondary structure of amyloid fibrils were observed by VandenAkker et al. [106] who monitored the amide I band in SFG measurements.

10.4 Exploiting Distinct Architectures in Solid Organised Films

In the Introduction we have emphasized that the control of molecular architecture was one of the main features exploited in organised films, with which a variety of applications could be envisaged. In this Chapter we shall restrict ourselves to sensing and biosensing, which has already led to a considerable bulk of contributions in the literature. Most importantly, applications in sensing provide beautiful examples of how molecular control can be explored. We consider organised films advantageous for three main reasons: (i) the films are ultrathin, with controllable thickness, which normally leads to higher sensitivity; (ii) in some cases the film-forming molecules may be arranged in a way to facilitate interaction with the analyte to be detected; (iii) more than one material can be adsorbed on the same sensing unit, seeking synergy in the combination.

Numerous papers have been published in sensors and biosensors built with LB films, LbL films and SAMs, for one or more of the reasons above. The diversity of materials used in the sensing units is immense because the film fabricating methods are complementary. These materials stem from organic compounds insoluble in water in the LB films to water soluble materials used in LbL films, and may include hybrid organic-inorganic structures. The principles of detection for such sensors and biosensors are equally diverse, mostly based on electrochemical techniques, electrical impedance and optical methods, where in the latter emphasis is placed on spectroscopic techniques. For the sake of space, we shall exemplify the use of organised films in two types of applications that have been extensively studied in our research group, namely electronic tongues and biosensors. Reviews on these subjects can be found in Refs. [21, 107].

10.4.1 Electronic Tongues

The concept of electronic tongues (e-tongues) has been created to mimic what a human tongue does when processing the signals from taste felt by the papillae, particularly with regard to the prevalence of a global selectivity [108] rather than a specific interaction with the substances giving rise to taste [109]. It so happens that the sensors in the human tongue are not able to discriminate specific substances; instead they send signals to the brain which are a composition of basic tastes, namely salty, sweet, sour, bitter and umami. Therefore, in order to mimic this system an e-tongue is made with a sensor array where the sensing units may be obtained in various architectures, while the principles of measurements have mainly been electrochemical methods and impedance spectroscopy [107].

In terms of organised films, the first e-tongue was presented in 2002 [110], in which LB and LbL films deposited onto interdigitated gold electrodes were used in the sensor array. Impedance spectroscopy was the principle of detection, and the sensor arrays in this type of e-tongue normally contain from 3 to 6 units of different materials, which may be judiciously chosen for the task at hand. For example, if one needs to distinguish similar liquids that vary in their acidity, a suitable choice is polyaniline and its derivatives, whose electrical properties are strongly dependent on pH. We shall return to this choice of materials later on. The sensing units are immersed into the liquid sample to be measured and the impedance spectrum is taken, being therefore the response to an electrical stimulus of a very complex system. Indeed, the electrical response depends on the liquid, on the interaction between the liquid and the film-forming molecules in each sensing unit and on the double-layer that is usually formed at the film-liquid interface. There is no theoretical model that can account for all of these phenomena at once, which is why equivalent electrical circuits have been used to model the impedance data [111]. Figure 10.9 shows an equivalent circuit with capacitors and resistors used to model typical impedance data also shown in the Figure in the form of capacitance versus frequency and loss versus frequency curves.

Because the electrical properties of interfaces are affected by any small changes in the liquid, on the double-layer and on the films of the sensing units, the sensitivity of e-tongues based on impedance spectroscopy is very high. Just by

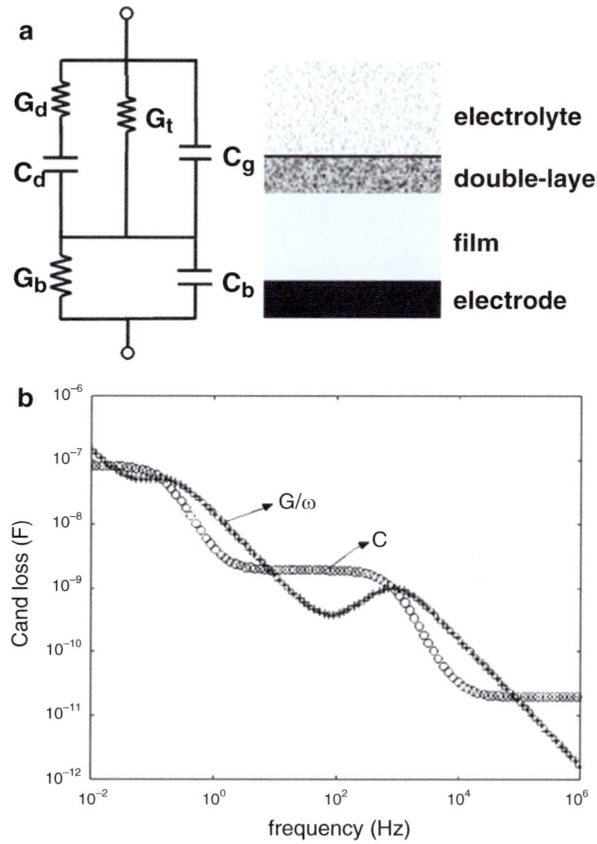

Fig. 10.9 (**a**) Typical sensing unit of an e-tongue showing the metal-film-solution interface and the equivalent circuit used to model its electrical properties. (**b**) Plots of capacitance and loss versus frequency for general metallic electrodes covered with low conductivity thin film and immersed in an electrolyte (Reproduced with permission from Ref. [111])

way of illustration, an e-tongue is at least 1,000–10,000 times more sensitive than the human tongue for the salty taste. This means that an e-tongue may be able to tell there is salt in water at concentrations from 1,000 to 10,000 lower than a human being could do. A demonstration of this capability is in the Principal Component Analysis (PCA) [112] plot in Fig. 10.10, where aqueous solutions at μM concentrations are represented. Note that instead of plotting impedance spectra for each sample, one resorts to the statistical PCA tool, which serves to reduce the dimensionality of the original dataset. PCA is a technique through which linear combinations of the data variance are used to obtain a new orthogonal basis of lower dimension; in Fig. 10.10 it is a 2D plot.

Considering the high sensitivity of e-tongues just mentioned, one could perhaps guess that they could apply to more sophisticated tasks such as classifying wines or other beverages. This turns out to be true, as indicated by the 3D PCA plot in Fig. 10.11 with various types of wine differing in the grapes, vintages and producers. Similar papers have been published with distinction among samples of complex liquids, including coconut water [110], orange juice [115], milk [116], tea and coffee [117].

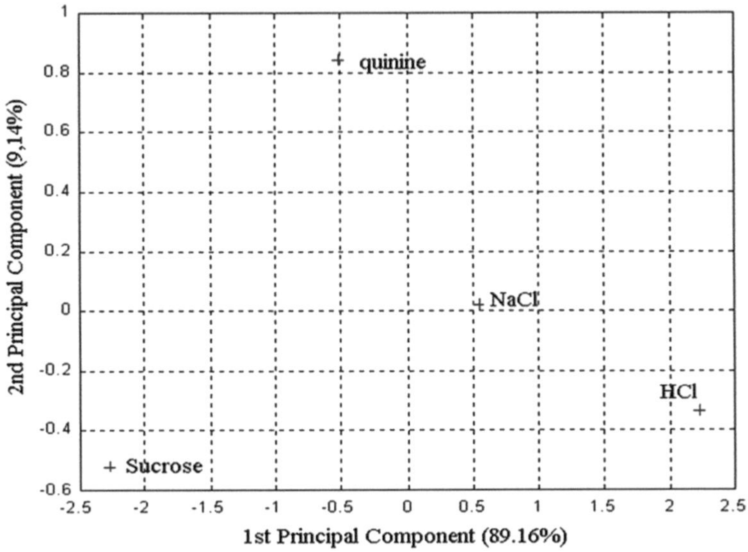

Fig. 10.10 2D PCA plot of capacitance values at 1 kHz data for several sensing units subjected to solutions representing salty (with NaCl), sour (HCl), sweet (sucrose) and bitter (quinine) tastes (Reproduced with permission from Ref. [113])

A high distinguishing ability is reached by considering several issues, including the frequency region for measuring the impedance, an adequate choice of materials and film architectures for the sensing units. According to our experience over many years, it is easier to distinguish similar samples using the impedance signal at low frequency (ca. 100 Hz), because the electrical response is governed by the double-layer in this frequency range [111]. As for the type of material and film architecture in the sensing units, there are many possibilities to optimise the performance of an e-tongue. In a systematic study of the influence of film thickness, LbL films from 1 to 20 bilayers of poly(o-ethoxyaniline) (POEA) and poly(styrene sulfonate) (PSS) were used as sensing units [118]. Thicker films were proven more sensitive to NaCl in solution, but specific sensors could be more sensitive depending on the range of NaCl concentration. The dependence on film thickness was related to the change in film roughness that varied with the thickness. This e-tongue was innovative in that the sensing units were made with LbL films of the same materials but differing in the number of layers. Differences in thickness providing distinct electrical responses were also made use of in an e-tongue made with 10-bilayer LbL films of POEA alternated with chitosan or chitosan nanoparticles in a matrix of poly(methacrylic acid) (CS-PMA) [119]. The film architecture is shown in Fig. 10.12, where it is indicated that POEA could either be doped or not [120]. The POEA layers co-deposited with chitosan nanoparticles were much thicker. Sensor arrays made with POEA and chitosan could detect copper ions in aqueous solutions.

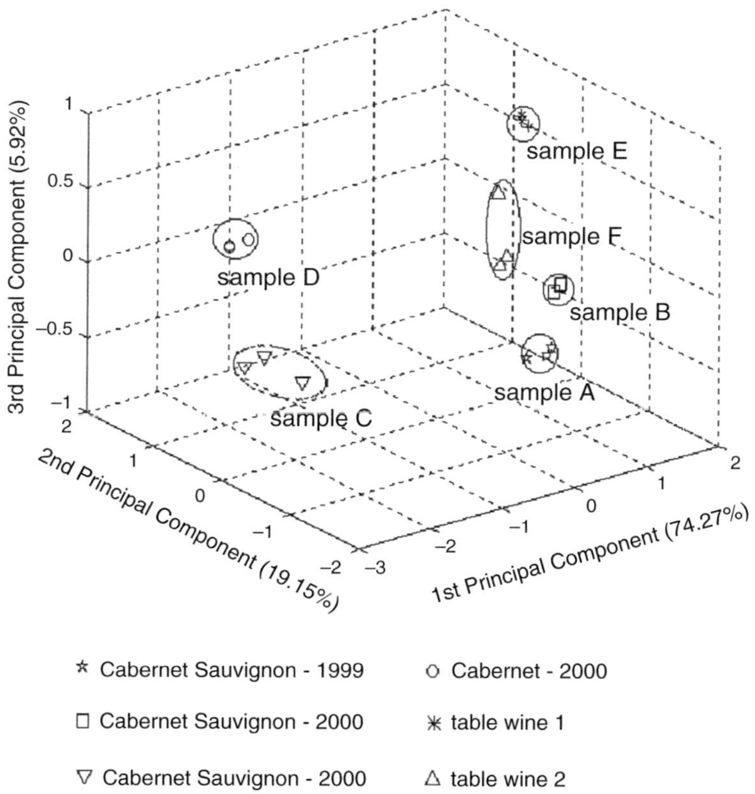

Fig. 10.11 PCA plot of impedance data, at 1 kHz, of the sensor array made of a 3-bilayer chitosan/PSS film and a bare gold electrode in contact with red wine samples (Adapted with permission from [114])

In another example of the same material being used in the sensing units, Volpati et al. [121] employed iron phthalocyanine to obtain an e-tongue with sensing units containing LbL, Langmuir-Blodgett (LB) and Physical Vapor Deposited (PVD) films. The e-tongue could detect copper ions (Cu^{2+}) in aqueous solutions prepared with ultrapure water, in addition to distilled and tap waters. The sensitivity, down to 0.2 mg/L, is sufficient for controlling the quality of water for human consumption.

It should be clear by now that in the applications mentioned the e-tongues were merely used to distinguish liquid samples from each other, which could either be prepared by adding an analyte to a given solution or be real liquid samples, such as those of wine. Nevertheless, upon employing artificial intelligence methods one can try and "teach" an electronic tongue what is good or bad in terms of taste, obviously from the human perception. This may be performed with machine learning techniques where samples tested by human experts are used as examples from which the e-tongue system may learn. Ferreira et al. [122] have demonstrated

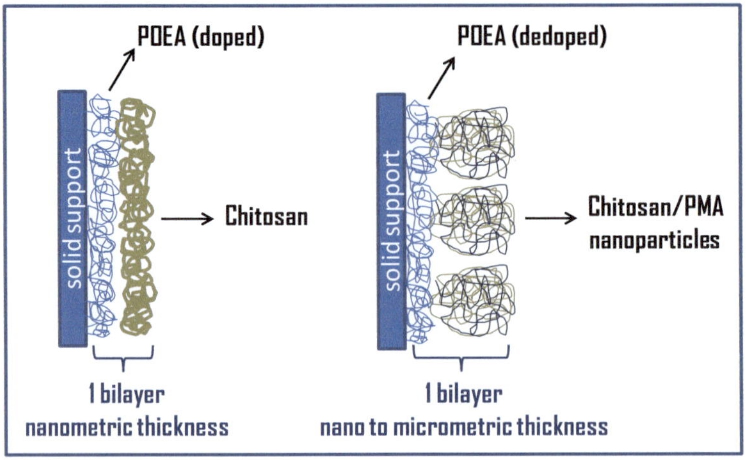

Fig. 10.12 Illustration of a bilayer, deposited on a solid substrate, of POEA and Chitosan (*left*) or POEA and Chitosan-PMA nanoparticles (*right*) (Reproduced with permission from Ref. [120])

the feasibility of such an approach by correlating the impedance spectroscopy data for a variety of coffee samples with the evaluation of human experts. With a combination of machine learning and linear regression techniques, they were able to predict the scores assigned to the coffee samples by the human experts with a precision of ± 0.3, for scores varying typically from 4 to 9. The correlation between the predicted and the assigned scores was indeed high, reaching ca. 0.96 in the Pearson coefficient (1.0 would be maximum) [122].

Last but not least, there is no reason why some of the sensing units in an e-tongue could not be made of materials capable of molecular recognition toward a given analyte, thus imparting selectivity to the sensor – rather than just the global selectivity inherent in e-tongues. In fact, this has been pursued since 2007 [123], whereby e-tongues can serve as biosensors. This topic is addressed in the next subsection.

10.4.2 Biosensors

There are four main challenges in research and development in biosensors, especially for clinical diagnosis, which are associated to enhancing performance and making them viable for the market place. The first major challenge is to ensure high selectivity – thus avoiding false positives and negatives – while reaching high sensitivity. The second challenge is in device engineering to reduce costs and fabricate robust, simple-to-operate devices, particularly when point-of-care use is envisaged. The third challenge is related to understanding the physicochemical principles behind sensing in the devices. Though this is basically a fundamental scientific problem, it has important implications for the design of novel biosensing devices, since the mechanisms inherent in molecular recognition are poorly known. In the fourth challenge, one has to go beyond exploring limited datasets, from a single laboratory or from a single patient in clinical diagnosis, to take advantage of the new scientific paradigm represented by e-Science (also referred to as "Big Data").

Here we shall not touch upon the problems in device engineering, but rather concentrate on the other challenges. Achieving high selectivity and sensitivity is in fact motivation for much of the work on biosensing based on organised films owing to the already-emphasized ability to control molecular architectures. Most of this section will be dedicated to using novel film architectures. Then we shall briefly dwell upon the experimental methods being used to unveil the mechanisms of molecular recognition, followed by a description of attempts to introduce statistical and computational methods to treat large amounts of data.

10.4.2.1 A Variety of Film Architectures for Biosensing

A generic biosensor has at least three important components, namely immobilized biomolecules capable of molecular recognition, signal transducers, and a measuring device that can be based on different detection principles. Organised films are used for the two first components, in the immobilization of a variety of biomolecules and with adsorption of so-called reporter molecules which may serve as signal transducers. The three methods discussed so far, i.e. the LB, LbL and SAMs techniques, are suitable for immobilization of biomolecules with preserved activity because film fabrication can be carried out under mild experimental conditions and the resulting films contain entrained water. The hypothesis of the presence of water as a requirement for activity preservation has not been proven, but this can be expected from the importance of water for proteins to keep their native state.

Biosensors made with organised films can be obtained from a long list of biomolecules, including antibodies, antigens, enzymes and DNA [124–127]. In many instances, a matrix is crucial for immobilizing the biomolecules, and again there are several possibilities. For biosensors containing LB films such a matrix may be obtained with lipids or phospholipids [128–130]. For LbL films, a larger number of materials have been employed. To our experience, dendrimers, lipid liposomes and chitosans are among the best matrices to preserve the bioactivity of enzymes, peptides and antigens. The organised films are commonly adsorbed on planar solid supports, but they can also coat particles and quantum dots [131, 132] that may act in sensing.

Owing to the variety of biosensors built with organised films, we shall describe some of the possibilities by classifying them according to the principle of detection. The examples from the literature will therefore be presented in subsections whose titles refer to the techniques for detection.

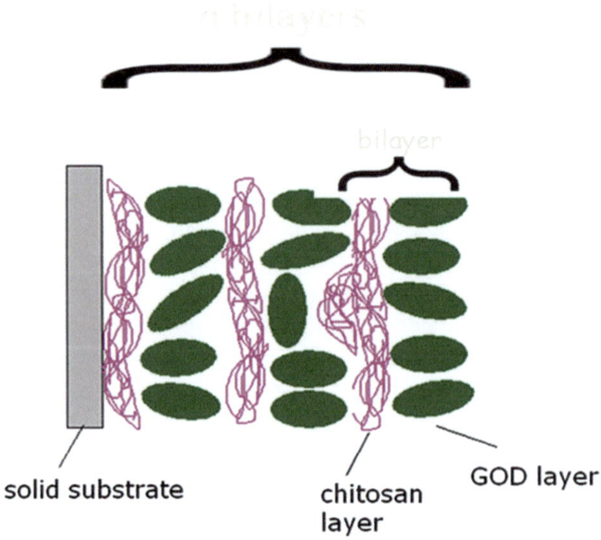

Fig. 10.13 Biosensor fabricated with a film architecture, which alternates glucose oxidase (Gox) (GOD in the figure) and chitosan layers (Reprinted with permission from Ref. [135])

Electrochemical Biosensors

In this type of biosensor, the organised films usually contain an electroactive material to interact with the analyte of interest, or electroactive species are generated from such an interaction. Cyclic voltammetry, amperometry, chronoamperometry, and electrochemical impedance spectroscopy are the specific techniques used, as discussed at length on reviews for the use of LbL films [133] and LB or LbL films [134].

Many of these biosensors contain immobilized enzymes, where the performance may be optimized by controlling the molecular architecture in at least two ways: by changing the matrix material and/or varying the number and order of deposition of the layers with the biomolecules. As one should guess, advantage is taken of the layer-by-layer fashion with which LB and LbL films can be deposited, from more than one material on the same unit. Consider for example the LbL film made with alternating layers of chitosan and glucose oxidase (GOx) in Fig. 10.13 [103]. Detection of glucose is indirect because GOx catalyzes a reaction of β-glucose with O_2, thus generating gluconic acid and H_2O_2, which is then actually detected. The scheme below shows the reaction.

$$\beta - \text{glucose} + O_2 \stackrel{\text{GOx}}{\rightarrow} \text{gluconic acid} + H_2O_2$$

One could ask whether the sensitivity of the biosensor would increase with the number of bilayers or even whether it should be best to adsorb the enzyme only at the top most layer.

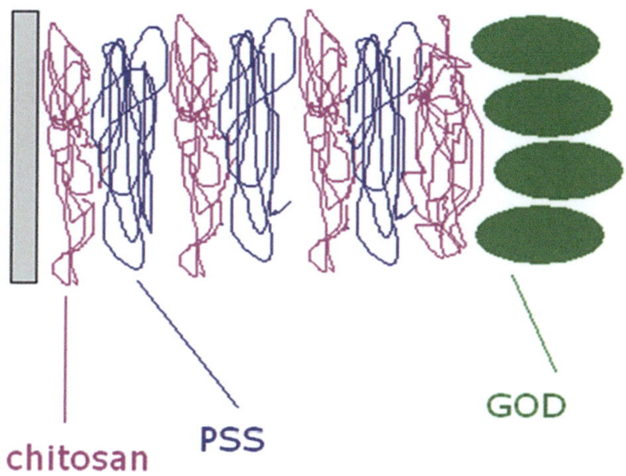

Fig. 10.14 Film architecture optimized for glucose detection. The LbL film was built with a few bilayers of chitosan and PSS and the enzyme GOx (GOD in the figure) was deposited only on the outer layer (Reprinted with permission from Ref. [135])

After checking all of the possibilities of film architectures, Caseli et al. [135] found that the highest efficiency was achieved with GOx only at the topmost layer, in the architecture shown in Fig. 10.14. A cushion of a few bilayers was built with chitosan and the polyelectrolyte poly(styrene sulfonate) (PSS) onto which the GOx layer was adsorbed. That this was the optimised architecture was attributed to the lack of contribution from the enzyme in inner layers to glucose oxidation, probably because the GOx molecules were arranged in a closely packed arrangement and hampered electron transport and analyte diffusion through the LbL film, or chitosan acted as a barrier for glucose [135]. It should be remarked, nevertheless, that this behaviour is not universal. There are cases in which the sensitivity increases with the amount of enzyme adsorbed and therefore with the number of bilayers deposited [136], thus reinforcing the influence from diffusion of small molecules through the film. As for an optimum number of layers, there is empirical evidence that optimised performance is reached with 3–5 bilayers, perhaps because thicker films would be less sensitive while very thin films may not warrant full coverage of the substrate.

In 2006, Crespilho et al. [137] conceived a biosensor to detect glucose seeking synergy among the various components that may be introduced with the LbL method. Three aims were targeted: (i) because amperometry was to be used, gold nanoparticles were incorporated into the dendrimer film to increase the electrical current measured; (ii) in order to detect H_2O_2 at a low potential, the gold nanoparticles were capped with a redox mediator; (iii) for the enzyme glucose oxidase (GOx) to be in a friendly environment, it was co-deposited with another protein, bovine serum albumin (BSA). The procedures for film fabrication are illustrated in Fig. 10.15, where the gold nanoparticle capped with polyamidoamine dendrimer

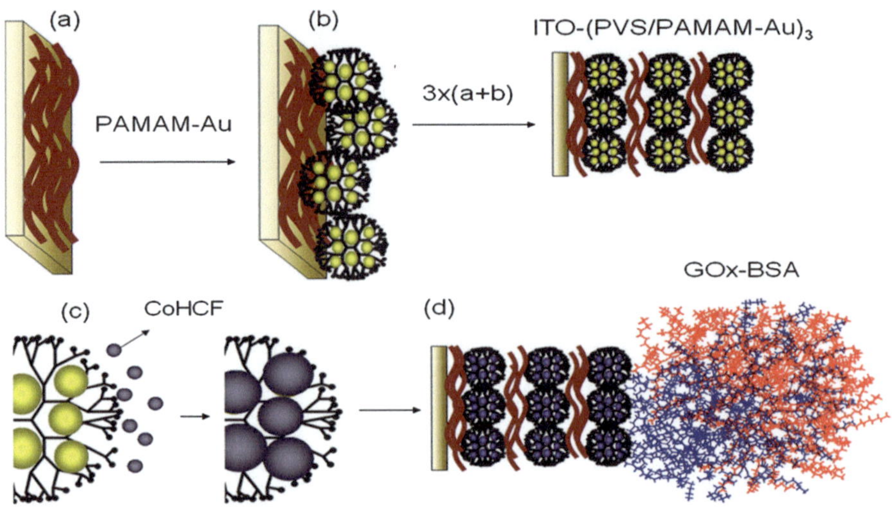

Fig. 10.15 Illustration of fabrication of LbL films of PVS and PAMAM–Au. The deposition of LbL multilayers was performed by immersing the substrates alternately into (**a**) PVS and (**b**) PAMAM–Au solutions for 5 min. The ITO–(PVS/ PAMAM–Au)3@CoHCF electrode (**c**) was obtained, after the deposition of 3 bilayers, by potential cycling. The enzyme immobilization (**d**) was carried out in a solution containing BSA, glutaraldehyde and GOx, producing the ITO–(PVS/PAMAM–Au)3@CoHCFGOx (Reprinted with permission from Elsevier Ref. [137])

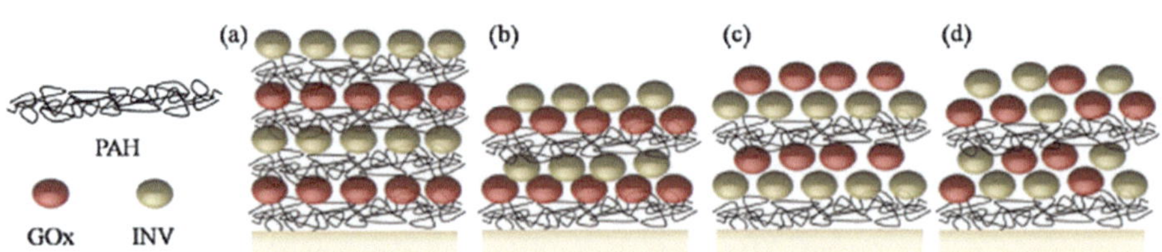

Fig. 10.16 Architectures of LbL films (**a**) PAH/GOx/PAH/INV; (**b**) PAH/GOx/INV (**c**) PAH/INV/GOx (**d**) PAH/GOx + INV (Reprinted with permission from Ref. [138])

(PAMAM-Au) layers were alternately adsorbed with poly(vinyl sulfonic acid) (PVS), the nanoparticles were capped with CoCHF redox mediator, and the enzyme was co-immobilized with BSA only on the topmost layer. Similarly to the study of Caseli et al. [135], various architectures were tested, e.g. with enzyme in all bilayers, but optimised performance was observed with the architecture shown in the figure.

We finish this section of electrochemical biosensors with an example in which more than one enzyme was immobilized in the same LbL film in order to exploit cascade reactions. Figure 10.16 shows the film architectures built with the enzymes glucose oxidase (GOx) and invertase (INV) for the simultaneous detection of glucose and sucrose. From the four architectures tested, only the one with GOx and INV co-adsorbed in the same layer (architecture d, (PAH/(GOx + INV))$_{10}$, in Fig. 10.16) provided good detection ability for both glucose and sucrose.

The reactions responsible for detection are as follows. For glucose,

$$\beta - \text{glucose} + O_2 \xrightarrow{\text{GOx}} \text{gluconic acid} + H_2O_2$$

On the other hand, the activity of INV may be monitored through sucrose catalysis

$$\text{sucrose} + H_2O \xrightarrow{\text{INV}} \alpha - \text{glucose} + \beta - \text{frutose}$$

Since no electroactive species is produced from the latter catalysis process, the two mechanisms are combined as

$$\text{sucrose} + H_2O \xrightarrow{\text{INV}} \alpha - \text{glucose} + \beta - \text{frutose}$$
$$\alpha - \text{glucose} + O_2 \xrightarrow{\text{GOx}} \text{gluconic acid} + H_2O_2$$

The simultaneous detection is shown in Fig. 10.17 featuring a large change in the amperometric signal for the addition of glucose, and small changes when sucrose was added. In spite of this small change due to sucrose, the changes increased with successive additions. The effect of sucrose was small because GOx had small activity toward α-glucose generated in the sucrose catalysis.

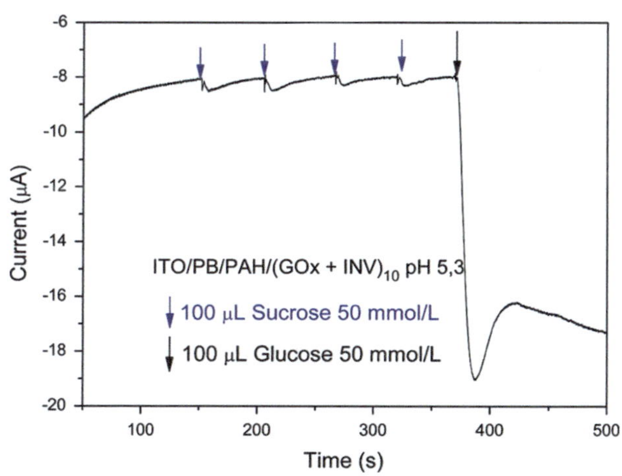

Fig. 10.17 Amperometric response, obtained at 0.0 V in a phosphate buffer at pH 6.3, of the electrode ITO/PB/PAH/(GOx + INV)$_{10}$ prepared at pH 5.3 for the additions of glucose and sucrose, as indicated (Reprinted with permission from Ref. [138] Copyright © 2011 American Chemical Society)

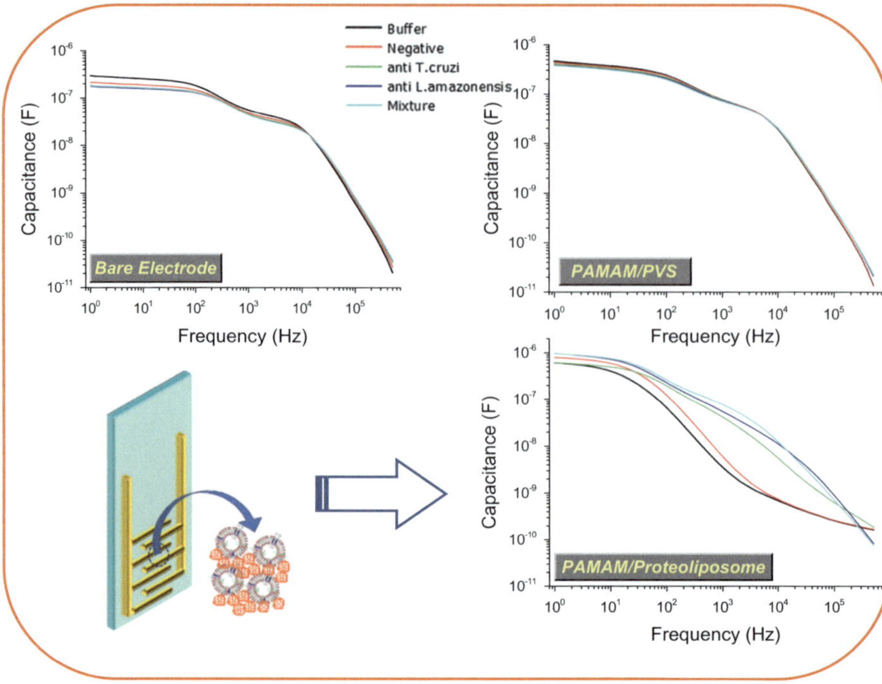

Fig. 10.18 Illustration of an interdigitated electrode coated with LbL films of antigens in proteoliposomes (*bottom left*). The graphs show capacitance vs frequency plots taken with three types of electrode for samples with 10^{-5} mg/mL antibody solutions: bare, coated with PAMAM/PVS LbL films and coated with PAMAM/proteoliposome LbL film (Reproduced with permission from Ref. [139])

Biosensing Based on Impedance Spectroscopy and Field-Effect Devices

This topic has been largely developed as an extension of the work on e-tongues, whose concept was adapted to distinguish similar diseases and to detect trace amounts of biologically-relevant analytes. In this type of biosensor, one may either use only one sensing unit with an immobilized biomolecule or an array of sensing units, some of which will not be specific.

Figure 10.18 illustrates the use of e-tongue in biosensing for distinguishing between Leishmaniasis and Chagas' disease, where four sensing units were employed [139]. Two of those contained antigens associated with the two diseases, in the form of proteoliposomes, while the other two sensing units were a bare electrode and an electrode coated with LbL films of polyelectrolytes. The capacitance spectra shown in the figure for samples containing antibodies for the two diseases, a mixture of them and a buffer solution, do indicate that – as expected – similar responses are obtained for the diseases which are of the same family. In fact, motivation for this study came from the difficulties in distinguishing these diseases. Also shown is a schematic drawing of the interdigitated gold electrode on which LbL films of proteoliposomes were adsorbed.

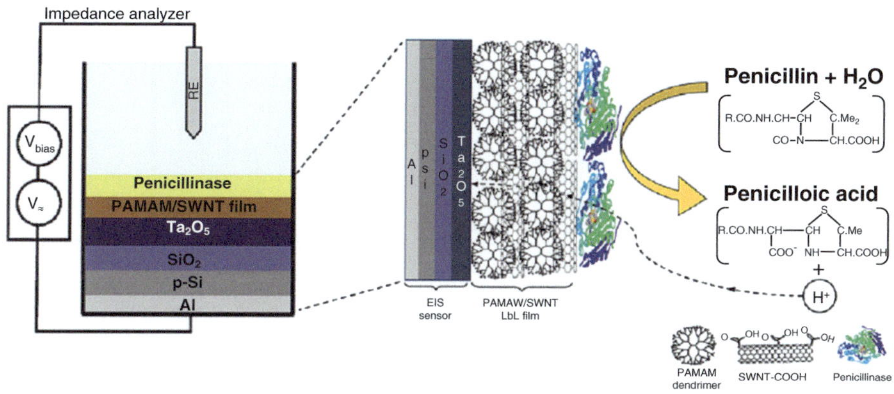

Fig. 10.19 Architecture of the sensing unit made with a semiconductor multilayer chip with an LbL film adsorbed, made of PAMAM/SWNT bilayers and one top layer of penicillinase (*left*), with a zoomed view of the LbL film (*center*). The reactions catalyzed by penicinillinase responsible for the detection are shown on the scheme (*right*) (Reproduced with permission from Ref. [21])

While the distinction of the few samples – with just one antibody concentration – can be made visually as in Fig. 10.18, the same does not apply when a large number of samples are used, especially if "real" samples extracted from blood serum need to be treated. In such cases, statistical and computational methods are required, and we shall return to this point later on.

Impedance measurements may be taken in field-effect devices for biosensing, where there is convergence of several technologies. On one hand, by using field-effect devices one may exploit the possibilities offered by the semiconductor industry for mass production of biosensors. On the other hand, with methods related to organised films, the devices may be functionalised for very specific goals. For example, Siqueira et al. [21, 140] obtained capacitance vs. voltage curves in a field-effect (bio-) chemical sensor to detect penicillin G. The biosensor was made of LbL films containing single-wall carbon nanotubes (SWNTs) and polyamidoamine (PAMAM) dendrimers, having penicillinase enzyme immobilized on the top layer. The penicillin concentrations tested ranged from 5.0×10^{-6} to 25×10^{-3} M, and the sensitivity achieved was 116 mV/decade, using a device of the capacitive electrolyte-insulator-semiconductor (EIS) type, whose structure in shown in Fig. 10.19.

A number of features in this biosensor are illustrative of the advantages of molecular control in film architectures. With the enzyme penicillinase on the top layer, it was possible to detect penicillin selectively and with high sensitivity because the LbL film was porous owing to the presence of PAMAM dendrimers. With such porosity and the increase in film roughness caused by deposition of SWNT layers, the signal in potentiometry increased. This is illustrated in Fig. 10.20 that compares the response of an EIS sensor containing the SWNT layers with that of a sensor without SWNT. The performance of the SWNT-containing sensor was superior not only in sensitivity, as is apparent in the graphs, but also in terms of higher stability, lower drift and shorter response times [140].

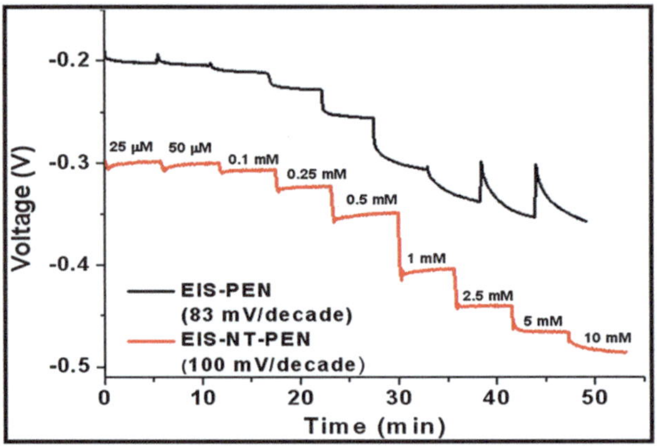

Fig. 10.20 Voltage signal in the EIS device containing penicinillase, either without (*Black*) or with carbon nanotubes (*Red*), whose evolution in time was dictated by addition of successive amounts of penicillin G. The final concentration of the detected solution is marked in the figure for the NT-containing device. One may note a considerably higher change for the NT-containing device (Adapted from Ref. [140])

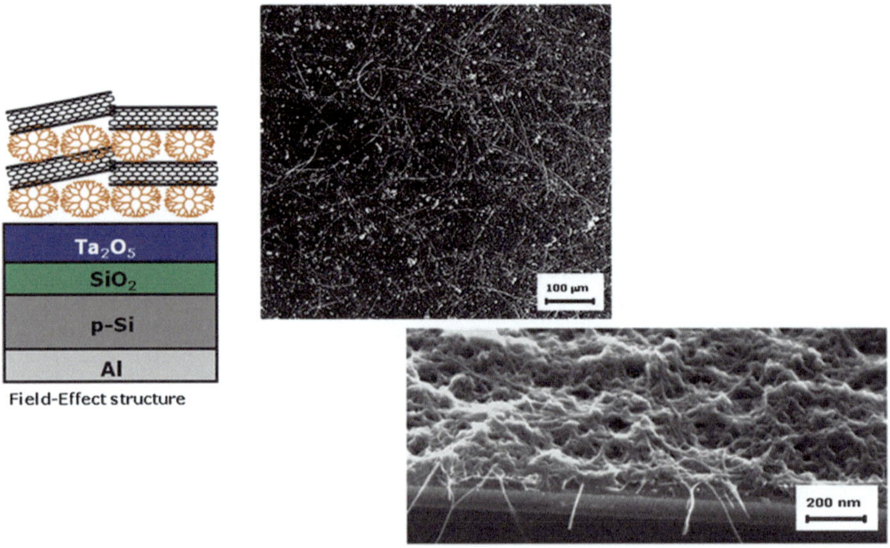

Fig. 10.21 Illustration of the LAPS structure functionalized with a PAMAM/SWNT LbL Film (*left*), FESEM images for a 6-bilayer PAMAM/SWNT LbL film: top view (*right-top*) and cross-sectional view (*right-bottom*) (Reproduced with permission from Ref. [141])

The effects from SWNT could in principle be attributed to a higher conductivity, but the data available were not sufficient to confirm it. On the other hand, there was strong evidence that a major effect from SWNT layers comes from the much larger surface area of the films, since film roughness increased considerably with deposition of such layers. This is indicated in the micrographs of Fig. 10.21.

10.4.2.2 Importance of Computational Methods for Data Analysis

For decades there have been predictions of a paradigmatic change in automation and use of machines toward replacing humans in ever increasingly-sophisticated tasks, e.g. with robots capable of performing many of the intellectual human activities. This is becoming a reality with the Internet of things and enhanced capabilities of artificial intelligence systems, particularly with development of machine learning methods that exploit the tremendous amount of electronic data available worldwide. Such developments are now reaching the field of sensing and biosensing, with prospects of expert systems being created for clinical diagnosis [142]. We shall discuss here the use of information visualisation techniques to treat data from biosensors, which allowed for an enhanced performance [143].

While describing contributions related to e-tongues in Sect. 10.4.1, we have already mentioned the need to treat the data with statistical methods, and results with PCA plots were introduced. When the amount of data increases significantly or if there is a highly non-linear response in the measurements, as one can expect for biosensors based on very specific interactions, then additional methods may be required. These methods of data mining and visualisation are also crucial owing to the limited capacity of interpretation by humans for a lot of data. Information visualisation (*InfoVis*) [144] may be closely related to multivariate or multidimensional data analysis, known from statistics. The generality of *InfoVis* is readily apparent in that representations can be built from abstract entities such as text, images and census data, which do not necessarily have a physical or geometric representation. Since these data instances are normally multidimensional, *InfoVis* methods are used to reduce dimensionality in order to obtain a 2D or 3D plot by mapping each data instance to a graphical marker. The aim with multidimensional projection techniques is to convey similarity relationships among data instances in a way that dissimilar instances are placed far from each other on the resulting plot.

Figure 10.22 shows a 2D plot obtained with the technique referred to as Sammon's mapping [146] for the impedance data taken with a 4-unit sensor array designed to distinguish between Chagas' disease and *Leishmaniasis*, two tropical diseases of the same family whose detection is fraught with false positives when conventional clinical analysis methods are used. The four sensing units were a bare interdigitated gold electrode and electrodes coated with LbL films of three types: containing only polyelectrolytes, and containing antigens specific for antibodies of Tripanosoma Cruzi (for Chagas' disease) and of *LeishmaniasisAmozenensis* [145]. The samples tested were synthetic samples made with buffer to which antibodies were added and real samples of the blood serum of infected animals (see [145] for details). The synthetic samples are the same as those of Fig. 10.18, but now including different concentrations as well. For this large number of samples, distinction using visual inspection of the capacitance spectra as in Fig. 10.18 is out of the question. PCA was also tried but the distinction was poor. We believe that the excellent distinction in Fig. 10.22 could only be achieved because Sammon's mapping is a non-linear technique, which is able to capture subtle changes inherent in the non-linear response expected in biosensing.

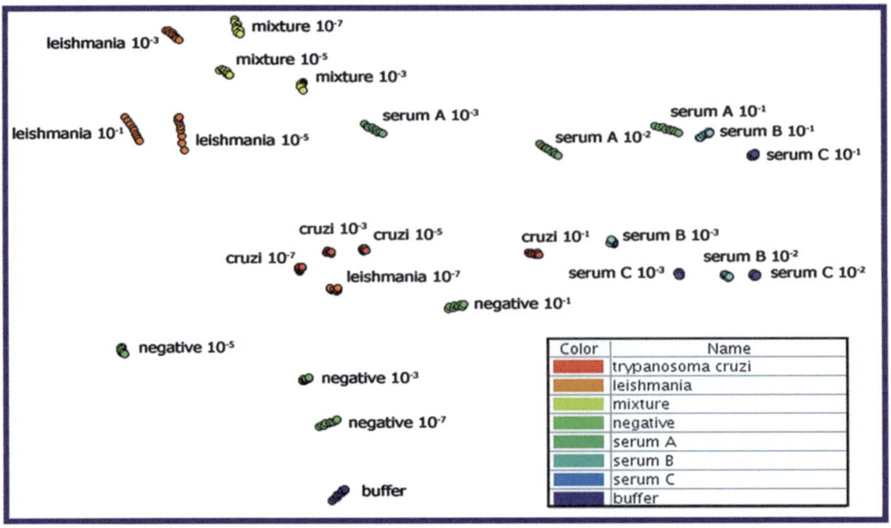

Fig. 10.22 Two dimensional plot obtained with the Sammon's mapping technique for the electrical impedance data taken with the four sensing units for synthetic and real samples. The synthetic samples contained TRIS buffer to which anti-Leishmania, anti-T. cruzi and negative antibodies (not related to these two diseases) were added. In addition, some samples contained mixtures of the two antibodies. The real samples consisted of different concentrations of blood serum, referred to as serum A (containing negative antibodies), serum B (containing anti-Leishmania antibodies), and serum C (containing anti-T. cruzi antibodies). For the data treatment, impedance was converted into capacitance. The axes in the plot are not labeled because what matters is the relative distance (i.e. dissimilarity) among the samples (Reproduced with permission from Ref. [145])

Multidimensional projections can be combined with other *InfoVis* methods, such as Parallel Coordinates, and genetic algorithms for optimization of biosensing performance [147]. They can also be used in speeding up analysis of thousands of spectra that may now be routinely acquired with modern equipment. One important application in this regard has been the use of information visualisation for identifying hot spots where single molecules were placed in LB films explored in single molecule detection via surface-enhanced Raman scattering (SERS) [148].

The examples above were limited to cases where only local data – from a single laboratory – were treated. However, this type of data treatment can be extended to distributed data treatment, which can be associated with yet other computational techniques for data mining and classification. This will open a new era where nanotechnology in general, and biosensing in particular, will be integrated into the e-Science (or "Big Data") paradigm.

10.5 Concluding Remarks

In this Chapter, we have tried to emphasize the growing importance of nanostructured, organised films in coating surfaces where molecular control can be exploited. After mentioning a number of possible applications, we concentrated on the use of such films in e-tongues and biosensors, precisely because the main features related to organisation in the films are taken advantage of for reaching high sensitivity and selectivity. We also highlighted the use of computational methods to treat sensing and biosensing data, with which one may now envisage the development of expert systems for clinical diagnosis, theranostic use, surveillance and environmental control. In all of these cases, the computational system will result from converging technologies that include the fabrication of tailored sensors and biosensors.

When referring to applications we included the modifier "possible" since products based on organised films apparently have not gone to mass production. We have already mentioned that for LB films the limitations include the relatively poor stability and the costly experimental procedures for film fabrication. LbL films, on the other hand, should not suffer from such limitations, at least not to the same extent. They can be made more stable and with the spraying techniques the films can be fabricated within very short times. Nevertheless, these procedures have not been integrated into industrial processes, perhaps because a "killer" application has not been identified, which would justify the huge investment required for mass production.

In spite of the absence in commercialised products, the technology inherent in organised films serves a testbed for characterising materials and identifying applications. Of particular importance, in this regard, is the research on Langmuir monolayers as cell membrane models, which has been covered in this Chapter. Though these films are surely not going to be incorporated in a product, the results from interaction with biologically-relevant molecules may be helpful for drug design and development of drug delivery systems.

Acknowledgments The authors acknowledge the financial support from FAPESP, CNPq, CAPES and nBioNet network (Brazil).

References

1. Whitesides GM, Mathias JP, Seto CT (1991) Molecular self-assembly and nanochemistry: a chemical strategy for the synthesis of nanostructures. Science 254:1312–1319. doi:10.1126/science.1962191
2. Decher G (1997) Fuzzy nanoassemblies: toward layered polymeric multicomposites. Science 277:1232–1237. doi:10.1126/science.277.5330.1232
3. Hoeppener S, Maoz R, Cohen S, Chi L, Fuchs H, Sagiv J (2002) Metal nanoparticles, nanowires, and contact electrodes self-assembled on patterned monolayer templates: a bottom-up chemical approach. Adv Mater 14:1036–1041. doi:10.1002/1521-4095(20020805)14:15<1036::AID-ADMA1036>3.0.CO;2-J
4. Decher G, Hong JD, Schmitt J (1992) Buildup of ultrathin multilayer films by a self-assembly process: III. Consecutively alternating adsorption of anionic and cationic polyelectrolytes on charged surfaces. Thin Solid Films 210–211:831–835. doi:10.1016/0040-6090(92)90417-A

5. Blodgett KB (1935) Films built by depositing successive monomolecular layers on a solid surface. J Am Chem Soc 57:1007–1022
6. Langmuir I (1920) The mechanism of the surface phenomena of flotation. Trans Faraday Soc 15:62. doi:10.1039/tf9201500062
7. Langmuir I (1917) The constitution and fundamental properties of solids and liquids. II Liquids. 1. J Am Chem Soc 39:1848–1906. doi:10.1021/ja02254a006
8. Blodgett KB (1934) Monomolecular films of fatty acids on glass. J Am Chem Soc 56:495. doi:10.1021/ja01317a513
9. Petty MC (1996) Langmuir-Blodgett films. Cambridge University Press, Cambridge, UK
10. Roberts G (1990) Langmuir-Blodgett films. Plenum Press, New York
11. Tredgold RH (1994) Order in thin organic films. Cambridge University Press, Cambridge, UK
12. Ulman A (1991) An Introduction to organic ultrathin films from Langmuir-Blodgett to self-assembly. Academic, Boston
13. Oliveira Jr ON, He J-A, Zucolotto V, Balasubramanian S, Li L, Nalwa HS, Kumar J, Tripathy SK (2002) Layer-by-layer polyelectrolyte-based thin films for electronic and photonic applications. In: Kumar J, Tripathy SK, Nalwa HS (eds) Handbook of polyelectrolytes and their applications. American Scientific Publishers, Los Angeles, pp 1–33
14. Park JY, Advincula RC (2011) Nanostructuring polymers, colloids, and nanomaterials at the air–water interface through Langmuir and Langmuir–Blodgett techniques. Soft Matter 7:9829. doi:10.1039/c1sm05750b
15. Miyashita T (1993) Recent studies on functional ultrathin polymer films prepared by the Langmuir-Blodgett technique. Prog Polym Sci 18:263–294
16. Tredgold RH, Winter CS (1982) Langmuir–Blodgett monolayers of preformed polymers. J Phys D Appl Phys 15:L55–L58. doi:10.1088/0022-3727/15/6/003
17. Moon GD, Lee T II, Kim B, Chae G, Kim J, Kim S, Myoung J-M, Jeong U (2011) Assembled monolayers of hydrophilic particles on water surfaces. ACS Nano 5:8600–8612. doi:10.1021/nn202733f
18. Mitzi DB (2001) Thin-film deposition of organic – inorganic hybrid materials. Chem Mater 13:3283–3298. doi:10.1021/cm0101677
19. Girard-Egrot AP, Godoy S, Blum LJ (2005) Enzyme association with lipidic Langmuir-Blodgett films: interests and applications in nanobioscience. Adv Colloid Interface Sci 116:205–225. doi:10.1016/j.cis.2005.04.006
20. Sassolas A, Blum LJ, Leca-Bouvier BD (2012) Immobilization strategies to develop enzymatic biosensors. Biotechnol Adv 30:489–511. doi:10.1016/j.biotechadv.2011.09.003
21. Siqueira JR, Caseli L, Crespilho FN, Zucolotto V, Oliveira Jr ON (2010) Immobilization of biomolecules on nanostructured films for biosensing. Biosens Bioelectron 25:1254–1263. doi:10.1016/j.bios.2009.09.043
22. Brezesinski G, Möhwald H (2003) Langmuir monolayers to study interactions at model membrane surfaces. Adv Colloid Interface Sci 100–102:563–584. doi:10.1016/S0001-8686(02)00071-4
23. Netzer L, Sagiv J (1983) A new approach to construction of artificial monolayer assemblies. J Am Chem Soc 105:674–676. doi:10.1021/ja00341a087
24. Evans SD, Urankar E, Ulman A, Ferris N (1991) Self-assembled monolayers of alkanethiols containing a polar aromatic group: effects of the dipole position on molecular packing, orientation, and surface wetting properties. J Am Chem Soc 113:4121–4131. doi:10.1021/ja00011a010
25. Maoz R, Sagiv J (1987) Penetration-controlled reactions in organized monolayer assemblies. 2. Aqueous permanganate interaction with self-assembling monolayers of long-chain surfactants. Langmuir 3:1045–1051. doi:10.1021/la00078a028
26. Richer J, Stolberg L, Lipkowski J (1986) Quantitative investigations of adsorption of tert-amyl alcohol at the gold(110)-aqueous solution interface. Langmuir 2:630–638. doi:10.1021/la00071a019

27. Tillman N, Ulman A, Penner TL (1989) Formation of multilayers by self-assembly. Langmuir 5:101–111. doi:10.1021/la00085a019
28. Zhao X-M, Wilbur JL, Whitesides GM (1996) Using two-stage chemical amplification to determine the density of defects in self-assembled monolayers of alkanethiolates on gold. Langmuir 12:3257–3264. doi:10.1021/la960044e
29. Arya SK, Solanki PR, Datta M, Malhotra BD (2009) Recent advances in self-assembled monolayers based biomolecular electronic devices. Biosens Bioelectron 24:2810–7. doi:10.1016/j.bios.2009.02.008
30. Gooding JJ, Darwish N (2012) The rise of self-assembled monolayers for fabricating electrochemical biosensors–an interfacial perspective. Chem Rec 12:92–105. doi:10.1002/tcr.201100013
31. Booth MA, Vogel R, Curran JM, Harbison S, Travas-Sejdic J (2013) Detection of target-probe oligonucleotide hybridization using synthetic nanopore resistive pulse sensing. Biosens Bioelectron 45:136–40. doi:10.1016/j.bios.2013.01.044
32. Iler RK (1966) Multilayers of colloidal particles. J Colloid Interface Sci 21:569–594. doi:10.1016/0095-8522(66)90018-3
33. Lvov Y, Haas H, Decher G, Moehwald H, Kalachev A (1993) Assembly of polyelectrolyte molecular films onto plasma-treated glass. J Phys Chem 97:12835–12841. doi:10.1021/j100151a033
34. Lvov Y, Decher G, Moehwald H (1993) Assembly, structural characterization, and thermal behavior of layer-by-layer deposited ultrathin films of poly(vinyl sulfate) and poly(allylamine). Langmuir 9:481–486. doi:10.1021/la00026a020
35. Cassagneau T, Fendler JH (1999) Preparation and layer-by-layer self-assembly of silver nanoparticles capped by graphite oxide nanosheets. J Phys Chem B 103:1789–1793. doi:10.1021/jp984690t
36. Fendler JH (1996) Self-assembled nanostructured materials. Chem Mater 8:1616–1624. doi:10.1021/cm960116n
37. He J-A, Valluzzi R, Yang K, Dolukhanyan T, Sung C, Kumar J, Tripathy SK, Samuelson L, Balogh L, Tomalia DA (1999) Electrostatic multilayer deposition of a gold – dendrimer nanocomposite. Chem Mater 11:3268–3274. doi:10.1021/cm990311c
38. Balasubramanian S, Wang X, Wang HC, Yang K, Kumar J, Tripathy SK, Li L (1998) Azo chromophore-functionalized polyelectrolytes. 2. Acentric self-assembly through a layer-by-layer deposition process. Chem Mater 10:1554–1560. doi:10.1021/cm9707418
39. Tripathy SK, Katagi H, Kasai H, Balasubramanian S, Oshikiri H, Kumar J, Oikawa H, Okada S, Nakanishi H (1998) Self assembly of organic microcrystals 1: electrostatic attachment of polydiacetylene microcrystals on a polyelectrolyte surface. Jpn J Appl Phys 37:L343–L345. doi:10.1143/JJAP.37.L343
40. Fou AC, Onitsuka O, Ferreira M, Rubner MF, Hsieh BR (1996) Fabrication and properties of light-emitting diodes based on self-assembled multilayers of poly(phenylene vinylene). J Appl Phys 79:7501. doi:10.1063/1.362421
41. Wu A, Yoo D, Lee J-K, Rubner MF (1999) Solid-state light-emitting devices based on the tris-chelated ruthenium(II) complex: 3. High efficiency devices via a layer-by-layer molecular-level blending approach. J Am Chem Soc 121:4883–4891. doi:10.1021/ja9833624
42. Linford MR, Auch M, Möhwald H (1998) Nonmonotonic effect of ionic strength on surface dye extraction during dye – polyelectrolyte multilayer formation. J Am Chem Soc 120:178–182. doi:10.1021/ja972133z
43. Tedeschi C, Caruso F, Möhwald H, Kirstein S (2000) Adsorption and desorption behavior of an anionic pyrene chromophore in sequentially deposited polyelectrolyte-dye thin films. J Am Chem Soc 122:5841–5848. doi:10.1021/ja994029i
44. Caruso F, Möhwald H (1999) Protein multilayer formation on colloids through a stepwise self-assembly technique. J Am Chem Soc 121:6039–6046. doi:10.1021/ja990441m
45. He J-A, Samuelson L, Li L, Kumar J, Tripathy SK (1998) Oriented bacteriorhodopsin/polycation multilayers by electrostatic layer-by-layer assembly. Langmuir 14:1674–1679. doi:10.1021/la971336y

46. Lvov Y, Ariga K, Ichinose I, Kunitake T (1995) Assembly of multicomponent protein films by means of electrostatic layer-by-layer adsorption. J Am Chem Soc 117:6117–6123. doi:10.1021/ja00127a026
47. Lvov Y, Ariga K, Kunitake T (1994) Layer-by-layer assembly of alternate protein/polyion ultrathin films. Chem Lett 23:2323–2326. doi:10.1246/cl.1994.2323
48. Oliveira Jr ON, Raposo M, Dhanabalan A (2011) Polymeric, Langmuir-Blodgett and self-assembled films. In: Nalwa HS (ed) Handbook of surfaces and interfaces of materials. Academic, San Diego, pp 1–63
49. Pavinatto FJ, Caseli L, Oliveira Jr ON (2010) Chitosan in nanostructured thin films. Biomacromolecules 11:1897–908. doi:10.1021/bm1004838
50. Izquierdo A, Ono SS, Voegel J-C, Schaaf P, Decher G (2005) Dipping versus spraying: exploring the deposition conditions for speeding up layer-by-layer assembly. Langmuir 21:7558–67. doi:10.1021/la047407s
51. Caruso F (2000) Hollow capsule processing through colloidal templating and self-assembly. Chem Eur J 6:413–419. doi:10.1002/(SICI)1521-3765(20000204)6:3{\mathsurround=\opskip$<$}413::AID-CHEM413{\mathsurround=\opskip$>$}3.0.CO;2-9
52. Wang Y, Angelatos AS, Caruso F (2008) Template synthesis of nanostructured materials via layer-by-layer assembly †. Chem Mater 20:848–858. doi:10.1021/cm7024813
53. Estephan ZG, Qian Z, Lee D, Crocker JC, Park S-J (2013) Responsive multidomain free-standing films of gold nanoparticles assembled by DNA-directed layer-by-layer approach. Nano Lett 13:4449–55. doi:10.1021/nl4023308
54. Raoufi M, Schönherr H (2014) Fabrication of complex free-standing nanostructures with concave and convex curvature via the layer-by-layer approach. Langmuir 30:1723–8. doi:10.1021/la500007x
55. Boudou T, Crouzier T, Ren K, Blin G, Picart C (2010) Multiple functionalities of polyelectrolyte multilayer films: new biomedical applications. Adv Mater 22:441–67. doi:10.1002/adma.200901327
56. Granicka LH (2014) Nanoencapsulation of cells within multilayer shells for biomedical applications. J Nanosci Nanotechnol 14:705–716. doi:10.1166/jnn.2014.9106
57. Skorb EV, Andreeva DV (2013) Layer-by-layer approaches for formation of smart self-healing materials. Polym Chem 4:4834. doi:10.1039/c3py00088e
58. Kerdjoudj H, Berthelemy N, Boulmedais F, Stoltz J-FJ-F, Menu P, Voegel JC (2010) Multilayered polyelectrolyte films: a tool for arteries and vessel repair. Soft Matter 6:3722. doi:10.1039/b920729e
59. De Temmerman M-L, Demeester J, De Smedt SC, Rejman J (2012) Tailoring layer-by-layer capsules for biomedical applications. Nanomedicine (Lond) 7:771–788. doi:10.2217/nnm.12.48
60. Wohl BM, Engbersen JFJ (2012) Responsive layer-by-layer materials for drug delivery. J Control Release 158:2–14. doi:10.1016/j.jconrel.2011.08.035
61. Yeagle PL (1993) The membranes of cells. Academic, San Diego
62. Krägel J, Derkatch SR (2010) Interfacial shear rheology. Curr Opin Colloid Interface Sci 15:246–255. doi:10.1016/j.cocis.2010.02.001
63. Hansen FK, Rødsrud G (1991) Surface tension by pendant drop. J Colloid Interface Sci 141:1–9. doi:10.1016/0021-9797(91)90296-K
64. Langevin D (2014) Rheology of adsorbed surfactant monolayers at fluid surfaces. Annu Rev Fluid Mech 46:47–65. doi:10.1146/annurev-fluid-010313-141403
65. Dynarowicz-Latka P, Dhanabalan A, Oliveira Jr ON (2001) Modern physicochemical research on Langmuir monolayers. Adv Colloid Interface Sci 91:221–293
66. Sadewasser S, Glatzel T (2012) Kelvin probe force microscopy. Springer Ser Surf Sci. doi:10.1007/978-3-642-22566-6
67. Dhanabalan A, Mello SV, Oliveira Jr ON (1998) Preparation of Langmuir – Blodgett films of soluble polypyrrole. Macromolecules 31:1827–1832. doi:10.1021/ma970606g
68. Helmholtz H (1902) Abhandlungen zur thermodynamik. 51

69. Demchak RJ, Fort T (1974) Surface dipole moments of close-packed un-ionized monolayers at the air-water interface. J Colloid Interface Sci 46:191–202. doi:10.1016/0021-9797(74)90002-2
70. Oliveira Jr ON, Taylor DM, Lewis TJ, Salvagno S, Stirling CJM (1989) Estimation of group dipole moments from surface potential measurements on Langmuir monolayers. J Chem Soc Faraday Trans 1 Phys Chem Condens Phases 85:1009. doi:10.1039/f19898501009
71. Oliveira Jr ON, Taylor DM, Morgan H (1992) Modelling the surface potential-area dependence of a stearic acid monolayer. Thin Solid Films 210–211:76–78. doi:10.1016/0040-6090(92)90172-8
72. Adamson AW (1967) Physical chemistry of surfaces. Wiley, New York
73. Taylor DM, Oliveira Jr ON, Morgan H (1989) The surface potential of monolayers formed on weak acidic electrolytes: implications for lateral conduction. Chem Phys Lett 161:147–150. doi:10.1016/0009-2614(89)85047-X
74. Brown JQ, McShane MJ (2005) Core-referenced ratiometric fluorescent potassium ion sensors using self-assembled ultrathin films on europium nanoparticles. IEEE Sens J 5:1197–1205. doi:10.1109/JSEN.2005.859252
75. Shi W, Lin Y, Kong X, Zhang S, Jia Y, Wei M, Evans DG, Duan X (2011) Fabrication of pyrenetetrasulfonate/layered double hydroxide ultrathin films and their application in fluorescence chemosensors. J Mater Chem 21:6088. doi:10.1039/c1jm00073j
76. Stubbe BG, Gevaert K, Derveaux S, Braeckmans K, De Geest BG, Goethals M, Vandekerckhove J, Demeester J, De Smedt SC (2008) Evaluation of encoded layer-by-layer coated microparticles as protease sensors. Adv Funct Mater 18:1624–1631. doi:10.1002/adfm.200701356
77. Hénon S, Meunier J (1991) Microscope at the Brewster angle: direct observation of first-order phase transitions in monolayers. Rev Sci Instrum 62:936. doi:10.1063/1.1142032
78. Honig D, Mobius D (1991) Direct visualization of monolayers at the air-water interface by Brewster angle microscopy. J Phys Chem 95:4590–4592. doi:10.1021/j100165a003
79. Weidemann G, Brezesinski G, Vollhardt D, Möhwald H (1998) Disorder in Langmuir monolayers. 1. Disordered packing of alkyl chains. Langmuir 14:6485–6492. doi:10.1021/la980188o
80. Popovitz-Biro R, Edgar R, Weissbuch I, Lavie R, Cohen S, Kjaer K, Als-Nielsen J, Wassermann E, Leiserowitz L, Lahav M (1998) Structural studies on Langmuir films of C50H102, nylon-6,6 polymer and its oligomeric analogue. Acta Polym 49:626–635. doi:10.1002/(SICI)1521-4044(199810)49:10/11{\mathsurround=\opskip$<$}626::AID-APOL626{\mathsurround=\opskip$>$}3.0.CO;2-8
81. Zhou X-L, Chen S-H (1995) Theoretical foundation of X-ray and neutron reflectometry. Phys Rep 257:223–348. doi:10.1016/0370-1573(94)00110-O
82. Kago K, Fürst M, Matsuoka H, Yamaoka H, Seki T (1999) Direct observation of photoisomerization of a polymer monolayer on a water surface by X-ray reflectometry. Langmuir 15:2237–2240. doi:10.1021/la981084g
83. Li ZX, Bain CD, Thomas RK, Duffy DC, Penfold J (1998) Monolayers of hexadecyltrimethylammonium p -tosylate at the air – water interface. 2. Neutron reflection. J Phys Chem B 102:9473–9480. doi:10.1021/jp9821432
84. Blaudez D, Buffeteau T, Cornut JC, Desbat B, Escafre N, Pezolet M, Turlet JM (1993) Polarization-modulated FT-IR spectroscopy of a spread monolayer at the air/water interface. Appl Spectrosc 47:869–874
85. Shultz MJ, Baldelli S, Schnitzer C, Simonelli D (2002) Aqueous solution/air interfaces probed with sum frequency generation spectroscopy. J Phys Chem B 106:5313–5324. doi:10.1021/jp014466v
86. Greenler RG (1966) Infrared study of adsorbed molecules on metal surfaces by reflection techniques. J Chem Phys 44:310. doi:10.1063/1.1726462
87. Golden W (1981) A method for measuring infrared reflection? Absorption spectra of molecules adsorbed on low-area surfaces at monolayer and submonolayer concentrations. J Catal 71:395–404. doi:10.1016/0021-9517(81)90243-8

88. Lambert AG, Davies PB, Neivandt DJ (2005) Implementing the theory of sum frequency generation vibrational spectroscopy: a tutorial review. Appl Spectrosc Rev 40:103–145. doi:10.1081/ASR-200038326
89. Williams CT, Beattie DA (2002) Probing buried interfaces with non-linear optical spectroscopy. Surf Sci 500:545–576. doi:10.1016/S0039-6028(01)01536-9
90. Volpati D, Aoki PHB, Alessio P, Pavinatto FJ, Miranda PB, Constantino CJL, Oliveira Jr ON (2014) Vibrational spectroscopy for probing molecular-level interactions in organic films mimicking biointerfaces. Adv Colloid Interface Sci 207C:199–215. doi:10.1016/j.cis.2014.01.014
91. Miranda PB, Du Q, Shen YR (1998) Interaction of water with a fatty acid Langmuir film. Chem Phys Lett 286:1–8. doi:10.1016/S0009-2614(97)01476-0
92. Nicholson GL (2014) The fluid-mosaic model of membrane structure: still relevant to understanding the structure, function and dynamics of biological membranes after more than 40 years. Biochim Biophys Acta 1838:1451–1466. doi:10.1016/j.bbamem.2013.10.019
93. Jones MN, Chapman D (1994) Micelles, monolayers and biomembranes. Wiley-Liss, New York
94. Berkowitz ML, Vácha R (2012) Aqueous solutions at the interface with phospholipid bilayers. Acc Chem Res 45:74–82. doi:10.1021/ar200079x
95. Blume A, Kerth A (2013) Peptide and protein binding to lipid monolayers studied by FT-IRRA spectroscopy. Biochim Biophys Acta 1828:2294–305. doi:10.1016/j.bbamem.2013.04.014
96. Chattoraj DK, Birdi KS (1984) Adsorption and the Gibbs surface excess. Springer, New York, pp 219–223
97. Galvez Ruiz MJ, Cabrerizo Vilchez MA (1991) A study of the miscibility of bile components in mixed monolayers at the air-liquid interface I. Cholesterol, lecithin, and lithocholic acid. Colloid Polym Sci 269:77–84. doi:10.1007/BF00654662
98. Caetano W, Ferreira M, Tabak M, Mosquera Sanchez MI, Oliveira Jr ON, Krüger P, Schalke M, Lösche M (2001) Cooperativity of phospholipid reorganization upon interaction of dipyridamole with surface monolayers on water. Biophys Chem 91:21–35. doi:10.1016/S0301-4622(01)00145-4
99. Hidalgo AA, Caetano W, Tabak M, Oliveira Jr ON (2004) Interaction of two phenothiazine derivatives with phospholipid monolayers. Biophys Chem 109:85–104. doi:10.1016/j.bpc.2003.10.020
100. Moraes ML, Bonardi C, Mendonça CR, Campana PT, Lottersberger J, Tonarelli G, Oliveira Jr ON, Beltramini LM (2005) Cooperative effects in phospholipid monolayers induced by a peptide from HIV-1 capsid protein. Colloids Surf B Biointerfaces 41:15–20. doi:10.1016/j.colsurfb.2004.10.026
101. Pickholz M, Oliveira Jr ON, Skaf MS (2006) Molecular dynamics simulations of neutral chlorpromazine in zwitterionic phospholipid monolayers. J Phys Chem B 110:8804–14. doi:10.1021/jp056678o
102. Torrano AA, Pereira ÂS, Oliveira Jr ON, Barros-Timmons A (2013) Probing the interaction of oppositely charged gold nanoparticles with DPPG and DPPC Langmuir monolayers as cell membrane models. Colloids Surf B Biointerfaces 108:120–6. doi:10.1016/j.colsurfb.2013.02.014
103. Caseli L, Pavinatto FJ, Nobre TM, Zaniquelli MED, Viitala T, Oliveira Jr ON (2008) Chitosan as a removing agent of beta-lactoglobulin from membrane models. Langmuir 24:4150–6. doi:10.1021/la7038762
104. Casal E, Montilla A, Moreno FJ, Olano A, Corzo N (2006) Use of chitosan for selective removal of β-lactoglobulin from whey. J Dairy Sci 89:1384–1389
105. Damalio JCP, Nobre TM, Lopes JL, Oliveira Jr ON, Araújo APU (2013) Lipid interaction triggering Septin2 to assembly into β-sheet structures investigated by Langmuir monolayers and PM-IRRAS. Biochim Biophys Acta 1828:1441–8. doi:10.1016/j.bbamem.2013.02.003
106. vanden Akker CC, Engel MFM, Velikov KP, Bonn M, Koenderink GH (2011) Morphology and persistence length of amyloid fibrils are correlated to peptide molecular structure. J Am Chem Soc 133:18030–3. doi:10.1021/ja206513r

107. Riul A, Dantas CAR, Miyazaki CM, Oliveira Jr ON (2010) Recent advances in electronic tongues. Analyst 135:2481–95. doi:10.1039/c0an00292e
108. Toko K (1996) Taste sensor with global selectivity. Mater Sci Eng C 4:69–82. doi:10.1016/0928-4931(96)00134-8
109. Dulac C (2000) The physiology of taste, vintage 2000. Cell 100:607–610. doi:10.1016/S0092-8674(00)80697-2
110. Riul A, dos Santos DS, Wohnrath K, Di Tommazo R, Carvalho ACPLF, Fonseca FJ, Oliveira Jr ON, Taylor DM, Mattoso LHC (2002) Artificial taste sensor: efficient combination of sensors made from Langmuir – Blodgett films of conducting polymers and a ruthenium complex and self-assembled films of an azobenzene-containing polymer. Langmuir 18:239–245. doi:10.1021/la011017d
111. Taylor DM, Macdonald AG (1987) AC admittance of the metal/insulator/electrolyte interface. J Phys D Appl Phys 20:1277
112. Gorban AN, Kégl B, Wunsch DC, Zinovyev A (2007) Principal manifolds for data visualisation and dimension reduction. Springer, Berlin
113. Ferreira M, Riul A, Wohnrath K, Fonseca FJ, Oliveira Jr ON, Mattoso LHC (2003) High-performance taste sensor made from Langmuir – Blodgett films of conducting polymers and a ruthenium complex. Anal Chem 75:953–955. doi:10.1021/ac026031p
114. Dos Santos DS, Riul A, Malmegrim RR, Fonseca FJ, Oliveira Jr ON, Mattoso LHC (2003) A layer-by-layer film of chitosan in a taste sensor application. Macromol Biosci 3:591–595. doi:10.1002/mabi.200350027
115. Liu M, Wang J, Li D, Wang M (2012) Electronic tongue coupled with physicochemical analysis for the recognition of orange beverages. J Food Qual 35:429–441. doi:10.1111/jfq.12004
116. Winquist F, Bjorklund R, Krantz-Rülcker C, Lundström I, Östergren K, Skoglund T (2005) An electronic tongue in the dairy industry. Sens Actuators B Chem 111–112:299–304. doi:http://dx.doi.org/10.1016/j.snb.2005.05.003
117. Riul A, Gallardo Soto AM, Mello SVV, Bone S, Taylor DMM, Mattoso LHC (2003) An electronic tongue using polypyrrole and polyaniline. Synth Met 132:109–116. doi:10.1016/S0379-6779(02)00107-8
118. Wiziack NKL, Paterno LG, Fonseca FJ, Mattoso LHC (2007) Effect of film thickness and different electrode geometries on the performance of chemical sensors made of nanostructured conducting polymer films. Sens Actuators B 122:484–492. doi:10.1016/j.snb.2006.06.016
119. Borato CE, Leite FL, Mattoso LHC, Goy RC, Filho SPC, de Vasconcelos CL, da Trindade Neto CG, Pereira MP, Fonseca JLC, Oliveira Jr ON (2006) Layer-by-layer films of poly(o-ethoxyaniline), chitosan and chitosan-poly(methacrylic acid) nanoparticles and their application in an electronic tongue. IEEE Trans Dielectr Electr Insul 13:1101–1109. doi:10.1109/TDEI.2006.247838
120. Brugnollo ED, Paterno LG, Leite FL, Fonseca FJ, Constantino CJL, Antunes PA, Mattoso LHC (2008) Fabrication and characterization of chemical sensors made from nanostructured films of poly(o-ethoxyaniline) prepared with different doping acids. Thin Solid Films 516:3274–3281. doi:10.1016/j.tsf.2007.08.118
121. Volpati D, Alessio P, Zanfolim AA, Storti FC, Job AE, Ferreira M, Riul A, Oliveira Jr ON, Constantino CJL (2008) Exploiting distinct molecular architectures of ultrathin films made with iron phthalocyanine for sensing. J Phys Chem B 112:15275–82. doi:10.1021/jp804159h
122. Ferreira EJ, Pereira RCT, Delbem ACB, Oliveira Jr ON, Mattoso LHC (2007) Random subspace method for analysing coffee with electronic tongue. Electron Lett 43:1138. doi:10.1049/el:20071182
123. Zucolotto V, Daghastanli KRP, Hayasaka CO, Riul A, Ciancaglini P, Oliveira Jr ON (2007) Using capacitance measurements as the detection method in antigen-containing layer-by-layer films for biosensing. Anal Chem 79:2163–7. doi:10.1021/ac0616153
124. Such GK, Johnston APR, Caruso F (2011) Engineered hydrogen-bonded polymer multilayers: from assembly to biomedical applications. Chem Soc Rev 40:19–29. doi:10.1039/c0cs00001a

125. Takahashi S, Sato K, Anzai J (2012) Layer-by-layer construction of protein architectures through avidin-biotin and lectin-sugar interactions for biosensor applications. Anal Bioanal Chem 402:1749–58. doi:10.1007/s00216-011-5317-4
126. Vannoy CH, Tavares AJ, Noor MO, Uddayasankar U, Krull UJ (2011) Biosensing with quantum dots: a microfluidic approach. Sensors (Basel) 11:9732–63. doi:10.3390/s111009732
127. Yan Y, Björnmalm M, Caruso F (2014) Assembly of layer-by-layer particles and their interactions with biological systems. Chem Mater 26:452–460. doi:10.1021/cm402126n
128. Caseli L, Crespilho FN, Nobre TM, Zaniquelli MED, Zucolotto V Jr, Oliveira Jr ON (2008) Using phospholipid Langmuir and Langmuir–Blodgett films as matrix for urease immobilization. J Colloid Interface Sci 319:100–108, doi:http://dx.doi.org/10.1016/j.jcis.2007.12.007
129. Caseli L, Moraes ML, Zucolotto V, Ferreira M, Nobre TM, Zaniquelli MED, Rodrigues Filho UP, Oliveira Jr ON (2006) Fabrication of phytic acid sensor based on mixed phytase – lipid Langmuir – Blodgett films. Langmuir 22:8501–8508. doi:10.1021/la061799g
130. Pavinatto FJ, Fernandes EGR, Alessio P, Constantino CJL, de Saja JA, Zucolotto V, Apetrei C, Oliveira Jr ON, Rodriguez-Mendez ML (2011) Optimized architecture for tyrosinase-containing Langmuir–Blodgett films to detect pyrogallol. J Mater Chem 21:4995. doi:10.1039/c0jm03864d
131. Lu F, Tian Y, Liu M, Su D, Zhang H, Govorov AO, Gang O (2013) Discrete nanocubes as plasmonic reporters of molecular chirality. Nano Lett 13:3145–51. doi:10.1021/nl401107g
132. Srivastava S, Kotov NA (2008) Composite layer-by-layer (LBL) assembly with inorganic nanoparticles and nanowires. Acc Chem Res 41:1831–41. doi:10.1021/ar8001377
133. Manickam A, Johnson CA, Kavusi S, Hassibi A (2012) Interface design for CMOS-integrated Electrochemical Impedance Spectroscopy (EIS) biosensors. Sensors (Basel) 12:14467–14488. doi:10.3390/s121114467
134. Marzo FF, Pierna AR, Barranco J, Lorenzo A, Barroso J, García JA, Pérez A (2008) Determination of trace metal release during corrosion characterization of FeCo-based amorphous metallic materials by stripping voltammetry. New materials for GMI biosensors. J Non Cryst Solids 354:5169–5171. doi:10.1016/j.jnoncrysol.2008.08.014
135. Caseli L, dos Santos DS, Foschini M, Gonçalves D, Oliveira Jr ON (2006) The effect of the layer structure on the activity of immobilized enzymes in ultrathin films. J Colloid Interface Sci 303:326–31. doi:10.1016/j.jcis.2006.07.013
136. Ferreira M, Fiorito PA, Oliveira Jr ON, Córdoba de Torresi SI (2004) Enzyme-mediated amperometric biosensors prepared with the Layer-by-Layer (LbL) adsorption technique. Biosens Bioelectron 19:1611–5. doi:10.1016/j.bios.2003.12.025
137. Crespilho FN, Emilia Ghica M, Florescu M, Nart FC, Oliveira Jr ON, Brett CMA (2006) A strategy for enzyme immobilization on layer-by-layer dendrimer–gold nanoparticle electrocatalytic membrane incorporating redox mediator. Electrochem Commun 8:1665–1670. doi:10.1016/j.elecom.2006.07.032
138. De Oliveira RF, de Moraes ML, Oliveira Jr ON, Ferreira M (2011) Exploiting cascade reactions in bienzyme layer-by-layer films. J Phys Chem C 115:19136–19140. doi:10.1021/jp207610w
139. Perinoto ÂC, Maki RM, Colhone MC, Santos FR, Migliaccio V, Daghastanli KR, Stabeli RG, Ciancaglini P, Paulovich FV, de Oliveira MCF, Oliveira Jr ON, Zucolotto V (2010) Biosensors for efficient diagnosis of leishmaniasis: innovations in bioanalytics for a neglected disease. Anal Chem 82:9763–8. doi:10.1021/ac101920t
140. Siqueira JR, Abouzar MH, Poghossian A, Zucolotto V, Oliveira Jr ON, Schöning MJ (2009) Penicillin biosensor based on a capacitive field-effect structure functionalized with a dendrimer/carbon nanotube multilayer. Biosens Bioelectron 25:497–501. doi:10.1016/j.bios.2009.07.007
141. Siqueira JR, Werner CF, Bäcker M, Poghossian A, Zucolotto V, Oliveira Jr ON, Schöning MJ (2009) Layer-by-layer assembly of carbon nanotubes incorporated in light-addressable potentiometric sensors. J Phys Chem C 113:14765–14770. doi:10.1021/jp904777t

142. Oliveira Jr ON, Iost RM, Siqueira JR, Crespilho FN, Caseli L (2014) Nanomaterials for diagnosis: challenges and application in smart devices based on molecular recognition. ACS Appl Mater Interfaces. doi:10.1021/am5015056
143. Siqueira JR, Maki RM, Paulovich FV, Werner CF, Poghossian A, de Oliveira MCF, Zucolotto V, Oliveira Jr ON, Schöning MJ (2010) Use of information visualization methods eliminating cross talk in multiple sensing units investigated for a light-addressable potentiometric sensor. Anal Chem 82:61–5. doi:10.1021/ac9024076
144. De Oliveira MCF, Levkowitz H (2003) From visual data exploration to visual data mining: a survey. IEEE Trans Vis Comput Graph 9:378–394. doi:10.1109/TVCG.2003.1207445
145. Paulovich FV, Maki RM, de Oliveira MCF, Colhone MC, Santos FR, Migliaccio V, Ciancaglini P, Perez KR, Stabeli RG, Perinoto AC, Oliveira Jr ON, Zucolotto V (2011) Using multidimensional projection techniques for reaching a high distinguishing ability in biosensing. Anal Bioanal Chem 400:1153–9. doi:10.1007/s00216-011-4853-2
146. Sammon JW (1969) A nonlinear mapping for data structure analysis. IEEE Trans Comput C-18:401–409. doi:10.1109/T-C.1969.222678
147. Paulovich FV, Moraes ML, Maki RM, Ferreira M, Oliveira Jr ON, de Oliveira MCF (2011) Information visualization techniques for sensing and biosensing. Analyst 136:1344–50. doi:10.1039/c0an00822b
148. Aoki PHB, Carreon EGE, Volpati D, Shimabukuro MH, Constantino CJL, Aroca RF, Oliveira Jr ON, Paulovich FV (2013) SERS mapping in Langmuir-Blodgett films and single-molecule detection. Appl Spectrosc 67:563–9. doi:10.1366/12-06909

MIX
Papier aus verantwortungsvollen Quellen
Paper from responsible sources
FSC® C105338

If you have any concerns about our products,
you can contact us on
ProductSafety@springernature.com

In case Publisher is established outside the EU,
the EU authorized representative is:
Springer Nature Customer Service Center GmbH
Europaplatz 3, 69115 Heidelberg, Germany

Printed by Libri Plureos GmbH
in Hamburg, Germany